World Sustainability Series

Series Editor

Walter Leal Filho, European School of Sustainability Science and Research, Research and Transfer Centre "Sustainable Development and Climate Change Management", Hamburg University of Applied Sciences, Hamburg, Germany

Due to its scope and nature, sustainable development is a matter which is very interdisciplinary, and draws from knowledge and inputs from the social sciences and environmental sciences on the one hand, but also from physical sciences and arts on the other. As such, there is a perceived need to foster integrative approaches, whereby the combination of inputs from various fields may contribute to a better understanding of what sustainability is, and means to people. But despite the need for and the relevance of integrative approaches toward sustainable development, there is a paucity of literature which addresses matters related to sustainability in an integrated way.

Notes on the quality assurance and peer-review of this publication

Prior to publication, the works published in this book are initially assessed and reviewed by an in-house editor. If suitable for publication, manuscripts are sent for further review, which includes a combined effort by the editorial board and appointed subject experts, who provide independent peer-review. The feedback obtained in this way was communicated to authors, and with manuscripts checked upon return before finally accepted. The peer-reviewed nature of the books in the "World Sustainability Series" means that contributions to them have, over many years, been officially accepted for tenure and promotion purposes.

Mona-Maria Narra · Satyanarayana Narra
Editors

African Green Transition Through Innovative Pathways

Springer

Editors
Mona-Maria Narra
Material and Energy Valorisation of Biogenous Residues—Waste and Resource Management
University of Rostock
Rostock, Germany

Satyanarayana Narra
Material and Energy Valorisation of Biogenous Residues—Waste and Resource Management
University of Rostock
Rostock, Germany

ISSN 2199-7373 ISSN 2199-7381 (electronic)
World Sustainability Series
ISBN 978-3-031-87042-2 ISBN 978-3-031-87043-9 (eBook)
https://doi.org/10.1007/978-3-031-87043-9

This Springer imprint is published by the registered company Springer Nature Switzerland AG
The registered company address is: Gewerbestrasse 11, 6330 Cham, Switzerland

Preface

Africa stands at a pivotal crossroads in its development journey. The pressing need for economic growth and industrialization is set against the backdrop of climate change, environmental degradation, and energy inequities. With a rapidly growing population and urbanization rates among the highest in the world, the continent faces immense challenges in meeting the demands for energy, waste management, and resource efficiency. At the same time, Africa is endowed with vast natural resources, abundant sunlight, and agricultural potential that offer unique opportunities to lead a green transition. This book, African Green Transition Through Innovative Pathways, is both a call to action and a roadmap for leveraging innovative solutions to address these challenges while fostering sustainable development.

The African green transition is not merely about adopting global sustainability trends but about tailoring solutions to the continent's unique contexts. From rural communities seeking basic energy access to sprawling urban centers grappling with waste management, Africa's challenges require a multifaceted approach that balances technological innovation, social inclusivity, and environmental stewardship. The articles in this volume, contributed by experts and researchers from diverse fields, illuminate these pathways, providing critical insights and actionable strategies for achieving a sustainable future.

A Focus on Innovation

Innovation lies at the heart of this book. Each chapter delves into practical and transformative solutions for some of Africa's most pressing issues. Whether it's through waste valorization, renewable energy systems, or data-driven energy optimization, the book highlights how science and technology can pave the way for sustainable development. For instance, strategies for recovering valuable materials from waste streams, such as magnetic particles from wastewater or energy from agricultural byproducts, demonstrate the untapped potential of Africa's resources. Similarly, the focus on hydrogen economy and advanced solar technologies underscores the continent's ability to harness cutting-edge approaches to energy transitions.

A Holistic Approach to Sustainability

This book is structured to address the interconnected nature of sustainability challenges. Waste management and resource recovery form the foundation, exploring how traditional burdens can be transformed into opportunities. Subsequent chapters build on this by investigating bioenergy potential, urban and rural energy solutions, and the integration of circular economy principles. By examining both technological advancements and systemic transitions, the book offers a comprehensive view of how Africa can chart a sustainable path forward.

A significant portion of the book also addresses the social and economic dimensions of sustainability. Chapters on energy access highlight the critical need for inclusivity, particularly for underserved communities. By exploring comparative studies of biofuels, renewable energy systems, and urban energy planning, the book provides insights into equitable solutions that prioritize people alongside the planet.

Looking Ahead

As you journey through the chapters of this book, we invite you to consider not only the challenges Africa faces but also the immense opportunities at hand. Each chapter presents a piece of the puzzle, contributing to a larger vision of a sustainable Africa. By the end of this book, we hope you will be inspired to take part in this transformative journey, whether as a researcher, policymaker, or advocate for change.

Together, we can chart a path toward an Africa that thrives through innovative pathways, ensuring a future where economic growth, social equity, and environmental stewardship go hand in hand.

Acknowledgments

This book is the culmination of the hard work and dedication of researchers, academics, and practitioners who have contributed their expertise to its pages. Their insights provide a wealth of knowledge for anyone committed to sustainability in Africa. The articles in this book were first presented at the "Second International

Conference on Circular Economy, Renewable Energy and Green Hydrogen" in Lomé, Togo in October 2024.

Rostock, Germany

Mona-Maria Narra
Satyanarayana Narra

Acknowledgements

The "Second International Conference on Circular Economy, Renewable Energy and Green Hydrogen" and the resulting book "African Green Transition Through Innovative Pathways" has been funded by the German Federal Ministry of Education and Research under the project "Verbundvorhaben Waste2Energy: Hybrid Energie aus Abfall als nachhaltige Lösung für Ghana".

SPONSORED BY THE

Federal Ministry
of Education
and Research

Contents

Innovations in Waste Management and Resource Recovery

Assessment of Energy Potential of Residential Solid Waste in the Asokore Mampong Municipality of the Ashanti Region of Ghana

Grant Degbe Dzitse, Helen M. K. Essandoh, and Lawrence Darkwah

Abstract Waste management is a key issue for various local authorities in Ghana, with the most common disposal method being landfilling. Sustainable waste management requires reliable data that could assist in the management process. To establish these data, waste types and generation rates, composition, separation efficiency of household waste, and calorific value of waste, were studied in Asokore Mampong Municipality over four weeks. The study population consisted of residents in the households of the selected study areas. Sample sizes of 100 households were selected purposively from three (3) residential areas in the Municipality including Sawaba (low-income area), Asokore Mampong (middle-income area), and Bouban (high-income area). Field observations, mini-questionnaire interviews, key informant interviews and secondary data were used to gather neighbourhood data. The calorific value of solid waste samples were determined using the Oxygen Bomb Calorimeter (Parr 6400). The results show a total of 5343.240 kg of waste sampled from the three socioeconomic areas with an average generation rate estimated to be 0.470 kg/capita/day and an average household size of 4.5 persons. The composition of the residential solid waste was dominantly organics 68%, plastic 15.4%, paper and cardboard 6.3%, metals 5%, glass 0.8%, textiles 2.0% and 2.5% "Others". The highest source-separated component was plastics and cardboard (94%) followed by organic

The original version of the chapter has been revised. A correction to this chapter can be found at https://doi.org/10.1007/978-3-031-87043-9_18

G. D. Dzitse (✉) · H. M. K. Essandoh
Department of Civil Engineering, Regional Water and Environmental Sanitation Centre, Kumasi (RWESCK), College of Engineering, Kwame Nkrumah University of Science and Technology, Kumasi, Ghana
e-mail: dzitsegrant@gmail.com

H. M. K. Essandoh
e-mail: hmkessandoh.coe@knust.edu.gh

L. Darkwah
Department of Chemical Engineering, College of Engineering, Kwame Nkrumah University of Science and Technology, Kumasi, Ghana
e-mail: ldarkwah.soe@knust.edu.gh

M.-M. Narra and S. Narra (eds.), *African Green Transition Through Innovative Pathways*, World Sustainability Series, https://doi.org/10.1007/978-3-031-87043-9_1

(92%) and the least, the "Others" component (86%). The mean calorific value of the major components of the waste: organic, plastic, textile, paper and cardboard were found to be 19.72 MJ/kg. The cumulative calorific values of combustible contents of residential waste in the municipality was determined as 7.5 GWh per day in the study which is potentially good for waste-to-energy recovery through incineration.

Keywords Waste Generation Rate · Waste to Energy · Residential Solid Waste · Incineration · Calorific value

1 Introduction

The Ghana Environmental Sanitation Policy (2010) describes solid wastes as all materials emanating from establishments, as well as healthcare waste from clinics industries, hospitals, households and commercial centres, discarded for collection. It also includes street sweepings, drain cleanings, dead animals, demolition and construction waste, and other waste materials. According to the Waste Framework Directive of the European Union (WFD, 2018/851), waste is characterised as an article or substance that the owner has disposed of or needs to throw away. Waste does not happen in nature; however, it emerges from human activities, and mishandling waste could represent a genuine danger to the climate and general well-being of the environment and public health.

On the other hand, Solid waste may be defined as any undesirable or unwanted solid materials generated from collective industrial, commercial, residential, agricultural, and other activities in a given jurisdiction (Adu-Boahen et al., 2014). Increases in the quantity and diversity of solid wastes produced by residential, industrial, mining, and agricultural operations have been attributed to population growth, increased urbanisation, and diverse life style standards as well as result of technological developments (Medina, M., 2010). These together with limited resources resort to extensive practice of improper waste disposal system of landfill and burning that are deteriorating the quality of the environment and posing a risk to sustainable development.

As stated by (Kaza et al., 2018; Mandal, 2019) globally, the world waste generation was 2.01 billion tonnes and was projected to 3.40 billion tonnes of waste to be produced by 2050. The constant increase in both the amount and type of solid waste necessitates the appropriate usage of these wastes by several countries as one of their solid waste management options, rather than relying on traditional forms such as collection, transportation, and disposal.

These sustainable waste management practices adopted by developed countries like the Netherlands, Singapore, and Germany have brought about socioeconomic development and improvement in environmental quality. According to Donkoh (2016) solid waste recycling as a means of improving environmental quality and resource recovery could be assessed based on how recycling activities either improve or degenerate environmental characteristics which may ultimately advance or deteriorate the health of humans and other organisms, hence, societies that effectively make use of a large amount of its solid waste are most likely to eliminate a significant proportion of these environmental and public health consequences. This is prevalent mostly in developed countries such as the Netherlands where solid waste is utilized effectively through recycling, re-use, generation of a renewable form of energy called biogas, and harnessing them into compost for improved agriculture among others (Adipah & Kwame, 2019). These solid waste utilization methods greatly minimize the accumulation of solid waste in towns and cities. Several European nations, including Belgium, Germany, Austria, the Netherlands, Norway, Sweden, and Switzerland, Sweden are known for their efficient municipal solid waste management, with some virtually facing off landfilling with more than 50% MSW recovery rate (EEA, 2016).

In addition, effective solid waste utilization brings about a clean and serene environment, revenue generation, employment creation, production of essential products, reducing government's expenditure on imported goods, and effecting favourable Terms of Trade (Donkoh, S., 2016).

In contrast, societies that utilize little amount of their generated solid waste are most likely to encounter solid waste disposal and accumulation problems. Also, people may be compelled to dispose of the generated solid waste at unapproved places since they may not have designated places to dispose of their waste. This situation is common in many developing countries whose solid waste is dumped largely at unapproved places such as marketplaces, near a public toilet, and final disposal sites. Municipal SWM in Ghana's capital, Accra, remains inadequate, similar to many developing countries. For nearly two decades, the private sector has been responsible for waste management, however, waste collection coverage is only 75%, with just 62% of the total waste generated being disposed of at regulated landfills (Oduro-Appiah et al., 2017) which not different from scenerios in other cities across the country.

For instance, in many cities in Ghana, people dump solid waste into large containers, placed near market centres and public toilets, and at times dump very close to residential areas and along beaches while others dispose of their waste into water bodies causing water pollution and adversely affecting the survival of aquatic ecosystem. These measures have, however, not solved the problem of waste disposal since with time, the level of solid waste accumulation increases and causes the spread of epidemics such as malaria, cholera, typhoid, and others among posing environmental challenges like risk of flooding.

Though the utilization of solid waste for manure production, biogas production, combustion as waste to energy, and recycling among others is carried out in developing countries like Ghana, little significant strides are made toward sustainable integrated solid waste management, utilising these processes to contribute significantly to minimizing solid waste accumulation, protect public health and the environment. The challenges of sustainable solid waste management in Ghana according to the World Bank (2019b) are key issues such as: (i) absence of effective collection, segregation and recycling systems; (ii) lack of community awareness, and (iii) inadequate disposal capacity. In Accra, Ghana for instance, it was revealed that recycling rate was only 5% (Oduro-Appiah et al., 2017) while (Wahabu et al., 2014) discovered that, the informal sector Kumasi is vigorously involved in recycling on a small-scale resulting is a low-recovery rate of 0.5%.

By reducing the need for "conventional" waste disposal and lowering Greenhouse gases, recycling prevents the loss of potentially valuable materials, reduces the consumption of fresh raw materials, reduces energy consumption, and reduces air pollution from uncontrolled burning and water pollution from landfill disposal. Local decision-makers routinely face the choice of whether or not to invest in Waste-to-Energy (WtE) technology in their effort to modernize their waste management systems. WtE technologies are increasingly being promoted as a viable solution for addressing not only urgent waste disposal issues, but also several other issues at the same time, including power shortages due to over dependence on hydro- energy and thermal fossil power plant landfill space constraints, and greenhouse gas emissions from improper waste disposal.

With about 1,500 tonnes of solid waste per day in the city of Kumasi, Ghana, and about 1,300 tonnes are received from industrial wastes, commercial, and domestic, including 300 tonnes of market waste daily (Oduro-Kwarteng et al., 2016), waste management is a topmost item of priority in the budget of Kumasi Metropolitan Area (KMA). Huge substantial amount of money not less than GH¢ 45,000 (8,333.33 US dollars) is spent daily by KMA for waste collection service (Sarfo-Mensah et al., 2019). These wastes are usually transported to the landfill site, which is made up of 78 percent combustible and 22 percent non-combustible.

Organic components of MSW are combustible and for this reason, can be incinerated to obtain thermal energy or generate electric power for use for domestic purposes. Incineration is the process of converting organic solid waste into other useful forms such as heat, steam, and ash residues through combustion which is at times referred to as Waste- to Energy (WTE) plants, which reduces solid waste by 70 to 80 percent and thus minimizes waste sent to landfills. Countries that have land space problems such as Japan and Singapore seriously resort to incineration (Magutu & Onsongo, 2011). European countries like Denmark and Sweden have resorted to incineration rather than landfilling due to bans and taxation on landfills (ISWA Report, 2013). Landfilling has the least priority on the SWM hierarchy because of the negative environmental impacts through greenhouse emissions and other forms of pollution that affect soils, groundwater, and land surfaces (Franca and Bassin, 2020). In Africa, the landfill method is characterised by the disposal of waste on open, non-structured areas of land without consideration for environmental impact, where structures for leachate and greenhouse containment are absent. The increased population could put a burden on landfill disposal sites, as more solid waste will be generated in the near future, shortening the life of landfills. The above-mentioned socio-economic and environmental implications, sustainable energy supply as well as the increased quantity of solid waste generated in the Greater Kumasi metropolitan Area (GKMA), have driven this research to develop sustainable means of handling MSW and, as a result, Waste to Energy.

2 Materials and Methods

Solid waste characterisation through sorting, rate of generation, and separation efficiency was conducted at the household level of the three socioeconomic communities low, middle, and high-income neighbourhood of the Asokore Mampong Municipality between May and June 2021 while calorific value analysis was conducted on the various samples using Parr6400 Oxygen Bomb Calorimeter.

2.1 Study Area

The study was conducted in Asokore Mampong, Bouban, and Sawaba in the Asokore Mampong Municipality as shown in Fig. 1. Asokore Mampong Municipal Assembly (AMMA) located in the Ashanti Region of Ghana, is part of the Eight Assemblies in the Greater Kumasi Metropolitan Area (GKMA). The Municipality covers a total land area of 23.91sq.km. Geographically, Asokore Mampong Municipal is situated on 6°42′N 1°36′W in the central part of the Region and shares boundaries with Kumasi Metropolitan Assembly (KMA) to the East, South and West is Kwabre East Municipal to the North-west and Ejisu Municipal Assembly to the South-East. According to the National Population and Housing Census (GSS, 2014), the population of the Municipality is approximately 304,815 with gender distribution of 159,036 females about 4.4% higher than that of male population of 145,779. Currently, the estimated population from the based population of 304,815 in 2010 and with a population growth rate of 8.68% which is projected to population of 791,956 as of 2021. The areas of employment consist of a web of marketplaces, financial institutions, wholesalers and retailers, airlines and transportation companies, hotels, and restaurants. Other commercial activities include petty trading, bread bakery, metallic pot making, masonry, dressmaking, carpentry, and vehicle repairs among others. The study approach involved an assessment of the physical characteristics (composition, generation rate, separation efficiency, and analysis of calorific value) of solid waste generated from 100 households in the study area over four weeks using the Parr 6400 Bomb Calorimeter. The stratified and purposive random sampling method was used in the selection of the study households both for solid waste composition analysis and questionnaire administration. By this method, households and study respondents were selected from low, middle, and high-income neighbourhoods within the study area to provide a general idea about the waste characteristics, rate of generation, and separation efficiency. The criteria used to classify the households were based on the residents' living standards, housing type, and access to basic essential services such as potable water, electricity, and toilet facilities, and population density among others, according to the Asokore Mampong Municipal Assembly Valuation Unit. Three communities from Municipality were therefore selected for study: Asokore Mampong, Bouban, and Sawaba as low, middle, and high-income communities respectively. Thirty-three households from each of middle and high-income

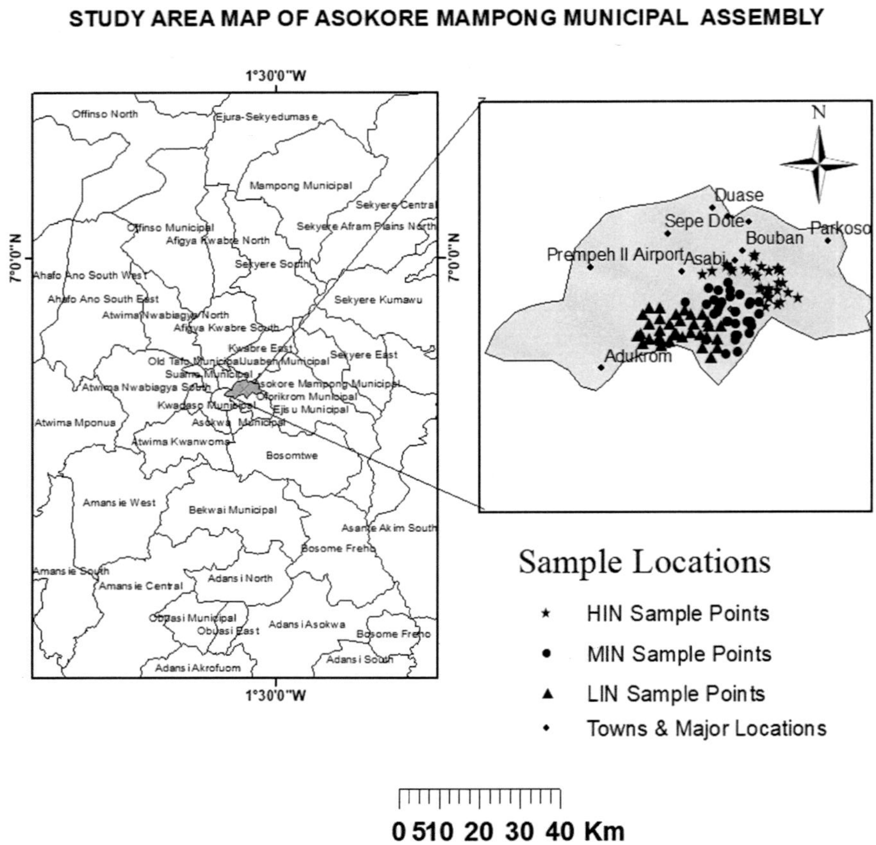

Fig. 1 A location map showing Asokore Mampong Municipal (*Source* The Authors' 2021 design using ArcGIS version 10.7.1)

neighbourhoods, while Thirty-four households represented low-income areas, were used for the study. Each of the participating households were provided with a three different bins/polythene after thoroughly sensitizing families on the study objectives for their total cooperation.

2.1.1 Classification of Socioeconomic Income Areas

Residential classification into three fundamental socioeconomic income statuses (low, middle, and high-residential neighbourhoods) was established by the AMMA. The AMMA classification which is grounded on socioeconomic development takes into consideration the type of road network, water supply, buildings, and other social amenities in the area similar framework espoused by (Oduro-Kwarteng et al., 2016; Zakarya et al., 2020). The three communities chosen for this research were

Sawaba, a low-income neighbourhood (LIN), Asokore Mampong, a middle-income neighbourhood (MIN), and Bouban Residential area a high-income neighbourhood (HIN).

2.1.2 Low-Income neighbourhood

These are places with inadequate facilities and social services. They are primarily found in slum areas of cities. The structures range from multi-story or detached structures to squatter shacks. The low-income neighbourhoods responded enthusiastically because they faced problems with solid waste management issues. The financial benefit of not paying disposal costs during the study duration encouraged the households to participate actively.

2.1.3 Middle-Income neighbourhood

These neighbourhoods are known for their Flats or bungalows characterised by these residential areas. The buildings are commonly occupied by multi-family households and are usually referred to as "compound houses". The structures are detached or semi-detached, with paved courtyards and, on rare occasions, backyard gardens. Settlers in this area may have improved social services and amenities to some extent.

2.1.4 High-Income neighbourhood

These are suburban communities with generally good roads, as well as dependable social amenities and facilities like water, security, power, well-designed houses, schools, and supermarkets. Houses are typically single-story, standalone structures with grassed and a large paved compound (Miezah et al., 2015; Anarfi et al., 2013). These areas are typically inhabited by single-family or small families. Despite the general perception that members in this area are well-off, no empirical research on income segmentation has been done to determine settlers' socioeconomic status.

2.2 *Data Sources*

The data for this study was gathered from both secondary and primary sources. The sub-sections below describe the details of each of these sources.

2.2.1 Primary Data

Mixed processes such as preliminary field surveys, direct observations, questionnaire administrations, and face-to-face interviews were used to solicit primary data, including gathering solid waste samples from households. Preliminary field observation entailed a visit to the nominated households to notify inhabitants regarding the survey and seek their involvement in the survey. Direct interviews were conducted to obtain information from some key stakeholders such as the Environmental Health and Sanitation Department and Planning Department, households within Asokore Mampong Municipal Assembly (AMMA), Zoomlion, and other small-scale waste collectors who play administrative roles in the waste sector.

A well-structured questionnaire was developed and administered randomly and purposively to selected households to gather relevant information. This was followed by field measurements of waste samples collected and categorised from households.

2.2.2 Secondary Data

A review of available information on the subject matter was grounded in secondary data. The sources such as journals, books, articles, newsletters, and other helpful online resources were the sources of secondary data gathered. Secondary information was also obtained from the AMMA's District Medium Term Development Plan (DMTDP), the Kumasi Metropolitan Assembly Waste Management Department (KMA-WMD), and Zoomlion. The secondary data allowed the researcher to become acquainted with the subject and to determine what has been done and what has not been done thus far.

2.2.3 Sample Size Determination

According to the Ghana Statistical Service (2014), there was a total of 72,478 households with an average household size of 4 people in the study area. The new household size is determined by the exponential growth method. This was sampled for this study in the municipality represented in survey's sample frame. The formula of sampling for continuous variable measurements reported by Taro Yamane's (1973) mathematical model has been widely applied by other researchers including; Lyida (2015), Addo et al. (2019), Mionne et al. (2019), and Amo-Asamoah et al. (2020) in similar studies as in Eq. (1) was used to establish the survey's sample size. The procedure is described in Eq. (1)

$$n = \frac{N}{1 + N(\alpha^2)} \tag{1}$$

where, n = size of the sample, N = sample frame and (α) characterised the margin of error of 0.1 with a confidence level of 90%.

Substituting 180,052 projected number of households and 0.1 into the formula, n = 99.9 ≈ 100. The result of survey's sample size was one hundred (100) people.

Similarly, regarding the method subscribed by Miezah et al. (2015); and Zakarya et al. (2020), a total of 100 households were selected for the study, which comprised 34, 33, and 33 each from three neighbourhoods using mixed methods such as questionnaire, purposive, direct, and stratified, random sampling procedures were employed in each study area to choose the number of households.

These questionnaires collected data on socio-economic factors, household size, demographics, educational levels, knowledge of waste management, and reasons for participation. The selected households pre-received education on waste sorting and separation over a day and personal interactions and engagement.

2.3 Sampling Techniques

Cluster, purposive, stratified, systematic, and random sampling strategies were used to choose the households for the investigation. The study area was separated into three income groups: high, middle, and low according to AMMA's residential classification and similar procedure adopted by (Miezah et al., 2015; Zakarya et al., 2020). This was to account for variation in waste creation among the studied subjects over the study periods to reflect the true characteristics of waste. Using the weight analysis, the researcher assessed the compliance to waste separation, physical and thermo-chemical parameters such as composition, production rate, moisture content (MC), and calorific value (CV).

According to the Asokore Mampong Assembly's Valuation Unit, the criteria used to categorise the homes into various classes were based on the people's living standards, housing, and access to basic vital utilities such as potable water, electricity, and toilet facilities, among others. A total of 100 families were selected from three communities: Sawaba, a low-income neighbourhood (LIN), Asokore Mampong, a middle-income neighbourhood (MIN), and Bouban Residential Area, a high-income neighbourhood (HIN) in research area as depicted in Table 1.

After thoroughly sensitising residents on the study aims, each of the participating families was given three different plastic bags/ bins to sort the waste into it as a similar method of waste segregation from a source of generation adopted by (Miezah et al., 2015; Monney et al., 2013; Oduro-Kwarteng et al., 2016).

Table 1 Number of households selected from the socioeconomic areas

S/N	Class of residential area	Residential area	Number of samples
1	LIN	Sawaba	34
2	MIN	Asokore mampong	33
3	HIN	Bouban	33

Table 2 Collection of waste samples from households

Plastic bag/bin	Colour description	Content description
Black		Organics
White		Plastic, paper, and cardboard
Yellow		Glass, metals, textiles, leather, and others

The determination of the sample size, and the number of households required is sampled randomly and purposively within the stratified socioeconomic neighbourhood.

2.3.1 Sampling and Collection of Waste

The households selected through stratified random and purposive sampling were educated on waste sorting and separation for one day. During this session, a three-way separation method was introduced, involving categorisation into Organics, Plastics and cardboard, and "Other wastes", as detailed in Table 2. The importance of the survey was emphasized to encourage full participation. Feedback was collected from households to assess their understanding of the sorting process. This feedback included verbal responses and questionnaire completion to gauge comprehension of waste separation into the specified categories. Each household was provided with three designated waste bins or polythene bags labelled accordingly ('Biodegradable waste, Plastics' and 'Other Wastes'). Waste from each household was collected daily or every two days per week for four weeks. Collected waste was further sorted into different fractions to determine its composition.

Households were assigned numbers and given one each of black, white, and yellow plastic bags/bins respectively. The waste was divided into the desired categories using the proper coloured bags/bins. The household waste was collected for one to two days for four weeks, taking into account the probable differences in waste production rates on different days of the month.

2.3.2 Sorting and Sampling

The three bins were retrieved from households either daily or every two days per week, and sent to a designated site for further sorting, taking, and recording measurements. The weight of the waste was the key parameter measured with the number of waste generators. Using a measuring scale, the weight of each of the three plastic bags from individual households was calculated and recorded. The total daily waste generated per household for each waste type was represented by the combined weight of the three plastic bags.

The groupings into which the waste stream was sorted during the waste characterisation research, according to Zakarya et al. (2020) and Miezah et al. (2015), were determined by the study's goal. Organic wastes, plastics, paper, glass, and metal were among the principal categories sorted from the waste stream. In this regard, depending on the goal of a specific study, a small number of key categories (also known as main components) and a large number of secondary, tertiary, and other categories (subcomponents) are used.

2.4 *Physical Composition of RSW Analysis*

The residential solid waste from households was sorted into various compositions and analysed by weight and percentage composition, following the processing methods described by (Robert Ekwule et al., 2020; Nwoke et al., 2020; Oduro-Kwarteng et al., 2016) for separating unprocessed waste (Table 3). The waste was collected using a tricycle, wheelbarrow, and wheel dustbin from residences and transported to the main sorting centre, where a YD Digital and Pricing Computing Scale (0.005 kg–40 kg) was used to weigh the sorted fractions. A plastic sheet was laid on the ground to facilitate further separation and weighing, ensuring compliance with effective sorting by participating households. The segregated waste into sub-fractions was further sorted and analysed by percentage composition and weighed, as described by (Nwoke et al., 2020;) and the ASTM D5231-92 (Standard Test Method for Determination of the Composition of Unprocessed Municipal Solid Waste) scheme developed by the American Society for Testing and Materials (ASTM, 2008), the segregated wastes by respondents were further sorted into several sub-fractions and examined by their percentage composition.

Table 3 Composition of MSW and their sub-fraction for further sorting

Major fractions	Sub-fraction
Organic waste	Yard waste (grass trimmings), food waste, animal droppings, and wood
Plastics	PVC, HDPE, LDPE, PET, PS, PP, etc.
Papers	packaging/cardboard/office print/sheet/newsprint and tissue/diaper
Metal/can	Cans/tins, scrap
Rubber and leather	
Textile	Rags and cloths
Glass	Plates, cups, bottles, etc.
Inert	Rocks, sand, ash, etc.
Others	leather, foam, and other miscellaneous

The composition of the various portions of various types of waste were ascertained using Eq. 2.

$$\% \text{ Composition of separated waste} = \frac{\text{Weight of separated waste}}{\text{Total mix waste sampled}} \times 100 \quad (2)$$

2.5 Determination of Moisture Content

The moisture content of the waste sample from each of the three income neighbourhoods was determined using the method outlined by (Nwoke et al., 2020; Zakarya et al., 2020). This involved heating the waste in an oven at 105 °C for 12 h until a constant weight was achieved. The moisture content was calculated as the difference between the initial weight before drying and the final weight after drying refer to (Eq. 3). The moisture content of biodegradable waste from different socioeconomic income areas was also measured.

The percentage moisture content was calculated using Eq. 3:

$$\% \text{ Moisture content} = \frac{\text{IW} - \text{DW}}{\text{IW}} \times 100 \quad (3)$$

where:

IW = Initial weight of the sample as delivered.

DW = Weight of the sample after drying at 105 °C.

2.6 Determination of Calorific Value

2.6.1 Sample Preparation for Bomb Calorimeter Testing

A combustible portion of MSW was employed for calorific value analyses of the sample. After separating non-combustibles, a 500 g sample of combustibles was obtained. The samples were air-dried for a week and desiccated in an oven for 24 h at 105 degrees Celsius (°C) to remove moisture. The Bomb Calorimeter analysis was done at the KNUST Chemical Engineering Laboratory-Kumasi, where the dried samples decreased in size with a blender, pestle, and mortar and sieved with a 1-5 mm mesh to achieve fine aggregates for combustion.

The calorific values of each sample (food waste, textile, plastic, paper, and miscellaneous) were determined in duplicates and evaluated. The samples were burned in a confined space in pure oxygen, and the energy released was measured as the temperature of the bomb and its surroundings increased.

2.6.2 Samples Preparation

Before being analysed using the Parr 6400 Bomb Calorimeter, the sampled wastes were shredded, pounded, and ground into a fine powder with a pestle and mortar. Combustion is used to determine the heat level of various waste components. The calorific value (CV) of the waste composition was displayed in as table, with different waste components arranged horizontally to their calorific value as indicated in Table 9.

2.6.3 Procedure for Determining the Calorific Value

United Kingdom Company Limited's oxygen bomb calorimeter (SCIMED Scientific & Medical Products LTD, model Parr 6400) was used for calorific analyses. The bomb calorimeter was first pro-tested (calibrated) with Benzoic acid to check its readiness and effectiveness at the maximum operating energy value of 26.454 MJ/kg. Solid samples weighing approximately 0.3 g were prepared out of bulk samples shredded and ground into smaller particles before being placed inside the bomb's metallic cup. The sample material was fed through a 10 cm fuse thread of filament. The fuse thread concentrated heat on the sample and functioned as a solid support for keeping the sample in place while handling the bomb before it was ignited.

The bomb head was placed on the cup, which was then pushed into the body sleeve and secured with a screw cap. The bomb in the bench socket was set and the screw cap tightened tightly with the octagon wrench included with the calorimeter after these pieces were clamped together by hand. During this operation, the setup was carefully handled to ensure that the sample was not disturbed. The volume of water supplied to the bath was sufficient to cover the top of the bomb. On the oxygen filling connection, the needle valve and pressure relief valve were closed, and then the oxygen tank valve was opened by a quarter turn from the closed position.

By squeezing the coupling into the valve cap, the filling hose was connected to the gas inlet tube on the bomb head. The connection was pulled down until it firmly rested against the valve cap's collar. As the pressure in the bomb and connecting hose rose, the filling connection's valve slowly opened, and the gauge recorded it. The control valve was instantly closed when the pressure reached 2–3 MPa, and the pin wrench was used to close the valve cap on the inflow tube. The filling connection's relief valve was utilized to release the filling hose's residual pressure.

The pin wrench was withdrawn and the filling line was lifted from the bomb head once the pressure gauge returned to zero. The initial temperature readings were taken when the bomb calorimeter was opened and allowed to stabilize for roughly 2–3 min. The sample was started by pressing the ignition unit's button. Following that, the chart showed a definite temperature spike, indicating that the sample had burned.

The analysed energy potential of the sample was then transferred into a computer system for printout or copied manually.

2.7 Data Analysis

Field observations, mini-questionnaire interviews, secondary data, and key informant interviews were used to gather neighbourhood data for the study. In contrast, data were analysed using Microsoft Excel and Parr 640 Calorimeter.

2.7.1 Rate of Waste Generation

The per capita waste generation rate was computed for each study area using the total amount of waste collected from each household and the number of persons in the households. The number of persons in each household was calculated from the responses provided in a questionnaire survey conducted before the start of the project. The per capita generation is usually measured in kg/capita/day. This would be determined as per the mixed or the total waste collected in a day and the separated fractions using Eq. 4.

$$\text{Waste generation per person}\frac{\textit{Weight of RSW generated at household}}{\textit{Total household size} \times \textit{Total number of generation days}} \quad (4)$$

2.7.2 Composition of Waste generated in the Selected Neighbourhoods

The characterisation of household waste in this study was limited to physical appearance. The common waste components in the waste stream were classified as; organic, plastics, paper and cardboard, rubber, glass, textile, metals, and others (inert, residues, miscellaneous). The composition of the household waste stream created within the municipality fluctuates greatly between households and settlements and between different income categories as depicted in Table 4. Organic, paper, cardboard, plastic, and glass, among other materials, contribute most of the content. Residential solid waste (RSW) typically includes materials like paper and paperboard, biodegradables, metals/cans, plastics, textiles/clothing, and others (UNEP, 2013). Therefore, the RSW collected in this study was examined. The waste generated consisted of items used and discarded by the residents. A total sample weight of 5343.24 kg was measured. Table 4 presents the comprehensive composition of household solid waste sampled across the study period in the three selected income communities.

2.7.3 Source Separation Efficiency (SSE)

The ability of participating families to accurately separate and sort waste components into their designated coloured bags is measured in this study's source separation

Table 4 Waste Generation per Income Groups

Materials	(% By mass on wet basis)			
	LIN	MIN	HIN	Average
Organic	72.5	67.8	63.7	68.0
Plastic	9.1	15.7	20.6	15.4
Paper & cardboard	3.3	6.7	8.9	6.3
Metal	2.6	4.9	7.5	5.0
Glass	0.3	0.9	1.2	0.8
Textile	0.8	3.0	2.2	2.0
"Others"	2.7	2.3	1.9	2.5
Total	–	–	–	100.0

efficiency and as an avenue for stakeholders to consider employing source sorting and separation at the waste generation points. As defined in Eq. 5, it is the proportion of waste correctly divided by households, as determined by the weight of waste in each bag (Oduro-Kwarteng et al., 2016).

$$\text{SSE} = \frac{\textit{Weight of sorted waste in the right bin}}{\textit{The total weight of all waste separated into the bin}} \times 100 \tag{5}$$

Initially, a survey was run to measure the preparedness of nominated families to participate in the study. Following this, participants were requested to sort and separate their waste using a three-way sorting system that categorized waste into biodegradable, plastics, and others. The level of compliance with the separation and sorting process was assessed by computing the percentage of appropriately sorted waste in respectively provided polythene relative to the overall weight of waste in the same polythene.

The willingness of participating families to accurately sort waste fractions into their designated labelled bags and containers is measured in this study's source separation efficiency.

3 Results and Discussion

3.1 Characteristics of Households

A total of 100 households, comprising 34 from the low and 33, respectively, from the middle and high-income areas within the study area, were recruited in this study. Table 5 shows the total respondents of 403 who participated in the exercise. The average household size was approximately 4.0 persons.

Table 5 Number of Households Selected from the Socioeconomic Areas

Neighbourhood status	Community	Household size	Waste sample (Kg)	Per capita generation (Kg/day)
LIN	Sawaba	123	1184.736	0.344
MIN	Asokore Mampong	146	1978.592	0.484
HIN	Bouban	134	2179.912	0.581
		403	5343.240	0.470

In this study, communities such as Sawaba, Asokore Mampong, and Bouban were stratified, purposively, and randomly sampled as representative of low, middle, and high-income neighbourhoods.

3.1.1 Waste Generation Rate

Sawaba was classified as a low-income residential area in the municipality. The sampled total household size was 123, and the total waste generation was 1184.736 kg from a sample size of 34 households. The per capita waste generation rate for the low-income residential area was 0.344 kg/capita/day, and the average household size was 3.6 persons.

Asokore Mampong was classified as a middle-income residential neighbourhood. It sampled a total household size of 146 and a total waste generation of 1978.592 kg from a sampled size of 33 households. The per capita waste generation rate was 0.484 kg/person/day and the average household size of 4.4 persons.

Bouban community was classified as a high-income class residential area. It had a total household size of 134 sampled and a total waste generation of 2179.912 kg from a sample size of 33 households. The waste generation rate of this residential area was 0.5178 kg/person/day with an average household size of 4.1 persons. The average per capita waste generated in the study area is estimated 0.470 kg/capita/ day. A separate study by (Miezah et al., 2015) noted waste generation variation among socioeconomic areas have been reported by as; 0.56 kg/capita/day in the high-class zones, middle-class zones of 0.49 kg/capita/day, and low-class areas, 0.47 kg/capita/day for the Regional capitals in Ghana. Ansah (2014) in a related study amongst socioeconomic classes in Tarkwa municipal one of the highest economic areas of Ghana documented a higher waste generation rate of 0.7970, 0.9920, and 1.0527 respectively for low, middle, and high-class areas with an average of 0.92 kg per capita per day which was higher than 0.344, 0.484 and 0.581 with an average of 0.470 reported in this study. The low trend documented in this report compared to other findings could be attributed to the influence of the economic development of these areas and vice versa.

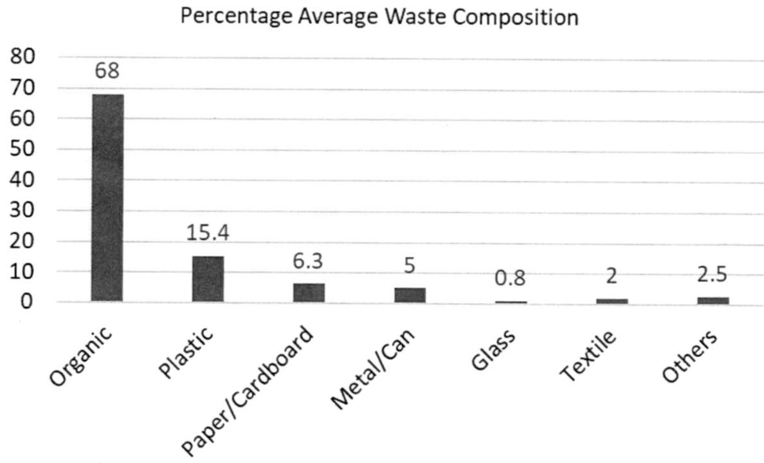

Fig. 2 Average per capita daily generation rate of household waste fractions in the Municipality

The high waste created in high-income communities is likely to be influenced by the socio-economic development of these areas. Authors such as Samuel et al., (2016) hinted that the per capita waste generation for different cities in Nigeria such as Lagos (0.63 kg/capita/day), Kano (0.56 kg/capita/day), Ibadan (0.51 kg/capita/day), Kaduna (0.58 kg/capita/day), Port Harcourt (0.60 kg/capita/day), Onisha (0.53 kg/capita/day), Nsukka (0.44 kg/capita/day) and Abuja (0.66 kg/capita/day) been obtained. Al Seadi et al. (2013) quoted 0.49 kg/capita/day for average Nigerian communities with household and market centers. Therefore, the per capita generation rate as revealed in the study is 0.470 kg/capita/ day inline with previous studies. Figure 2 depicts the sample size of the household population of the selected income areas and their generation rates during the period of study.

Other findings by Miezah et al. (2015) documented that, except for Tamale, the average domestic waste generation rate in the metropolises in Ghana was high, at 0.72 kg per person per day. Metropolises produced more garbage, 0.63 kg per person per day on average than 0.40 kg per person per day for municipalities and the least 0.28 kg per person per day in less developed districts. Combustible waste (organics 68%, papers and cardboard 6.3% plastic 15.4%, and textiles 2%) accounted for 91.7% of daily waste generation. In comparison, recyclables or non-combustible components (glass 0.8%, metals 5%) accounted for 5.8% and some marginal waste in "others" 2.5% contributed to total incombustible elements in the waste stream.

According to Robert Ekwule et al., (2020), analysis of the field data gathered from Nigeria Makurdi Metropolis indicates samples of SW generated were in order as follows; 81.30% (organic waste) > 5.86% (ash/sand) > 2.78% (Paper) > 2.03% (Metal/Tin) > 1.84% (Bag/Shoes) > 1.74% (Leather) > 1.47% (textiles) > 1.24% (Plastic) > 0.41% (glass/ceramics) and further stated per capita mean generation rate to be 0.45 kg/ capita/ day.

In a related study, Zakarya et al. (2020) discovered that the generation rate was highest in the first-class income areas, at a rate of 0.76 kg/capita/day, compared to the second and third-class income areas, which had rates of 0.66 kg/person/day and 0.69 kg/person/day, correspondingly, in the Sekondi-Takoradi Metropolis with a 0.70 kg/capita/day mean generation rate. There were no substantial changes in the generation rate per capita among the three socioeconomic income areas at the 5% significance level. The average waste generation rate from this study, 0.470 kg/capita/ day, is slightly lower than the per capita generation of 0.52 ± 0.26 and 0.65 ± 0.27 kg/per capita/ day for the low-income and middle income community respectively for metropolitan and municipal areas reported by (Oduro-Kwarteng et al., 2016).

Evaluation of waste generated in developing countries recorded a range of 0.4–0.6 kg/person/day (Chandrappa and Das 2012), and the values obtained in this research fall within this range. The variations in generation rates among the income groups indicate that household income level influences the waste generation rate. This could be attributed to the lower income level's relatively low standard of living, as the amount of waste generated upsurges with lifestyle changes that influence lifestyle towards goods and services and consumption, which is associated with the household disposable income level (Hoornweg & Bhada-Tata, 2012). They further noted that, in terms of per capita waste generation, North America had a much higher rate (2.21 kg/capita/day) than Central Asia and Europe (1.18 kg), Caribbean and Latin America (0.99 kg), Middle East and North America (0.81 kg), East Asia Pacific (0.56 kg), South Asia (0.52 kg), and Sub-Saharan Africa (0.46 kg). The characterization of SW from the Indian region of Tricity studied by (Rana et al., 2018) denotes that the physical composition of the MSW from Mohali, Panchkula and, Chandigarh, indicate a high percentage of 50% (organic matter) and 29% (inorganic fraction) in the waste stream of various socio-economic groups.

In sub-Saharan Africa, (Orhorhoro and Oghoghorie, 2019), noted that per capita waste production is mostly low in the province, but spans a wide range, from 0.09–3.0 kg per person per day, with a mean of 0.65 kg per capita per day which was similar to the average production rate discovered in this research of 0.47 kg daily. Hence, the low trend documented in this report compared to other findings could be attributed to the influence by economic development of these areas and vice versa.

3.1.2 Composition of Waste generated in the Neigbourhoods

The characterisation of household waste in this study was limited to physical appearance. The common waste components in the waste stream were classified as; organic, biodegradable materials, plastics, paper and cardboard, rubber, glass, textile, metals, inert or residues, and miscellaneous. The household waste stream within the municipality fluctuates greatly between households and settlements and between different income categories. As presented in Table 4 is the percentage rate of waste generation of solids for each of the nominated income neighbourhoods. High-income areas consume more packed products which gives rise to a higher percentage of non-organics like metals, glass/ceramics, and plastics as a higher percentage of inorganic materials were influenced by their income rate.

The study revealed that the major composition of residential solid waste is organics constituting about 68%, plastic 15.4%, papers, and cardboard (6.3%), glass (0.8%), metals (5%), textiles (2%), and "Others" 2.5%). In this study, the main fractions recorded across the three income areas were organic (72.5%, 67.8%, 63.7%), plastics (9.1%, 15.7%, 20.6%), and paper and cardboard (3.3%, 6.7%, 8.9%) for the respective low, middle, and high-income areas. The average organic composition amongst the income groups was 68% for all the classes. Similarly, organic waste was highest in all the income areas followed by plastics and paper. The high-income area had the lowest fraction of organic waste, 63.7%, whereas the highest was recorded in low-income community 72.5%. Organic waste is the most prevalent component of household waste in the composition of household waste in developing nations such as Ghana.

In a similar study by Hidalgo et al. (2019) who estimated household solid waste (HSW) composition and generation in Ecuador, Guayaquil revealed that about 72,6% SW is organic which is slightly above the average of 68% noted in this study.

According to Wilson et al. (2015), the organic fraction of MSW is higher in low-income countries, where organic waste typically constitutes 50% to 70% of the total waste, compared to high-income countries, where it accounts for 20% to 40%.

However, the average paper and cardboard waste composition in the waste stream was 6.3% in the study area. This high proportion of paper and cardboard discovered in this study could be attributed to the commercial activities that dominate the municipality.

Contrastingly, plastic waste was second highest in the component recorded 15.4% in the study. Metals, textiles, leather, and rubber percentage fractions were minimal in all the income communities. Inert also formed about 2.5% of all the waste in all the income areas with the highest recorded in the low-income neighbourhoods compared to what has been discovered in high and middle-income neighbourhoods. The moderate inert materials in the household waste from low-income neighbourhood area could be soil particles originating from untiled premises and ash resulting from use of wood and charcoal for cooking. According to the Global Waste Management Outlook (2015), Paper waste tends to correlate with income levels, making up 23% of MSW in high-income countries, 19% to 11% in middle-income countries, and 7% in low-income countries which agrees with discovery in this study. Plastic waste levels are relatively high across all income levels, ranging from 8 to 12%, and do not show as much variation based on income. 'Dry recyclable' materials, such as metals, glass, and textiles, represent 12% of MSW in high-income countries, 12% and 9% in middle-income countries, and 6% in low-income countries. Household hazardous waste (HHW) is estimated to constitute less than 1% of MSW across all income levels, but its presence complicates certain waste management options.

However, Ozcan et al., (2016), reported similar values where the organic content was found to be highest in low-income communities. 65.41, 57.69, 61.16 59.13% for low, middle, high, and downtown communities in the Kartal district, situated on the southwest part of the Kocaeli peninsula, which is found on the East coast of the jurisdiction of Istanbul. This is slightly lower than 72.5,67.8 and 63.7 revealed in this study. These differences could be attributed to factors such as seasonal change and feeding habits of people, differences in evaporation of water within solid waste aim high temperatures through the summer period, and higher fruit feeding in these calendar months.

In similar study (Otoo, 2013a), documented high percentages of putrescible waste in all the towns, with 69%, 54%, and 49% representing Fumesua, Ejisu, and Kwamo in a jurisdiction of study closer to the study area of this work. The high organic content amongst the low-to-middle and high-income neighbourhoods was attributed to the mode of feeding where people within low income normally prepare most of their food from local sources in their houses, while the people within the intermediate to higher income circle spend less time at home during the weekdays and patronize the street more.

According to Oduro-Kwarteng et al. (2016), there was a lower organic content amongst the lower income groups compared to the other income groups, and they attributed this to a possible diversion of organic waste in this income setting to aid as feedstuff for livestock. The plastic waste production in all the classes was analogous to 9.8%, 13.2%, and 23.2% respectively for the low, middle, and high-income classes, with a daily average generation rate of 15.4%. This was contrary to what has been discovered by Asare et al., (2020) who discovered a high compositional waste sample of plastic waste in their study area of 42.96% and attributed the large quantities of plastic waste as a result of the economic influence of household dynamics on the generation of solid waste.

The components of the waste generated were sorted manually and weighed and the results of the seven main waste fractions against household income levels are presented in Fig. 2. The varying composition of plastic waste from different areas from this study is higher that 1.24% discovered by (Ekwule et al., 2020) who studied per capita SW production and classification in Nigeria, Makurdi Metropolis, while similar to 16% reported by Ansah, (2014).

3.1.3 Source Separation Efficiency

The results of the average separation efficiency of households over the study period in the three income communities are presented in Table 6. The source separation efficiency in this study assesses the ability of participating households to correctly sort their waste component into designated bins. The results showed that separation efficiencies for all categories were higher in the middle-income group than in the low-income group. This may be due to a higher level of education in the middle-income group. However, in both groups, it was noted that separation efficiency was highest for organic waste with plastics and paper components having the lowest separation efficiency. It can be anticipated from the high-organic separation efficiency that high recovery of organic waste can be achieved if they are collected separately. The highest separation efficiencies of greater than 92, 87, and 79 percent, respectively for organic, paper, and plastics were 94, 81, and 73 percent while "others" were 86, 69, and 44 percent among the low, middle, and high-income areas. High wastes separation efficiency in this research could be attributed to the participating households were educated before the study. This underscores the importance of intensive education in improving source separation programmes.

The willingness of households to separate waste at source was reported as 92.4, 94, and 86% on average in low, middle, and high-income areas respectively, according to the findings in this study. This aligns with (Anarfi et al., 2013) findings which recorded 73.3 and 86.7% willingness in low and middle-income areas, and with Oduro-Kwarteng et al. (2016), where over 70% of respondents were willing to separate household waste at the source. The least willingness to separate waste was observed in high-income neighbourhoods, where residents cited concerns about potential extra costs and time delays as reasons for their reluctance. The success of source separation of waste depends on community participation, which can be gauged by their willingness to separate waste at source. However, this measure may not fully reflect actual participation due to possible changes of mind or the influence of neighbours' actions. Separation efficiency for organic waste was 92, 87, and 79% in low, middle, and high-income areas, respectively, while the efficiency for sorting cartons, papers and plastics of waste was 94, 81, and 73% in the same order. Separation effectiveness remained high from day 1 to day 20 of the study, especially for organic waste, with most sorting days achieving above 90% effectiveness across all income areas, except for occasional dips below 90% but above 85% in middle- and high-income areas. For other waste types, the low income area achieved the highest separation effectiveness, with middle income area ranging from 86 to 69%, and high-income areas recording between 56%. While separation effectiveness fluctuated slightly as the study progressed, ongoing education helped maintain participation across all socioeconomic groups. Lower separation effectiveness for other waste types in high-income areas was partly due to busy schedules, which limited household members' availability for education on sorting and separation processes. Additionally, some residents felt that a motivation package should be offered to encourage participation.

As indicated in Table 6, separation efficiencies were higher in the low and middle-income strata for all categories compared to the high-income stratum. However,

Table 6 Separation efficiency for the source separated wastes

Income	Efficiency of separation		
	Black bag (organics)	White bag (cartons & plastics)	Yellow bag ("others")
Low	92% pure	94% pure	86% pure
	8% impurity	6%	14% impurity
Middle	87% pure	81% pure	69% pure
	13% impurity	19% impurity	31% impurity
High	79% pure	73% pure	56% pure
	21% impurity	27% impurity	44% impurity

findings revealed that separation efficiency was highest for organic components in the low and middle-income group categories (92, 87%), followed by "plastics and cardboard" wastes with separation efficiency of (94, 81, and 73%), and "Others" having the least separation efficiency of (86, 69, and 56%), with overall organic waste separation efficiency among the income levels being highest at 86%, and plastics while "Others" was lowest at 70.33%.

These findings were slightly higher than what was discovered by (Oduro-Kwarteng et al., 2016) who examined source separation and recycling potential of MSW in Ghana's second-largest city, Kumasi, stated the maximum source separation efficiency (SSE) above 70% in the organic fraction, 69% inorganic and residue was over and 60% paper and plastics respectively. The high SSE achieved in this study, could be ascribed to the fact that contributing households, were dedicated, regularly educated, and motivated with incentives in the form of free bins/ bags. This highlights the effect more education and incentives could have on improving source separation programs and schemes, which are supported by (Oduro-Kwarteng et al., 2016), that the level of source separation attained in any ISWM scheme is a function of mutually the ability and particularly the motivation of householders. The SSE and level of compliance were generally more advanced in low and middle-income neighbourhoods than in high-income neighbourhoods. There was low diverting of organics among the participants compared to findings made by Oduro-Kwarteng et al. (2016) that usage of putrescible wastes such as plantain and cassava peels as feed for animals in low and middle-income areas could explain why low-income households have a smaller proportion of organics in their waste stream than middle-income families, which is contrary to the findings in this study.

A similar evaluation by (Miezah et al., 2015) discovered that domestic separation of waste proficiently at an average of 80%, but regarding sorting into the bin labelled organic, 84% efficiency was obtained, while 76% effectiveness was obtained for sorting into the bin labelled "other waste" which was similar to 86 and 70% in this study for both organic and "other" wastes.

Effective source separation of waste needs the participation of communities, and this could be measured by the willingness of people to sort waste at the source. This measurement might not be the same as the actual number participating in the segregation of waste at the source due to a change of mind or the mere fact that neighbours are separating their waste (Zakarya et al., 2020).

There was a slight fluctuation in the effectiveness of the separation as the survey proceeded, but constant education and motivation of some segments of households sustained the separation across the socioeconomic areas. The low separation efficiency noticed among participants could be attributed to the fact that in most households, not all household members were met during the education hence it has led to a low level of understanding of the separation process.

The outcomes from the evaluation enumerated above, although limited to this study, indicate the desire of homes in developing nations to sort their waste at the generation point is averagely high, and could be achieved through motivations and careful arrangement design, considering local conditions, household wastes source separation in developing nations might be successful. As stated by (Oduro-Appiah

Table 7 Moisture content of RSW

Neighbourhood/component	LIN	MIN	HIN
% Moisture content	63.50	57.40	51.70

et al., 2013b) sorting putrescible waste from MSW stream signifies a chance to decrease the amount of waste going to developing nations' landfills by nearly 50% by weight and further added that source separation upsurges the worth of MSW and encourages cost recovery structures adding to extending the lifecycle of the landfills.

3.1.4 Determination of Moisture Component

The moisture content of the organic wastes from all three income neighbourhoods of residence ranged from 51.70 to 63.50%, with waste from Sawaba having the highest moisture content and Bouban the lowest (Table 7). Organic waste contributes higher amounts of moisture to the waste stream. Although moisture content is important in bioconversion process, it affects the weight making the collection and transportation processes herculean tasks. Often, moisture and inert content in waste increase the weight of waste that gets to the disposal sites therefore increases the leachate flow rate from the disposal site. From the study, two of the three socioeconomic areas had an average moisture content in their waste 57.73% slightly below 64.3-78% revealed by (Ozcan et al., 2016) in the Kartal district of the province of Istanbul, Turkey which is in the optimum region for bacteriological activities. The optimal moisture content often ranges from 50 to 60% for most metabolic activity to occur (Zavala and Funamizu, 2005). Below this range, microbial activity decreases and biological conversion processes are slowed. The moisture contents obtained in this study compared fairly with work conducted by (Zakarya et al., 2020) as 59.05, 61.87 and 44.06% among low, middle, and high income neighbourhoods respectively.

3.1.5 Calorific Value of RSW

The calorific value of the different waste components is presented in Table 8. From Table 8, it was clear that plastic constituents had the maximum calorific value with a mean calorific value of 33.979 MJ/kg. The plastic consisted of a mixture of 50% low-density and high-density plastics. This means that a high amount of energy would be required to convert the sample into energy, which could also reduce calorific value of the plastic sample. This was similar to the findings made by (Nwoke et al., 2020a) where plastic constituents had the maximum calorific value with a mean calorific value of 26.948 MJ/kg. In a similar study by Rezaei et al. (2020), plastics component had the highest heating potential of 46.5 MJ/kg.

The textile ranked second with a mean value of 18.340 MJ/kg next by "others which comprised of foam, leather" among others recorded 17.219 MJ/kg, and organic and paper with values of 15.835 and 13.229 MJ/kg correspondingly. Other findings by Kamuk et al. (2013), discovered that the average calorific value of organic, plastic, textile, paper, and “others”, were 3, 35, 19, 16, and 11 MJ/Kg respectively, which was similar to findings in this study of 33.9789 MJ/Kg for plastic and 13.2286 for paper and cardboard respectively but contrary to what has been found for organic 3 against 15.8 MJ/Kg respectively. In other studies (Kuleape et al., 2014), cited the CV of Ghana’s solid waste ranges from 14 MJ per kg in low income settlements to 20 MJ/kg in high income settlements as reported by the Ministry of Power which is similar to the findings of this research of 19.72 MJ/Kg as mean CV.

Metals, glass, and “others” require high heating energy. Unfortunately, the oxygen bomb calorimeter available cannot be used to determine their calorific value.

The Calorific value of components of the waste generated was analysed with a Parr 640 Calorimeter as illustrated in Table 8 results of five samples analysed.

The Table 9 presents average calorific value in MJ kg^{-1} and the daily generation of waste components such as organic, plastic, paper, textile and Other was estimated as 227,738.80, 51,576.14, 20,429.51, 3349.10, and 8372.75 respectively. Multiplying the calorific values in MJ kg^{-1} (Table 8) with the daily waste generation in kg day^{-1} (Table 9) gives the calorific values in MJ day^{-1} for these components. The CV for the components: organic, paper, plastic, textile, and other were obtained as 360,068.90, 1,752,500.50, 270,253.82, 164,208.19, and 144,167.87 MJ day^{-1} respectively.

Using the conversion factor of 1 kWh = $3.60*10^6$ J = 3,600 kJ = 3.6 MJ. This translated into the overall daily gross energy from MSW in Asokore Mampong municipality is about $7.5*10^4$ MJ per day. Therefore, the electric energy that could

Table 8 Calorific value of MSW

Components	CV1 (MJ/Kg)	CV2 (MJ/Kg)	Mean CV (MJ/Kg)
Organic	15.6667	16.0033	15.8350
Plastic	35.2293	32.7284	33.9789
Paper & Cardboard	13.2567	13.2005	13.2286
Textile	18.3395	18.3395	18.3395
Other	17.2449	17.1925	17.2187

Source Field date analyses 2021

Table 9 Daily quantity of waste generation and energy estimate

Components	% Composition	Weight (kg)	Daily energy estimate (MJ/day)
Organic	68.0	227,738.80	360,068.90
Plastics	15.4	51,576.14	1,752,500.50
Paper	6.3	20,429.51	270,253.82
Metals	5.0	16,745.5	0
Glass	0.8	6,698.20	0
Textiles	2.0	3,349.10	164,208.19
Other	2.5	8,372.75	144,167.87

be generated from MSW in Asokore Mampong municipality is 27375000.0 MJ per year.

It could be deduced from Table 8 that, the mean energy level of RSW is estimated to be 19.72 MJ/Kg which was higher than the recommended standard, according to the ISWA Report (2013), the minimum energy content of MSW needs to be 7 MJ/kg to ensure sustainable energy generation which should not be less than 6 MJ per kg with the annual waste flow not less than 50,000 metric tons, and per week variations in waste availability to plant not surpassing 20 percent to justify incineration. Plastic, paper, cardboard, and organic have the greatest energy per kilogram. The energy produced mostly by plastic constituents in the waste stream is considerably large, demonstrating that additional energy may be retrieved from waste containing a higher percentage of plastic fractions. From Table 9, the daily generation of organic, plastic, paper, textile, and Other was estimated to be 227,738.80, 51,576.14, 20,429.15, 3349.10, and 8372.75 kg/day, respectively.

Determining the overall CV by multiplying the average energy contribution in MJ/kg with the daily waste generation of the various components in kg/day as in Table 9 gives contribution of the various CV in MJ/day. The gross calorific values for organic, plastic, paper, textile, and Other as obtained in Table 9 were obtained as $7.5*10^5$ MJ/day.

The gross calorific value obtained is low compared to that obtained from cities in advanced countries. The waste in undeveloped nations contains so much moisture from the organic origin that has to be fortified to maintain combustion compared to high plastic and paper composition in developed nations. For instance, Ozcan et al., (2016) documented 601,543.725 MJ/ kg in a similar study in the Istanbul province of Kartal district similar to $7.5*10^5$ MJ/day energy production in this study.

In other studies, by Moya et al., (2017) in Quito, the capital of Ecuador revealed MSW of composition of 69.31% organic and 30.69% inorganic has the energy generation potential which is favourable for biochemical and thermochemical methods. The energy potential of thermochemical has been estimated at around 5970 kWh per ton of MSW. In comparison, energy capacity of biochemical processes was evaluated at 62 kWh per ton of MSW, which could aid in determining the best methodology for producing energy from solid waste.

The above analysis and comparisons show incineration has higher option to recover energy in the municipal waste stream. In another study (Rockson et al., 2017) discovered that the CV of the waste in Accra regarding the characterisation of waste at the final disposal site gives an energy value range, 630–2,460 kcal/kg (6,822.85–10,297.068 kJ/Kg) and further stated that the compositional analysis showed technically, energy recovery from the waste is possible due to the presence of moderate to high energy value materials. This was similar to the average calorific value of 19.27 MJ/Kg revealed in this study which is slightly lower than an average of 19.93 kJ/Kg noted by (Mushimba, 2018) and high calorific value of 21.43 kJ/Kg and low value of 18.43 kJ/Kg for different periods in the year.

3.1.6 Forecasting Solid Waste Generation in the Municipality and Energy Estimate

Various factors drive the quantity of waste produced in our communities, towns, and cities according to Nwoke, et al. (2020). Generally, it is agreed that the following are important factors to consider when projecting future population for estimating waste amounts are;

The current population of the area under study,
Rate of Population Growth, and.
Waste generation per capita.

Other elements that contribute to the development of municipal waste are assumed to be steady in principle; only population dynamics remain considered relevant Nwoke, et al. (2020). The exponential growth formula is used for long-term population projection as depicted in Eq. 6.

$$Pn = [Poe]^{rn} \tag{6}$$

where Pn is the population in the dominant (estimate) period, P_O is the base population producing current waste amounts, r means yearly city population growth rate (8.68), n the projection period of 11 years, and e is 2.718.

Using Eq. 6 gives Pn = P 2021 = 304,815 $(2.718)^{0.0868*11}$

Pn = 791,955.367≈. 791,956.

The 2021 projection population for the study area is about 791,956.

This result shows that the average waste production rate in Asokore Mampong Municipality of 0.470 kg/capita/day transformed into an estimated 372.219 tons of waste per day per the current population of 791,956 projected for the study area.

This indicates a per capita waste generation rate of 0.470 and assumes that about 90% of daily household waste is combustible, translating to 334.997 tons. The calorific value is estimated to be 19.72 MJ/Kg daily household solid waste component of MSW.

4 Conclusion

The municipality has a projected population of 791,956 people, with an estimated municipal solid waste generation of 372.219 tonnes daily. This results in a waste generation rate of 0.470 kg per capita per day across all three income neighbourhoods, with the highest waste generation of 0.581 kg per person per day observed in households from the high-income neighbourhoods.

The composition of municipal solid waste (MSW) has been classified into seven categories: organic, plastics, paper and cardboard, textiles, metals, glass, and others. The average composition of this waste, organic constitutes the largest portion at 68%, followed by plastics at 15.4%, while paper and cardboard account for 6.3%. Textiles comprise 2.0%, metals contribute 5.0%, glass represents 0.8%, and the remaining "others" 2.5% comprise various other materials.

The total combustible element in the waste stream made of organics, papers cardboard, textiles, others (leather and rubber), and plastic contributed more than 90%, while less than (10%) proportion of the waste is non-combustible.

The highest source-separated component was cartons, paper and plastics (94%) followed by organic (92%) and the least, "Others" components (86%). This indicated that an effective separation at source program could significantly enhance the waste-to-energy recovery potential of MSW with the high organic, plastic, and cardboard content forming the major combustible wastes in the waste stream and diverging high amount of waste disposed at the landfill.

The calorific values of the major components of Residential Solid Waste were organic, plastic, paper, textile, and paper and cardboard were found to be reliable energy sources. The highest contributing components of CV were plastic, textile, others, organic and paper, and cardboard with their respective values of 33.9789, 18.3395, 17.2187, 15.8350 and 13.2286 MJ/Kg. The average calorific values of combustible contents in residential waste of MSW investigated was reported as 19.72 MJ/Kg as the energy content of household waste content in some communities within Asokore Mampong municipality, which is potentially good for waste to energy recovery through incineration. The gross energy that could be generated from Residential Solid Waste in the Municipality is about 7.5*10^5 kWh per day and 273.932 GWh/year which translates to 35.47 million/year using a conversion factor of 1 kWh to US$0.13.

However, it is recommended that a thorough economic analysis is conducted before deciding on this option.

Acknowledgements This study was partially funded by the Regional Water and Environmental Sanitation Centre Kumasi (RWESCK) at the Kwame Nkrumah University of Science and Technology (KNUST), with funding from the Ghana Government, through the World Bank under the Africa Centres of Excellence (ACE) Project. The opinions expressed in this paper do not reflect those of the Ghana Government, KNUST, and World Bank.

Author Contribution

1. Grant Degbe Dzitse: Writing—Original Draft, Methodology, Formal analysis, Investigation.
2. Helen Michelle Korkor Essandoh: Conceptualization, Supervision, Writing—Reviewing and Editing, Visualization.
3. Lawerence Darkwah: Supervision, Writing—Reviewing and Editing.

Declaration of competing interest The authors of this article have no competing interests.

References

Addo, G. A. & Wahabu, A. (2019) 'Quantification of municipal solid waste-the essential basis in waste management in tamale: A Ghanaian City', *8*(6), pp. 109–114. Available at: https://doi.org/10.11648/j.ajep.20190806.11.

Adipah, S. & Kwame, N.O. (2019). A novel introduction of municipal solid waste management. *Journal of Environmental Science and Public Health. 3*:2.

Adu-Boahen, K., Atampugre, G., Antwi, K. B., Osman, A., Osei, K. N., Mensah, E. A., & Adu-Boahen, A. O. (2014). Waste management practices in Ghana: challenges and prospect, Jukwa Central Region. *International Journal of Development and Sustainability, 3*(3), 530–546. Accessed 5th November 2020.

Agyei, G. Y. (2020). Characteristics of solid waste generated in Techiman, Ghana (Doctoral dissertation, University of Cape Coast). https://ir.ucc.edu.gh/xmlui: Accessed on 17th March, 2021.

Al Seadi, T., Owen, N., Hellström, H., & Kang, H. (2013). Source separation of MSW: An overview of the source separation and separate collection of the digestible fraction of household waste, and of other similar wastes from municipalities, aimed to be used as feedstock for anaerobic digestion in biogas plants. *IEA Bioenergy, 37*, 4–50.

American Society for Testing and Materials. (2008). Standard test method for determination of the composition of unprocessed municipal solid waste, In: ASTM D5231–92–2008, US.

Amo-Asamoah, E., Asumadu, G. & Edwards, D. J. (2020). 'Potential for waste to energy generation of municipal solid waste (MSW) in the Kumasi metropolis of Ghana', *14*(6), pp. 1315–1331. Available at: https://doi.org/10.1108/IJESM-12-2019-0005.

Anarfi, K. P. (2013). *"Source Separation, Characterization and Recycling Potential of Household Solid Waste. A case study of Kumasi, Ghana", Unpublished MSc* (p. 121). Thesis Report.

Ansah, B. (2014). Characterization of municipal solid waste in three selected communities in the Tarkwa Township of Tarkwa Nsuaem Municipality in Ghana (Doctoral dissertation). Accessed on 16th October 2020.

Asare, W. et al. (2020). 'Recovery of municipal solid waste recyclables under different incentive schemes in Tamale, Ghana', Sustainability (Switzerland), 12(23), pp. 1–19. Available at: https://doi.org/10.3390/su12239869.

Asase, M. A. D., & Oduro-Kwarteng, S. (2010). Potential for source separation of municipal solid waste: Case study of Ghana. *Unpublished doctoral dissertation, Department of Chemical Engineering, Kwame Nkrumah University of Science and Technology, Kumasi, Ghana.*

Chandrappa, R., & Das, D. B. (2012). *Solid waste management: Principles and practice.* In: Environmental science and engineering (pp. 48–50). Springer Nature. Accessed 6 November 2020.

Donkoh, S. (2016). Assessment of solid waste recycling as a means of improving environmental quality in Accra Metropolitan Area (Doctoral dissertation, University of Cape Coast). https://ir.ucc.edu.gh/xmlui; Accessed on 11th October, 2020.

EEA. (2016). *Municipal waste management across European countries.* Published on 14th November 2016. Accessed 5th November, 2020.

Ekwule, O., Okechukwu, E., & Ufedo, M. (2020). Per capita solid waste generation and characterization in makurdi metropolis, Nigeria. *Kastamonu University Journal of Engineering and Sciences, 6*(1), 33–40.

Franca, L. S., & Bassin, J. P. (2020). The role of dry anaerobic digestion in the treatment of the organic fraction of municipal solid waste: A systematic review. *Biomass and Bioenergy, 143*, 105866.

Ghana Statistical Service. (2014). 2010 population and housing census, summary report of final results, Accra, Ghana. Accessed on 12th August, 2020.

Global Waste Management Outlook. (2015). United Nations Environment Programme, ISBN: 978-92-807-3479-9

Hidalgo, J., Amaya, J. L., Jervis, F., & Moreira, C., (2019). Influence of socio-economic factors on household solid waste (HSW) generation of the city of Guayaquil, Ecuador. In *The 17th LACCEI International Multi-Conference for Engineering, Education, and Technology: "Industry, Innovation, and Infrastructure for Sustainable Cities and Communities"*. Latin American and Caribbean Consortium of Engineering Institutions.

Hoornweg, D., & Bhada-Tata, P. (2012). What a waste: A global review of solid waste management. *Urban Development Series Knowledge Papers, 15*, 10.

Howes, P., Warren, K., & Task, I.B. (2013). Integration of Thermal Energy Recovery into Solid Waste Management. *Paris. Obtained from:* http://www.ieabioenergytask36. org/vbulletin/attachment.php. Accessed on 10th October, 2020.

Kamuk, B., & Haukohl, J. (2013). *ISWA Guidelines: Waste to energy in low and middle income countries*. Austria.

Kaza, S. & Yao, L. (2018). At a glance: a global picture of solid waste management. Accessed on 1st October, 2020.

Kuleape, R., Cobbina, S. J., Dampare, S. B., Duwiejuah, A. B., Amoako, E. E., & Asare, W. (2014). Assessment of the energy recovery potentials of solid waste generated in Akosombo. *Ghana African Journal of Environmental Science and Technology, 8*(5), 297–305. https://doi.org/10.5897/AJEST2014.1663

Lyida, M. N. (2015). Solid waste management in Enugu Metropolis: Implications for urban administration. *IOSR Journal of Humanities and Social Science (IOSR-JHSS), 20*(1), 18–24.

Mandal, K. (2019). Review on evolution of municipal solid waste man- agement in India: Practices, challenges and policy implications. *Journal of Material Cycles and Waste Management, 21*, 1263–1279. https://doi.org/10.1007/s10163-019-00880-y

Magutu, P. O., & Onsongo, C. O. (2011). Operationalising municipal solid waste management. *Integrated Waste Management Journal, 2*, 1–9.

Medina, M. (2010). Solid wastes, poverty and the environment in developing country cities. *World Institute for Development Economic Research (UNUWIDER) Working Papers*. Accessed 4 Oct 2020.

Miezah, K. et al. (2015). Municipal solid waste characterization and quantification as a measure towards effective waste management in Ghana. Waste Management, 46, pp. 15–27. Available at: https://doi.org/10.1016/j.wasman.2015.09.009.

Mionne, M., Maniragaba, A., & Nshimiyimana, F. X. (2019). *Impact of solid waste management practices on environmental protection case of Kicukiro District*. Accessed 5 Nov 2020.

MLGRD. (2010). National environmental sanitation strategy and action plan (NESSAP). Unpublished, MLGRD, Accra.

Monney, I., Tiimub, B. M., & Bagah, H. C. (2013). Characteristics and management of household solid waste in urban areas in Ghana: The case of WA. *Civil and Environmental Research, 3*(9), 10–21.

Moya, D. et al. (2017). 'ScienceDirect sciencedirect waste-to-energy technologies: an opportunity of energy recovery from Municipal Solid Waste, using Quito—Ecuador as case study. *Energy Procedia, 134*, pp. 327–336. Available at: https://doi.org/10.1016/j.egypro.2017.09.537.

Mushimba, B. C. (2018). Municipal Solid Waste (MSW) characterization for possible waste-to-energy (WtE) conversion in Zambia. *Open Science Journal, 3*(4).

Nwoke, O. A., et al. (2020). Determination of the calorific value of municipal solid waste in Enugu, Nigeria and its potential for electricity generation. *Agricultural Engineering International: CIGR Journal, 22*(2), 86–97.

Oduro-Appiah, K., & Aggrey, B. E. (2013). Determinants of source separation of municipal solid waste in developing countries: The case of Ghana. *Journal of Sustainable Development in Africa, 15*(3), 15–20.

Oduro-Appiah, K., Scheinberg, A., Mensah, A., Afful, A., Boadu, H. K., & de Vries, N. (2017). Assessment of the municipal solid waste management system in Accra, Ghana: A 'Wasteaware 'benchmark indicator approach. *Waste Management & Research, 35*(11), 1149–1158.

Oduro-Kwarteng, S., Anarfi, K. P. & Essandoh, H. M. K. (2016). Source separation and recycling potential of municipal solid waste in Ghana. *Management of Environmental Quality: An International Journal, 27*(2), pp. 210–226. Available at: https://doi.org/10.1108/MEQ-03-2015-0038.

Orhorhoro, E. K., & Oghoghorie, O. (2019). Review on solid waste generation and management in sub-Saharan Africa: A case study of Nigeria. *Journal of Applied Sciences and Environmental Management, 23*(9), 1729–1737.
Otoo, W. (2013) Evaluation of households solid waste generation and disposal: A case study in Ejisu, Kwamo and Fumesua in the Ejisu–Juaben Municipality of Ghana (Doctoral dissertation). Accessed on 13th October 2020.
Ozcan, H. K., et al. (2016). Municipal solid waste characterization according to different income levels: A case study. *Sustainability, 8*(10), 1044.
Rana, R., Ganguly, R., & Gupta, A. K. (2018). Physico-chemical characterization of municipal solid waste from Tricity region of Northern India: A case study. *Journal of Material Cycles and Waste Management, 20*, 678–689.
Rezaei, H. et al. (2020). Pelletization of Refuse-Derived fuel with varying compositions of plastic, paper, organic and wood', pp. 1–11.
Rockson, G. N. K., Acheampong, I. B., & Yanful, E. (2017). Incineration as an option for municipal solid waste in developing countries: the case of greater accra metropolitan area. Accessed 6.10.201. *Sixteenth International Waste Management and Landfill Symposium.*
Sarfo-Mensah, P., Obeng-Okrah, K., Arhin, A. A., Amaning, T. K., & Oblitei, R. T. (2019). Solid waste management in urban communities in Ghana: A case study of the Kumasi metropolis. *African Journal of Environmental Science and Technology, 13*(9), 342–353.
UNEP. (2013). Municipal solid waste comparative analysis study in Jubah. *South Sudan, United Nations Environmental Programme.* Retrieved 28th January, 2015 at http://unep.org/SouthSudan/.
Wahabu, A., Oduro-Kwarteng, S., Monney, I. & Kotoka, P. (2014). Characteristics of diverted solid waste in Kumasi, a Ghanaian city. *American Journal of Environmental Protection. 3*(5), pp. 225–231. Williams, P.T. (2005) Waste treatment and disposal. John Wiley & Sons.
Wilson, D. C., Rodic, L., Modak, P., Soos, R., Carpintero, A., Velis, K., Iyer, M. & Simonett, O. (2015). Global waste management outlook. UNEP.
World Bank (2019). Solid waste management. Retrieved 10th August 2019, https://www.worldbank.org/en/topic/urbandevelopment/brief/solid-waste-management.
Yamane, T. (1973). *Statistics: An introductory analysis*, 2nd Ed. London: John Weather Hill, Inc. Accessed 5th October 2020
Zakarya, I. A., et al. (2020). 'Municipal solid waste characterization and quantification as a measure towards effective solid waste management in UniMAP. *In IOP Conference Series: Earth and Environmental Science, 616*(1). Available at: https://doi.org/10.1088/1755-1315/616/1/012047
Zavala, M. A. L., & Funamizu, N. (2005). Effect of moisture content on the composting process in a biotoilet system. *Compost Science & Utilization, 13*(3), 208–216. Accessed 6th November 2020.

Toxicity, Nutritional Quality, and Microbial Loads of Source-Separated Dried Food Waste as Poultry Feed

Nana Yaa Korankyewa Ayim, Sampson Oduro Kwarteng, and Jacob Alhassan Hamidu

Abstract On a global scale, a third of food prepared, manufactured and distributed is wasted. This costs the global economy $2.6 trillion annually. In Ghana, daily, 20 tonnes of waste is sent to the landfill. Food waste comprises 60%. This study aims to depict how food waste generated from sources can be collected, heat-treated and included in poultry feed diets confirming its potential with sensory and market analysis. Two heat treatment alternatives (oven vs rotary drying) were incorporated into food waste treatment to ascertain which is more economical and efficient. Food waste collected was categorised into Cereals & Grains (CW), Roots and Tuber (R&T) and Vegetable (VW). The general feed formulation of FW was 50% CW, 40% R&T and 10% VW. A total number of 250 Ross broilers were assigned five treatments in a completely randomised design for 7 weeks. Treatments design consisted of control (T1), 50% FW (T2), 50% FW with 2% soybean oil (T3), 100% FW (T4) and 100% FW with 3% soybean oil (T5). The remaining portions of the diet ratios were a standardised poultry diet formulated on most Ghanaian poultry farms (maize, wheat bran, oysters, soybean, fish meal, calcium). Data were analysed using one-way analysis of variance (ANOVA) using the generalized linear model procedure of SAS Statistical software. A comparison of means was performed using the Tukey's Studentized Range (HSD) test at a probability of 5%. The dehydrated food waste had 8.49% - 14.538%, 2.85% - 16.42% crude protein, fat 0.35% - 1.45% and fibre of 20.3% -30.96% respectively and average metabolizable energy of 2421.67 kcal/kg. The results indicate the treatment group, T2 (50%FW) had the highest body weight with 2743.32 g. The next treatment by body weight was a FW diet with

N. Y. K. Ayim (✉)
Regional Water and Environmental Sanitation Centre Kumasi (RWESCK) Centre of Excellence, KNUST, Kumasi, Ghana
e-mail: korankyewaayim@iss.nl

S. O. Kwarteng
Department of Civil Engineering, Regional Water and Environmental Sanitation Centre, Kumasi (RWESCK), Ghana

J. A. Hamidu
Animal Science Department, Faculty of Agriculture, KNUST, Kumasi, Ghana

M.-M. Narra and S. Narra (eds.), *African Green Transition Through Innovative Pathways*, World Sustainability Series, https://doi.org/10.1007/978-3-031-87043-9_2

50% maize as well however inclusive of soybean oil (T3) with a body weight of 2361.53 g. Corresponding with body weight and FCR (1.76), feed intake of T2 was 3.39 kg. Subsequent treatments (T1, T3 and T5) conformed to this positive relationship except for T4 which had a poor FCR of 2.3 despite a high feed intake of 2.9 kg. Abdominal fat was within the expected range of 17.75 g—33 g with the highest amount of fat being 50% FW with 2% soybean oil as expected. Internal and visceral organs were discovered to be within range, indicating a level of good health amongst birds. Haematological parameters of WBC, RBC, HCT were in low levels, suggestive of anaemic conditions yet to manifest. The microbial analysis confirmed the presence of bacteria and fungi active on feed, indicating feed to be of lesser quality. Toxicological analysis did not bear much influence on growth performance due to their minute amounts. Sensory analysis performed by 10 participants resulted in food waste inclusion meat dry matter 50% FW being selected as the overall best.

1 Introduction

There are close to 570 million smallholder farmer households worldwide. The International Fund for Agricultural Development (IFAD) notes that over 2 billion people depend on these farmers for their income sources. The small farms these smallholder farmers work on produce a combined 80 per cent of the food consumed in Asia and sub-Saharan Africa alone (IFAD, 2011). However, the Food and Agriculture (FAO) estimates that smallholder farmers living in rural communities in low-income countries consist of more than 70% of the world's impoverished and food-insecure population (FAO, 2014). Poultry rearing contributes significantly to these smallholder households around the world, most especially amongst low income families. Through the keeping of poultry, these households can provide food, energy and fertilizer for themselves ensuring sustainability to the lives in the families. The Food and Agricultural Organization estimates poultry rearing supports over 80 per cent of rural households across the world (Sonaiya & Swan, 2004).

In developing countries, many of these families involved in small-scale poultry production struggle with several challenges such as access to market, access to affordable input products and services, lack of knowledge, appropriate skills, techniques and technologies. Coupled with the increasing threat from climate change and global warming, changing rain patterns and quantity as well as reducing soil quality, the livelihood of many farmers are under threat (Alfani et al., 2019; Chamberlin, 2007; Ervin & Ervin, 2019; FAO, 2018). A drastic and adverse effect on the food support from poultry could increase the number of people sleeping hungry every day.

To strengthen the food security of these 570 million household families, there is a need to provide more sustainable means of supporting their current livelihoods (Lowder, Jakob, & Raney, 2016; The World Bank, 2016). There is a larger number of these small-holder farmers engaged in poultry rearing also into crop cultivation in a means to diversify their income (Senadza, 2012). So, there is an emphatic need to effectively utilise the scarce resources usually described as scavengable feed resources (SFR) within the urban, traditional and rural household settings to ensure the sustainability of poultry rearing.

In Ghana, a developing country in the West African Region, there are an estimated 5 million farmers (Ghana Statistical Service, 2010) with over 60% being rural farmers identified as the largest and most vulnerable component in the rural sector (Chamberlin, 2007). Agriculture supports 54% of the Gross Domestic Product (GDP) in the country, however, it is estimated 40% of farmers live on 40 Ghana cedis ($ 7) daily (FAO, 2018). This is further evidenced by a World Bank report on a decline of people working in the agricultural sector from 45% in 2013 to 40% in 2017 whiles Ghana's food imports increase (The World Bank, 2017). In a study conducted by Veihe (2000), 72% number of farmers were discovered to engage in a mixed farming model where they reared animals along with their cultivation of crops, typically poultry.

The feedstuffs from the scavengable feed resource base (SFRB) are usually made of by-products of local food processing and agro-processing settings within the scavenging poultry value chain. As well, there is a similar composition from food waste from schools, social events and markets, emphasizing the potential of food for poultry feed. Although much research has been conducted on the content and nutrition of the SFRB of developing countries such as Ghana, it is imperative to determine the quality of the feed by analysing current collection and treatment methods for food waste that is used for the poultry feed. Currently, food waste variability is high and food waste feed may contain toxics and microbes without heat treatment. Nutrition of the food waste is also affected by heat.

As well, the quality of the feed needs to be verified by analysing the toxicity, nutritional quality and microbial loads together to understand how the feed processing activity can be made more sustainable.

To ensure sustainability as well, the existing collection and treatment methods of the scavengable feed resource base must be closely reviewed and improved to ensure a standard process that submits that quality poultry feed can be developed.

1.1 Aim of Study

The main objective of this study is to determine the efficacy of source-separated dried food waste as poultry feed.

To achieve the main objective, the following specific objectives must be carried out:

1. Characterize source-separated food waste from restaurants, hotels and local food joints by determining variations in drying the food waste by analysing differing heat treatment methods
2. Determine the quality of the feed with parameters such as toxicity content levels, nutritional content and the microbial load

Fig. 1 Timeline of food waste collection, transport and treatment before formulation into feed

3. Determine the effect of adjusting different proportions of food waste additives on broiler birds by analysing their growth rate
4. Assess the meat quality and taste acceptability to consumers
5. Assess farmer acceptability for an alternative feed

2 Methodology

2.1 Sampling, Collection and Characterization of Food Waste

2.1.1 Sampling

Food waste (FW) was collected from various sources, including households, restaurants, and markets in the Oforikrom municipality of the Kumasi metropolis. This was based on previous study indicating the average percentage of organic material is 55%, suggesting the availability of food waste required for the study (Ketibuah et al., 2005; Opoku, 2011). The study took place from November 2020 to May 2021.

2.1.2 Collection

The collected food waste was sorted at source (restaurants, cafeterias) based on methods discussed by Amo (2020) and Oduro-Kwarteng et al., (2016), and categorized into three groups: Cereals & Grains (CW), Roots & Tubers (R&T), and Vegetables (VW). Collected food waste was transported from food source sites within 48 h to avoid extreme decomposition and odour irritation. Figure 1 shows the stages the food waste was taken through in order to prepare the formulations for treatment.

2.1.3 Characterization of Food Waste

Further separation was required to extract physical impurities (spoons, plastic food containers etc.). Characterisation of food involved:

a. Determining the percentage of composition of food waste of cereals, roots and tubers and vegetables
b. Determining the moisture content

2.1.4 Heat Treatment of Source Separated Food

Two heat treatment methods were employed: oven drying and rotary drying. For oven drying, food waste samples were dried at 70 °C for 48 h. For rotary drying, samples were processed at 100 °C for 24 h. The drying rates, nutritional composition, and microbial loads of the treated food waste were evaluated to determine the most effective method.

2.1.5 Determination of Nutritional Quality of Dried Food Waste

The nutritional quality of the dried food waste was analysed using standard procedures (proximate analysis). The parameters measured included moisture content, crude protein, crude fibre, crude fat, ash content, and metabolizable energy. These analyses were conducted in accordance with AOAC (1995) standards to ensure accuracy and comparability.

2.1.6 Experimental Design

A total of 250 Ross broilers were assigned to five dietary treatments in a completely randomized design, with each treatment replicated in a deep litter pen containing 24 broilers. The treated food waste was formulated into five different poultry diets: Control (T1), 50% FW (T2), 50% FW with 2% soybean oil (T3), 100% FW (T4), and 100% FW with 3% soybean oil (T5). The general feed formulation consisted of 50% CW, 40% R&T, and 10% VW. The remaining portions of the diet ratios were a standardised poultry diet formulated on most Ghanaian poultry farms (maize, wheat bran, oysters, soybean, fish meal, calcium). Table 1 summarises the feed formulation for each treatment.

The experiment lasted for seven weeks, during which growth performance, feed intake, body weight gain, feed conversion ratio (FCR), and carcass quality were monitored. Birds were provided with ad libitum access to feed and water, and the environmental conditions were maintained according to standard poultry rearing practices.

2.1.7 Microbial Analysis

Microbial loads in the treated food waste and the broiler intestines were assessed using standard microbiological techniques. This was to determine the effect of the FW in feed and their formulations on their welfare characteristics. The parameters measured included aflatoxins and total coliforms, E. coli, faecal enterococci, Salmonella, anaerobic plate count, mould, yeast, and Staphylococcus. Samples were collected and analysed at the beginning, mid-point, and end of the experiment to monitor changes in microbial contamination.

Table 1 Feed formulation for every 100 kg of each treatment

Ingredients/Kg	Control(T1)	T2	T3	T4	T5
Maize	63	31.5	30.5	0	0
Soya bean cake	13	13	13	13	13
Fish meal	16	16	16	16	16
Wheat bran	6	6	6	6	6
Oyster shell	0.5	0.5	0.5	0.5	0.5
Salt	1	1	1	1	1
Vitamin/premix	0.5	0.5	0.5	0.5	0.5
Dried Food Waste	0	31.5	30.5	63	63
Soybean Oil	0	0	2	0	3

* *T1: 0% food waste, T2: 50%food waste, T3: 50% food waste with soybean oil, T4: 100% food waste, T5: 100%food waste with soybean oil*

2.1.8 Toxicological and Haematological Analysis

Toxicological analysis of the feed (beginning and termination) and faeces (termination) was performed to detect the presence of harmful substances, including heavy metals (Cadmium, Arsenic, Mercury and Lead) and mycotoxins (to determination levels of cross-contamination of feed). Haematological parameters, including white blood cells (WBC), red blood cells (RBC), haematocrit (HCT), haemoglobin (HGB), mean corpuscular volume (MCV), and mean corpuscular haemoglobin concentration (MCHC), were measured to assess the health status of the broilers.

2.1.9 Sensory Analysis

Sensory evaluation of the broiler meat was conducted by 10 participants to determine the acceptability of meat from broilers fed different diets The chart used for evaluation was designed based on sensory analysis studies and guides by Lyon et al. (2004), Miller (2017) and Opoku (2018). The parameters assessed included appearance, taste, tenderness, texture, and mouth feel. A ten-point Likert scale was used for the evaluation, with scores ranging from 1 (dislike extremely) to 10 (like extremely).

2.2 Economic Analysis and Farmer Perception and Reception of Formulated Feed

2.2.1 Economic Analysis

An economic analysis was conducted on the cost of producing a 100 kg bag of each treatment to note the most affordable treatment to produce as well. The cost prices were included in a survey of 15 smallholder poultry farmers.

2.2.2 Farmer Perception and Reception of Formulated Feed

Two groups of poultry farmers were interviewed in the Ashanti and Bono Region for their views on the possibility of feeding their broilers with dried food waste. Some farmers were interviewed individually in local parlance and then the design questionnaire forms were filled out on their behalf by the interviewer. Some farmers were interviewed in groups. A total of 17 farmers and poultry farm managers were interviewed.

2.2.3 Statistical Analysis

The collected data were subjected to the one-way analysis of variance (ANOVA) using the generalized linear model procedure of SAS Statistical Software. A comparison of means was performed at 5% significance level using the Tukey's Studentized Range (HSD) test. This was done to determine additional patterns that could provide more meaning to the earlier identified objectives of this study.

3 Results

3.1 Generation and Characteristics of Source Separated Food

3.1.1 Characterization of Food Waste

The study collected source-separated food waste from six sources, finding that cooked leftovers made up 70% of the waste, with cereals being the most common component (50.29%), followed by roots and tubers (27.41%), and vegetables (22.29%). Figure 2 indicates the composition of the food waste varied slightly each week due to changes in menu demands and the seasonal availability of food crops, with cereals consistently being the highest fraction over the six weeks. This trend aligns with global patterns, particularly in developing regions where diets are heavily dependent on cereals, roots, and tubers, contrasting with global figures where vegetables are the leading

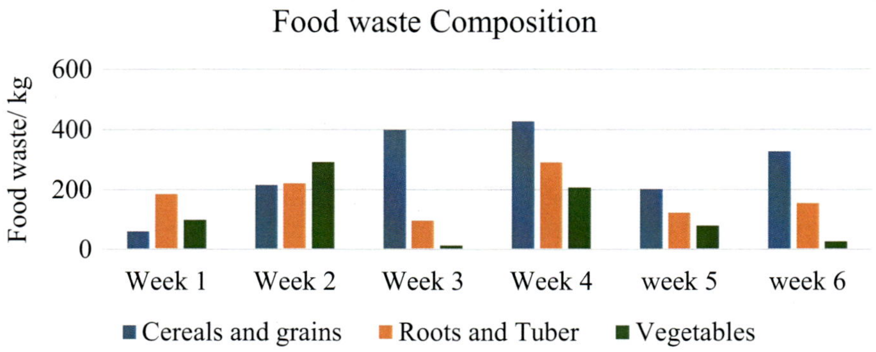

Fig. 2 Composition and generation of source-separated food waste (kg)

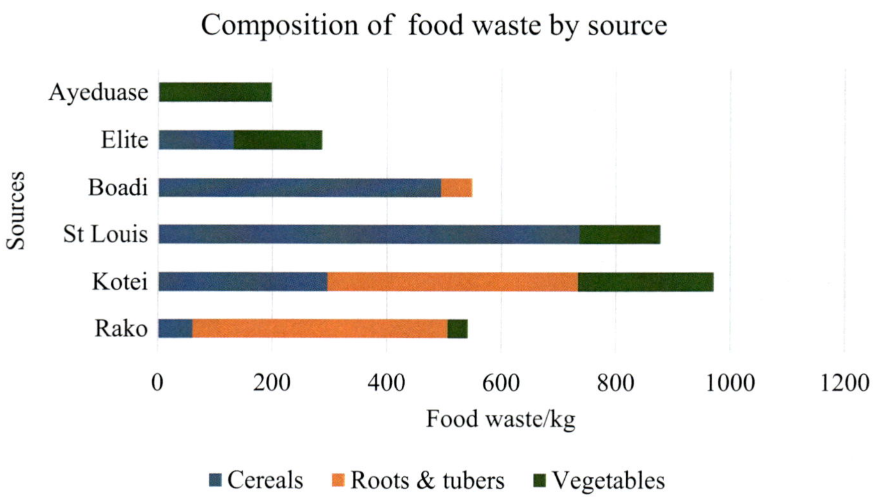

Fig. 3 Composition of waste generated by source (kg) locations within Oforikrom municipality of Kumasi

component of food waste (Gustavsson et al., 2011; Parfitt et al., 2010). Figure 3 also shows the amounts gathered from various sources.

3.1.2 Generation of Food Waste by Source

By the global average, an individual wastes 0.18 kg of food daily, the generation amount rates by source did not tally with the estimated number of individuals recorded to have discarded their food from the identified sources (Sharma et al., 2009). It was identified that a total count of 1,267 people on average contributed to the daily generated food waste. However, by a global average of 0.18 kg, this works

Table 2 Count number of people wasting food

Source	Daily generation rate (kg)	Number of estimated people generating waste by global average index	Count number
Rako chop bar	10.6551	59.195	151
Kotei	19.81347	110.0748333	200
St. Louis SHS	17.89796	99.43311111	600
Boadi	11.74694	65.26077778	118
Elite	5.873469	32.63038333	112
Ayeduase	4.302041	23.90022778	86
Total		390.4943333	1267

out to a third of the count number. From Table 2, the study shows sources where food waste was gathered and the daily generation rates.

3.1.3 Food Waste Drying Rates Variability (Sun Drying vs Oven vs Rotary Dryer)

Upon arrival, the wet food waste was weighed, categorized (cereal, root & tuber, vegetable), sun-dried for 24 h, and then further dried using either an oven or a rotary dryer. The rotary dryer, which operated at higher and more variable temperatures (169 °C to 267 °C) compared to the oven (105 °C), was more effective at reducing moisture content quickly but could only successfully dry cereals and roots and tubers, not vegetable waste. This is in line with the use of rotary dryers by other studies where particle size, weight and shape play a role in successful drying (Iguaz et al., 2003; Thanaraj et al., 2004). In Table 3 is a comparison of moisture content depicting the rates at which these two treatment options. The high variability in the temperatures of the Rotary Dryer can be attributed to the rapidly changing temperatures from the firewood which could not be maintained at one temperature like the oven (105 °C).

3.2 Nutritional Quality of Dried Food Waste

3.2.1 Proximate Analysis of Dried Food Waste Composition

The moisture content of dried food waste (DFW) ranges from 8.489% to 14.538%, with vegetable waste (VW) having the highest moisture content and root and tuber waste (RTW) the lowest. The preferred MC for maize in previous studies identified by Amo (2020) was between the range of 8% to 11%. A study that compared samples of poultry feed maize containing MC as high has 14.1% discovered traces of aflatoxins.

Table 3 Food Waste drying rate (oven versus rotary drying)

Food waste	Week 1			Week 2			Week 3		
	BSD	OD	RD	BSD	OD	RD	BSD	OD	RD
Cereal	33.8	9.9	26.6	39.2	25.7	11.6	38.6	36.3	4.4
Root & Tuber	60.3	35.7	21.9	50.9	35.8	12.4	49.1	12.2	3.7
Vegetable	51.4	36.8	–	52.1	21.8	–	50.4	17.3	–
Food waste	Week 4			Week 5			Week 6		
	BSD	OD	RD	BSD	OD	RD	BSD	OD	RD
Cereal	42.1	27.6	11.6	48.9	19.7	13.4	44.4	37.2	1.4
Root & Tuber	45.4	31.3	11.2	48.5	28.9	12.7	49.7	40.9	1.2
Vegetable	55.9	37.7	–	53.9	27.0	–	58.7	47.3	–

*** *BSD - Moisture Content % Before Solar Drying, OD – Moisture Content % after oven drying, RD- Moisture Content % after Rotary Dryer*

Table 4 Nutritional composition of dried food waste components

Parameters	Cereal	Maize	Vegetable	R&T
MC (%)	10.539	10.750	14.538	8.489
AC (%)	6.812	5.235	5.285	8.565
CP (%)	8.975	9.632	16.418	2.846
CF (%)	20.30	21.92	30.962	21.002
EE (%)	1.350	3.385	0.350	1.450
NFE (%)	60.999	58.713	48.865	60.493
ME (Kcal/Kg)	2577.458	2688.222	2346.385	2341.159

This is in significance links MC beyond 11% to possible aflatoxin contamination (Kana et al, 2013).

Table 4 presents nutritional composition of the dried food waste components. Crude protein (CP) values are 8.975% for cereal waste (CW), 2.846% for RTW, and 16.418% for VW. Crude fibre content is high across all DFW types, with CW at 20.30%, RTW at 21.0%, and VW at 30.962%, compared to the typical 0.9% to 2.5% for maize; high fibre can negatively impact poultry digestion and body weight (Enyisi et al., 2014).

3.2.2 Nutritional Composition

Table 5 presents the nutritional contents of the formulated diets from the dried food waste and other supplements (standard poultry diet, soybean oil). The moisture content of the experimental diets ranged from 7.617% to 9.772%, with the highest in treatment five (T5) due to added soybean oil, and all diets were within the standard moisture content for broiler feed. Crude protein (CP) content ranged from 17.013%

Table 5 Nutritional composition of formulated feed diets

Parameters	T1	T2	T3	T4	T5
MC (%)	9.114	9.563	8.483	7.617	9.772
AC (%)	6.685	5.716	7.539	7.267	3.995
CP (%)	17.013	22.547	22.110	23.861	22.985
CF (%)	2.765	5.647	10.14	9.214	12.70
EE (%)	9.013	8.063	17.955	7.589	18.574
NFE (%)	72.424	71.011	55.880	68.312	54.962
ME (Kcal/Kg)	3901.566	3979.191	4242.583	3894.581	4293.419

** *T1: 0% food waste, T2: 50%food waste, T3: 50% food waste with2% soybean oil, T4: 100% food waste, T5: 100%food waste with 3% soybean oil*

to 23.861%, with the highest in T4, confirming that food waste inclusion increased CP, while fat content varied from 7.589% to 18.574%, highest in T5, correlating with added soybean oil, and metabolizable energy ranged from 3901.566 to 4293.419 kcal/kg.

According to Amo (2020), Islam et al. (2015) indicated 10 – 12% as the conventional moisture content for prepared broiler feed. This is mainly because low moisture content meant reduced water activity and higher stability and quality associated with the feed as microbial activity cannot take place within unfavourable conditions. As protein serves a significant role in enhancing growth and immunity, being second in diet serving carbohydrates, the inclusion of food waste served as an alternative for increased CP content (Amo, 2020; Muduli et al., 2018). Esmail (2016), states that an acceptable broiler diet shall feature CP levels between 19–23%.

3.3 *Effect of Dried Food Waste on Broiler Growth Rate and Meat Quality*

3.3.1 Bodyweight Gain and Feed Efficiency

The growth performance of the broilers varied significantly across the dietary treatments. Figure 4 highlights the weight gain over a period of 7 weeks. The treatment group fed with 50% FW (T2) achieved the highest body weight (2743.32 g), followed by the group fed with 50% FW and 2% soybean oil (T3) (2361.53 g). Feed intake for T2 was 3.39 kg, and the FCR was 1.76. T4 exhibited poor FCR (2.3) despite high feed intake (2.9 kg), indicating inefficiencies in nutrient utilization when food waste constituted 100% of the diet. Mendes et al., 2011 conducted a study with 240 Cobb broilers in wood shaving litter, confirming that for birds without uniform initial body weight, lighter birds are unable to recover their weights during the six-week experimental period. This is established in the results obtained as T4 had the lowest initial body weight of 780.20 amongst all five treatments. By week seven (7), T4 still

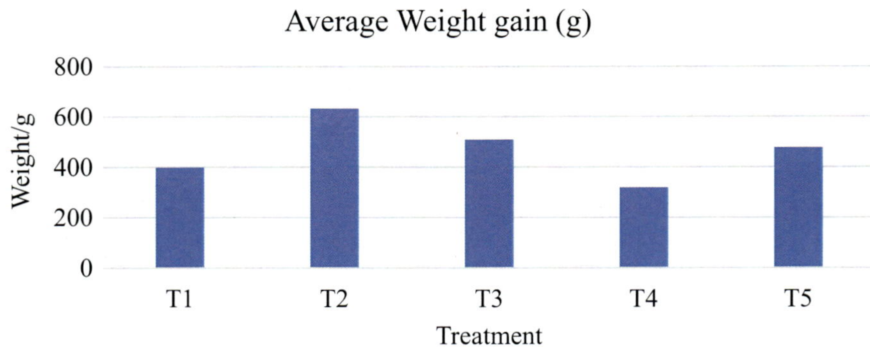

Fig. 4 Average weight gain for each treatment for 7 weeks

was the treatment with the lowest body weight gain (2008.98 g). The best performing treatment by body weight gain was T2 with the 50% food waste replacement diet of dry matter. This confirms the conclusion of Amo (2020) and Kangah (2019) that observed that diets with cereal waste and maize performed best.

3.3.2 Feed Intake/Consumption

Figure 5 highlights the average feed intake of each control group or treatment over the duration of 7 weeks. During week 4, T3 had the highest feed intake at 0.45913 kg, lower than the 0.700 kg observed in Amo's study, while the lowest was in T1 at 0.3880 kg, also lower than Amo's 0.515 kg for T1. Feed consumption in the following weeks varied slightly, peaking in week 7 at 1.4927 kg (T2) and lowest at 0.89097 kg (T1), with T2 having the highest total feed intake at 3.3891 kg. Significant variations in feed consumption were noted in weeks 4 and 6, but overall intake was below the 4 kg standard expected by the 6th week, possibly due to non-pelletized feed, the form of the feed (dry versus wet), and microbial growth affecting broiler performance.

3.3.3 Feed Conversion Ratio

The feed conversion ratio (FCR) measures the efficiency of feed conversion to body weight in broilers, with a lower FCR indicating higher efficiency. From Table 6, the best FCR among the treatment groups was T2 at 1.7577, followed by T5, T1, T3, and T4 with FCRs of 1.7667, 1.7891, 2.0011, and 2.3353 respectively; all were within acceptable ranges except for T4, which suggests poor feed quality.

Studies confirm that broilers with high feed intake have a high feed conversion ratio (FCR) converting much feed consumed into body weight gain (Scott, 2005). This is confirmed with the results displayed in Table 7, indicating the highest feed intake T2, had the highest FCR and highest body weight gain. However, the subsequent

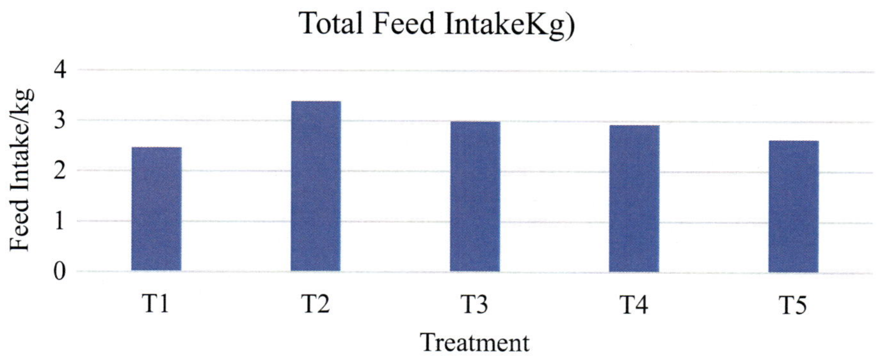

Fig. 5 Average feed intake for each treatment for 7 weeks

Table 6 Feed Conversion Ratio

Feed conversion							
Week	T1	T2	T3	T4	T5	SEM	P-value
4	0.8079	0.4870	0.7390	0.9889	0.7772	0.091	0.0378
5	3.3067	1.8364	1.1373	2.1334	1.9360	1.410	0.8796
6	2.1884	1.3618	1.4561	1.7948	1.0490	1.1727	0.9087
Overall FCR	1.7891	1.7577	2.0011	2.3353	1.7667	0.1968	0.1261

** *T1: 0% food waste, T2: 50%food waste, T3: 50% food waste with x soybean oil, T4: 100% food waste, T5: 100%food waste with 2 × soybean oil*

treatments do not conform to this as T3 and T4 with the next highest overall feed intake had poor FCRs. T3 broilers had the second-best body weight (2361.53 g) amongst all five despite the high FCR of 2.0. T4 broilers had the worst observably low weight of 2 kg. T5 and T1 had similar feed intakes of 2.4623 and 2.6332 as well as high FCRs of 1.7891 and 1.7667. However, their total body weights were closer to the 2 kg mark (2203.62 g for T1 and 2317.05 g for T5). Mortality was unusually high for all treatments. This was attributed to limited pen space coupled with vaccination failure identified by veterinary reports and microbial growth on feed weakening broiler performance to the level of intoxication.

3.4 *The Effect of the Feed on the Broiler Growth*

3.4.1 Caracass Analysis

Carcass quality analysis showed that abdominal fat ranged from 17.75 g to 33 g, with the highest fat content observed in T3. Internal and visceral organs were within

Table 7 Summary of Growth Rate Results

Summary of Growth rate Results							
Parameter	T1	T2	T3	T4	T5	SEM	P-value
Initial FCR	0.8079	0.4870	0.7390	0.9889	0.7772	0.091	0.0378
Final FCR	2.1884	1.3618	1.4561	1.7948	1.0490	1.1727	0.9087
Overall Weight gain	1385.75	1925.58	1543.75	1191.11	1499.26	156.83	0.23
Total Feed Intake	2.4623	3.3891	2.9896	2.9291	2.6332	0.2422	0.1576
Overall Body weight	2203.62	2743.32	2361.53	2008.98	2317.05	156.863	0.1723
Overall FCR	1.7891	1.7577	2.0011	2.3353	1.7667	0.1968	0.1261
Mortality	14	22	22	32	13	3.067	-

acceptable ranges across all treatment groups, suggesting that the inclusion of food waste did not adversely affect organ development.

Overall, broilers in T3 and T5 had the heaviest visceral body parts, suggesting good health conditions of the birds within those treatments. Again, it was observed that visceral organ weight, the diets of these two treatments were diets that included soybean oil added to their feed formulation. A study by Baião & Lara, (2005) did link oil and fat additions to food diets to enlarged hearts. This could explain the addition of soybean oil treatments having larger visceral organs.

3.4.2 Haematological and Microbial Analysis

Haematological parameters indicated low WBC, RBC, HGB and HCT levels in broilers fed 100% FW diets (T4 and T5), suggesting the potential onset of anaemia as Adeyemo et al., (2018) suggest that a range between $40.7 \times 10^3/\mu l$ to $43.2 \times 10^3/\mu l$ indicates healthier birds. However, no significant differences were observed in haemoglobin levels across the treatments.

Values of the treatment for the MCV was above the indicated healthy range of 81.60 Fl to 89.10 fL as treatment values ranged between 117.15 fL to as high as 679.75 fL in the exceptional case of T5. Broilers fed with food waste inclusion diet recorded the highest value while the control group recorded the lowest value. High MCV values are suggestive of a number of triggers most especially anaemia (Campbell, 2013). MCH and MCHC were within the expected range for healthy broilers.

3.4.3 Microbial Analysis on Feed and Intestinal Load

Microbial analysis revealed the presence of bacteria and fungi in both the feed and broiler intestines, with higher microbial loads detected in the rotary-dried samples compared to the oven-dried samples, however below threshold levels save T5. Fungal contamination according to Sultana et al, (2017) should not exceed 1×10^5 cfu/g. T3 and T5 exceeded acceptable threshold levels for fungal contamination.

Aflatoxin B1 levels were significantly higher ($P = 0.0176$) in broilers fed diets including food waste, with no aflatoxins detected in the control group fed 0% DFW, indicating the maize used was of good quality. Additionally, high bacterial concentrations of Staphylococcus, Salmonella, and E. coli were noted in four out of five treatments, especially in T5, T2, and T4 (however, below threshold levels), suggesting contamination and poor handling practices. Figures 6 and 7 provide an overview of the microbial load found for the five treatments. Additionally, Fig. 8 highlights, microbial contamination found in the intestines of the birds in the treatment groups after the 7-week period.

The high microbial loads within the intestines of broilers of the various treatments are consistent with an independent veterinary analysis conducted to ascertain the reason for the high mortality amongst the birds from week 5 to week 7. Veterinary analysis diagnosis concluded the main cause of death to be vaccine failure which drastically reduced the birds' immunity contributing to a gradual and systematic breakdown of the immune system of the birds. This was identified by several physical conditions of the birds. 25% of the birds were initially identified to have suffered from coccidiosis, respiratory disease and microbial infections evident in their liver, gizzard and other internal organs in the initial analysis (Ali et al., 2018). This was evidenced by scouring of the gizzard, lacerations on the gut lining, the gradual (lightening) whitish colour of the liver.

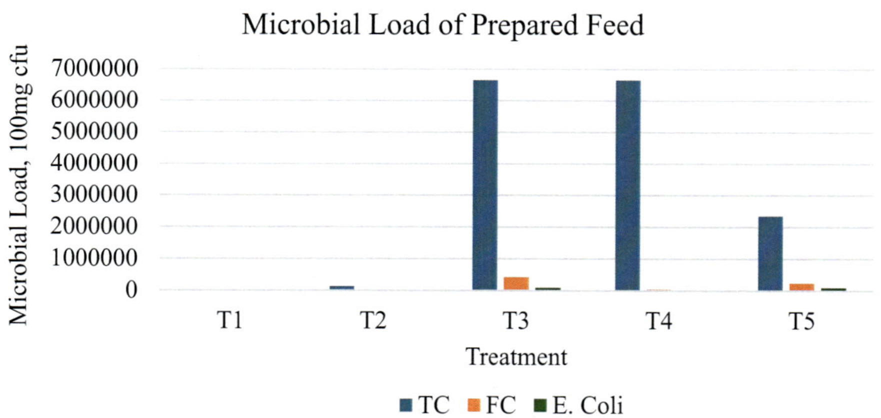

Fig. 6 Microbial load of prepared feed

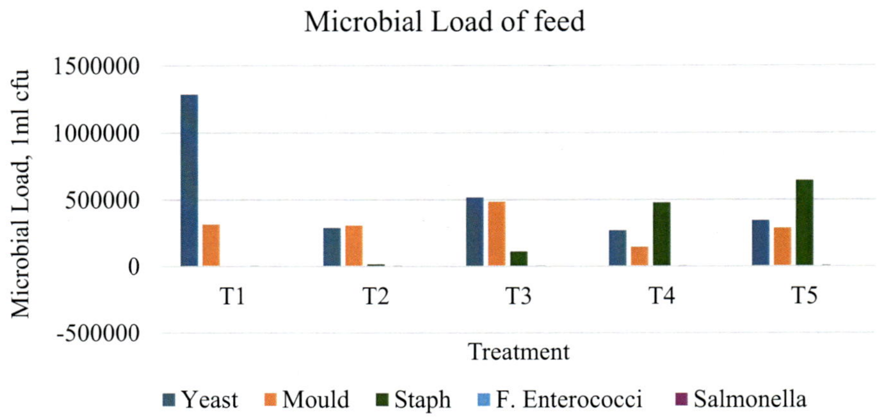

Fig. 7 Microbial load of feed

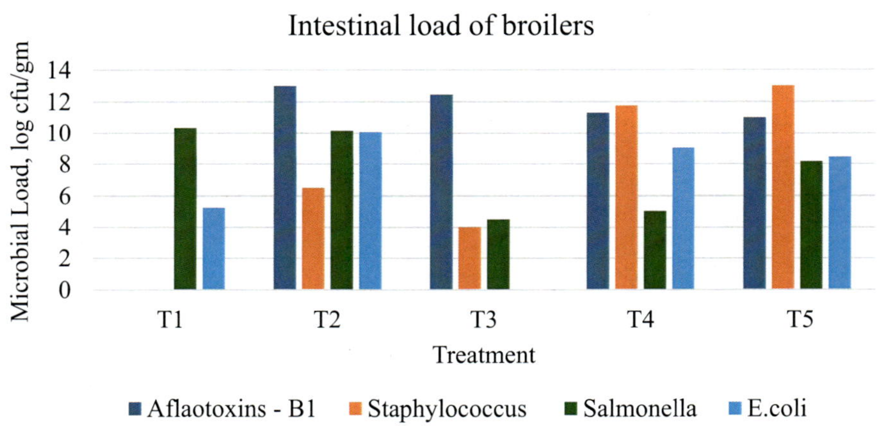

Fig. 8 Intestinal load of broilers

3.5 *Toxicology Analysis*

3.5.1 Toxicological Analysis on Feed

Heavy metal concentrations within the feed overall were minimal and usually far below the threshold guidelines set by FAO and WHO (1996). Despite variations within concentration values for all heavy metal parameters ($P < 0.0001$), they do not surpass the significant threshold mark. This can be evidenced in Fig. 9.

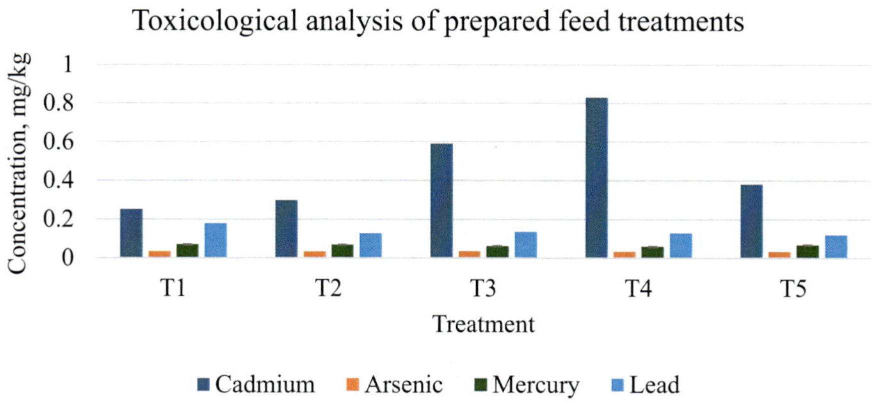

Fig. 9 Toxicological analysis on prepared feed

3.5.2 Toxicological Analysis on Faeces

In keeping with the set standards for the FAO (1996) for toxicology, it is observed that for all parameters of Cadmium, Arsenic, Mercury and Lead, almost all treatments were within acceptable thresholds. The thresholds for Cadmium, Arsenic, Mercury and Lead were not exceeded and below that of the concentrations within the formulated feed, suggestive of absorption of minute concentrations by the birds. This is moderately acceptable as it is within guidelines. Figure 10 visualises the concentrations in the faeces of the treatments.

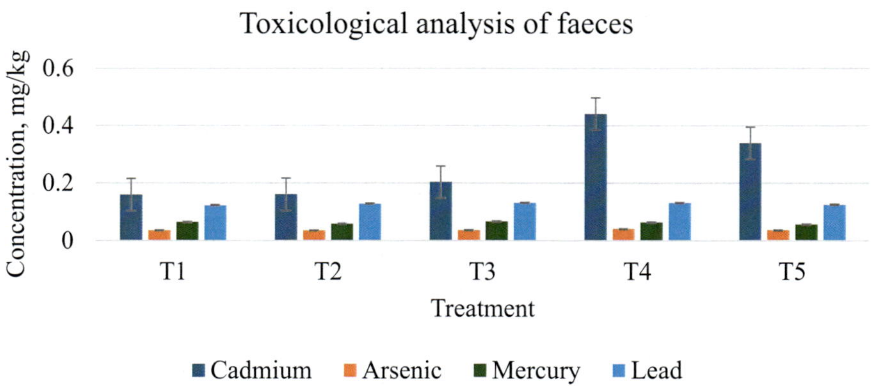

Fig. 10 Toxicological concentrations within faeces

Table 8 Meat sensory test

	Appearance	Texture	Taste	Tenderness	Mouth-feel
T1	7.5	8.4	7.1	8.5	6.9
T2	7.5	8.7	8.1	8.2	6.7
T3	7	8	8.20	8.4	7.5
T4	7.4	8.8	6.70	7.7	7.1
T5	6.8	8.20	6.20	8.6	8.1
SEM	0.3556	0.3581	0.2793	0.2944	0.3310
P-value	0.4533	0.4873	0.0003	0.2285	0.0364

3.6 *Sensory Analysis*

3.6.1 Meat Quality Analysis for Consumer Market

The sensory evaluation favoured meat from broilers fed a 50% FW diet, indicating good acceptability among participants. From Table 8, the meat from T2 and T3 received higher scores for taste, tenderness, and overall acceptability compared to the control and other treatments. Ranking the preferability of meat according to samplers is T2, T1, T3, T5 and T4. This suggests that dried feed food waste can be accepted by Ghanaian consumers as in the case in Asian countries (Lyon et al., 2004; Sealey et al., 2011). However, the inclusion of the soybean oil in the feed of T3 and T5 indicates there is a limit to which the greasiness in the flesh of the broiler is accepted as they gained minimal scores.

3.6.2 Feed Acceptability by Farmers

A total number of 17 poultry farmers and poultry farm managers were successfully interviewed to assess their interest in DFW feed. The results are depicted by their education, bird type they rear and the number of birds they manage, their interest in dried food waste at an affordable price and their concerns were noted.

Figure 11 highlights the responses the interviews. Out of 17 respondents, 7 breed up to 2,000 birds, another 7 breed between 2,001 to 10,000 birds, and 3 breed over 10,000 birds. All respondents expressed interest in DFW feed to reduce costs despite some cultivating their maize or having support from associations; their main concerns were the feed's side effects on bird health and growth, application methods, affordability, suitability for other bird types, and government approval in that order. There were follow up questions on accessibility, self-production, delivery indicating strong interest in the feed. As confirmed by previous studies, the DFW does have the interest of poultry farmers and has the potential for commercialisation(Amo, 2020; Dao et al., 2020; Hard, 2002; S. Salemdeeb et al., 2016; Tunde Adegoke Amole, 2019).

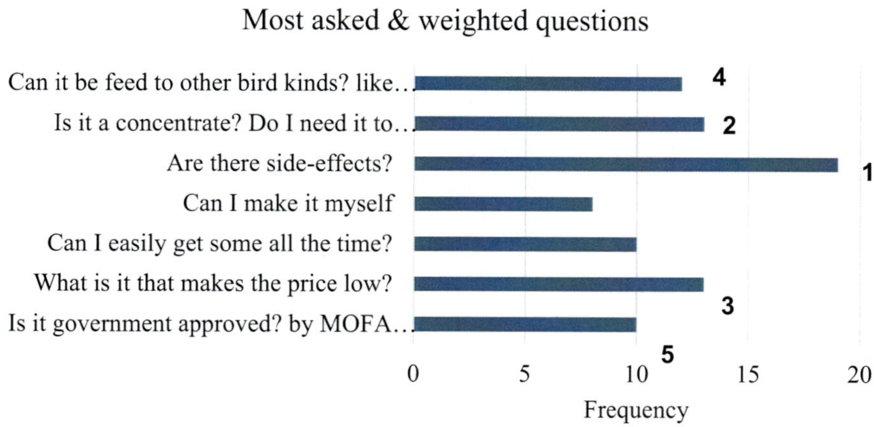

Fig. 11 Farmer perceptibility about feed

3.6.3 Cost Analysis of Feed Production

The unit cost of dried food waste which entails the cost of collection (incentive), transport and drying. At the time of this study in February–March 2021, the cost of producing 0.0011ton of dried food waste was GHS 0.70 and that of maize was GHS 2 per 0.0011ton.

The cost of maize is thrice the cost of producing DFW. Amo (2020), at the time of her study, noted the cost of processing DFW at GHS 0.65 per 0.0011ton and the market price of maize at GHS 1.30. The cost of maize was twice the cost of the feed with food waste and within a year, the market price has increased by 100% (GHS 0.70).

4 Discussion

The study demonstrated that source-separated dried food waste could serve as a viable poultry feed ingredient without adverse effects on broiler health or meat quality. The nutritional composition of the food waste, particularly the crude protein and metabolizable energy content, supported adequate growth performance. However, microbial contamination remained a concern, highlighting the need for stringent quality control measures during food waste processing (Amo, 2020).

The comparison between oven drying and rotary drying indicated that both methods effectively reserved the nutritional quality of the food waste, with slight variations in drying rates and microbial loads. The inclusion of soybean oil in the diets (T3 and T5) improved body weight gain and FCR, suggesting that additional fat sources could enhance the energy content of food waste-based diets.

Despite the positive growth performance, the low haematological parameters suggested a need for further investigation into the long-term health effects of food waste-based diets on poultry as analysis indicated an underlying health condition consistent with an overall low body immunity status. However, it could not be confirmed that this condition was caused from the feed formulation or a possible contamination. The presence of microbial contaminants also emphasized the importance of monitoring and controlling feed hygiene to prevent potential health risks.

Sensory Analysis performed indicated a market for DFW broiler meat. This is because the top choices selected by the panel of 10 tasters included meat fed on food waste diets (T2 and T3). The soybean oil DFW diets could not be conclusively affirmed however, there is the indication that the greasiness from high fat is acceptable to a limit for the Ghanaian consumer market.

The market study of the potential of the DFW for the Ghanaian poultry market indicated positive results with all 17 respondents indicating interest to switch to the more sustainable option. There was considerable concern for the side effects, applicability, regulation and accessibility of the feed. It is evident if these barriers can be resolved, commercial potential can be unlocked.

Economic analysis indicates low-cost margins associated with DFW waste, costing three (3) times less than producing with maize feed formulations.

Operation costs of rotary drum dryer were more affordable than oven as it required only firewood and an electricity GHS 50 lower than that of the oven. For this reason, a rotary drum dryer could be used in drying food waste more quickly and efficiently.

5 Conclusion and Recommendation

This study provides evidence that source-separated dried food waste can be effectively used as poultry feed, offering a sustainable solution to food waste management and reducing reliance on conventional feed ingredients. The results indicate that a 50% inclusion of food waste in poultry diets can achieve optimal growth performance without compromising meat quality. However, addressing microbial contamination and ensuring the long-term health of poultry are critical to the successful implementation of this approach.

Future research should focus on improving the safety and quality of food waste-based feeds, exploring additional dietary supplements to enhance nutritional balance, and conducting long-term studies to assess the overall impact on poultry health and productivity.

References

Adeyemo, G., Bolarinwa, M., & Ehiabhi, O. (2018). Haematology and external egg quality parameters of three Nigerian indigenous chicken genotypes. *International Journal of Molecular Biology, 3*(4), pp. 197–201. https://doi.org/10.15406/ijmboa.2018.03.00077.

Alfani, F., Aslihan, A., McCarthy, N., Cavatassi, R., & Sitko, N. (2019). Climate-Change Vulnerability in Rural Zambia The Impact of an El Nino-Induced Shock on Income and Productivity. *In FAO Agricultural Development Economics Working Papers* 21 June, No.2019/02, pp. 44. UN-iLibrary. Retrieved from https://doi.org/10.18356/9b6d387a-en.

Ali, A., Danbappa, R., Alhassan, K. A., & Shah, M. M. (2018). Isolation and identification of microbial contaminants associated with commercial poultry feed. https://doi.org/10.21839/jaar.2018.v3i5.231.

Amo, J. (2020). Conversion of food waste to broiler feed and its effects on growth and meat quality. KNUST.

Amole, T. A., & S. A. and I. O. (2019). A guide to the production of High-Quality Cassava Peel® mash as a feed for livestock. Nairobi, Kenya. Baião, N., & Lara, L. (2005). Oil and Fat in Broiler Nutrition. *Brazilian Journal of Poultry Science, 7*(3), 129–141.

Chamberlin, J. (2007). Defining smallholder agriculture in Ghana: Who are smallholders, what do they do and how are they linked with markets? Accra.

Dao, T. H., Boyle, N., & Rahman, M. (2020). Potential to produce poultry feed from food wastes. *Zootencia International.*

Ervin, P., & Ervin, L. (2019). Household Vulnerability to Food Security in the Face of Climate Change in Paraguay. *In FAO Agricultural Development Economics Working* Papers, 28 October, No.2019/04 (p. 48). Retrieved from https://doi.org/10.18356/9b6d387a-en

Enyisi, I. S., Umoh, V. J., Whong, C. M. Z., Alabi, O., & Abdullahi, I. O. (2014). Chemical and nutritional value of maize and maize products obtained from selected markets in Kaduna state Nigeria. *African Journal of Food Science and Technology, 5*(4), 100–104.

FAO. (2014). The state of food insecurity in the world, the multiple dimensions of food security. *Rome.*

FAO. (2018). Global food waste and its environmental impact. Retrieved from https://en.reset.org/knowledge/global-food-waste-and-its-environmental-impact-09122018#:~:text=Anestimated 1.3 billion tonnes, FAO) of the United Nations.

Ghana Statistical Service. (2010). Census.

Gustavsson, J., Cederberg, C., U, U. S., Otterdijk, R. van, & Meybeck, A. (2011). Global food losses and food waste. Extent, causes and prevention. Rome.

Hard, D. L. (2002). Innovative developments in the production and delivery of alternative protein sources. *In Protein Sources For The Animal Feed Industry Expert Consultation and Workshop.*

IFAD. (2011). Viewpoint: Smallholders can feed the world. Retrieved from https://www.ifad.org/documents/38714170/40706188/Smallholders+can+feed+the+world_e.pdf/460ca6c2-7621-40d8-9f79-a56f6f8fa75e

Iguaz, A., Esnoz, A., Martiınez, G., Lopez, A., & Vorseda, A. (2003). Mathematical modelling and simulation for the drying process of vegetable wholesale by-products in a rotary dryer. *Journal of Food Engineering, 8774*(September), 151–160. https://doi.org/10.1016/S0260-8774(02)00451-X

Islam, M. S., Moinul Haque, M., & Shakhawat Hossain, M. (2015). Effect of corn moisture on the quality of poultry feed. *Journal of Poultry Science and Technology, 3*(July), 24–31.

Kana, et al. (2013). Assessment of aflatoxin contamination of maize, peanut meal and poultry feed mixtures from different agroecological zones in cameroon. *Toxins, 5*(5), 884–894. https://doi.org/10.3390/toxins5050884

Kangah, A. B. (2019). Conversion of food waste into poultry feed for hubbard broiler. KNUST.

Ketibuah, E., Asase, M., Yussif, S., Mensah, M. Y., & Fischer, K. (2005). Comparative analysis of household waste in the cities of Stuttgart and Kumasi. *Options for Waste Recycling and Treatment in Kumasi Institute for Sanitary Engineering.* KNUST.

Lowder, S. K., Jakob, S., & Raney, T. (2016). The number, size, and distribution of farms, smallholder farms, and family farms worldwide. *World Development*, 87, 16–79. https://doi.org/10.1016/j.worlddev.2015.10.041

Lyon, B. G., Smith, D. P., Lyon, C. E., & Savage, E. M. (2004). Effects of diet and feed withdrawal on the sensory descriptive and instrumental profiles of broiler breast fillets. *Poultry Science, 83*, 275–281. *1*, pp. 1–3

Mendes, A., Paixão, S., Restelatto, R., Reffatti, R., Possenti, J., Morello, D., Carvalho, T. (2011). Effects of initial body weight and litter material on broiler production. *Brazilian Journal of Poultry Science, 13*(3) pp. 1–3. https://doi.org/10.1590/S1516-635X2011000300001Author.

Miller, R. K. (2017). The eating quality of meat: sensory evaluation of meat. in lawrie's meat science (pp. 461–499). Texas A&M University, College Station, TX: Elsevier Ltd. https://doi.org/10.1016/B978-0-08-100694-8.00015-7

Muduli, S., Champati, A., Popalghat, H. K., & Animal, M. (2018). Poultry waste management : An approach for sustainable development Poultry waste management: An approach for sustainable development.

Oduro-Kwarteng, S., Anarfi, K. P., & Essandoh, H. M. (2016). Source separation and recycling potential of municipal solid waste in Ghana. *Management of Environmental Quality: An International Journal, 27*(2). https://doi.org/10.1108/MEQ-03-2015-0038

Opoku Jnr, M. (2011). Biogas production from kitchen waste generated on knust campus by. KNUST.

Opoku, O. (2018). Black Soldier Fly Larvae Meal as a dietary protein source for broiler chickens: effect on growth performance, carcass and haematological parameters. Kwame Nkrumah University of Science and Technology (KNUST).

Parfitt, J., Barthel, M., & MacNaughton, S. (2010). Food waste within food supply chains: Quantification and potential for change to 2050. *Philosophical Transactions of the Royal Society B: Biological Sciences*, 3065–3081. https://doi.org/10.1098/rstb.2010.0126

Salemdeeb, S., Ermgassen, E. K. J., & z., Kim, M. H., Balmford, A., & Ali-Tabbaa, A. (2016). Support amongst UK pig farmers and agricultural stakeholders for the use of food losses in animal feed. *Journal of Cleaner Production, 5*, 49.

Scott, T. (2005). Variation in feed intake of broiler chicken. *Biology.*

Senadza, B. (2012). Non-farm income diversification in Rural Ghana: patterns and determinants. *African Development Review, 24*(3), 233–244. Retrieved from https://doi.org/10.1111/j.1467-8268.2012.00322.x

Sealey, W. M., Gaylord, T. G., Barrows, F. T., Tomberlin, J. K., McGuire, M. A., Ross, C., & St-Hilaire, S. (2011). Sensory analysis of rainbow trout, Oncorhynchus mykiss, fed enriched Black soldier fly prepupae, Hermetia Illucens. *Journal of World Aquaculture Society, 42*(1), 34–45.

Sharma, A., Chen, C., & Vu Lan, N. (2009). Solar-energy drying systems: A review. *Renewable and Sustainable Energy Reviews, 13*(6–7), 1185–1210.

Sonaiya and Swan. (2004). Small-Scale Poultry production.

Sultana, et al. (2017). Microbiological quality of commercially available poultry feeds sold in Bangladesh. *Asian Journal of Medical and Biological Research, 3*(1), 52–60.

Thanaraj, T., Dharmasena, D. A. N., & Samarajeewa, U. (2004). Development of a Rotary Solar Hybrid Dryer for Small Scale Copra Processing. Tropical Agriculture Research. Peradeniya.

The World Bank. (2016). A Year in the Lives of Smallholder Farmers. Retrieved from https://www.worldbank.org/en/news/feature/2016/02/25/a-year-in-the-lives-of-smallholder-farming-families

The World Bank. (2017). Economic diversification through productivity enhancement. Washington, DC. Retrieved from https://openknowledge.worldbank.org/handle/10986/33067

Veihe, A. (2000). Sustainable farming practices: ghanaian farmers ' perception of erosion and their use of conservation measures. *Environmental Management, 25*(4), 393–402.

Bioenergy from Agricultural and Organic Byproducts

Methanogenic Potential of Palm Oil Mill Effluents

Kossia Ouattara, Augustin A. Adima, and Félix A. Adjé

Abstract The processing of palm fruit into crude oil results in the production of various residues, of which liquid effluents are the most significant, with 0.5–0.8 m^3 of effluents produced per ton of processed seed bunches. The systems put in place to treat this waste have revealed some shortcomings and still entail several environmental risks. To remedy this situation, an alternative is being considered: anaerobic digestion, which is recognized as a sustainable process. In addition to treating palm oil mill effluent, this technology offers the prospect of making this activity profitable through the production of energy (gas, heat, electricity). However, given the high acidity (pH = 4.74), load (COD = 60,716 mg O_2/ L) and VFA content (VFA = 5000.33 mg/L) of palm oil mill effluent, it will be important to monitor these parameters during digester tests to reduce the risk of acidosis, a limiting factor in methane production. BMP tests carried out over 29 days at a constant temperature of 37°C, using sludge from a stationary digester as inoculum, revealed a methane yield of 170.26 Nml CH_4/g VS. This shows that palm oil mill effluent is an excellent source of renewable energy. To stabilize the reaction medium and achieve these yields on a larger scale, we need to find more suitable and optimal parameters.

Keywords Methanization · Palm oil mill effluent · Methanogenic potential

K. Ouattara (✉) · A. A. Adima · F. A. Adjé
Sciences agronomiques et procédés de transformation/INP—HB, Abidjan, Côte d'Ivoire
e-mail: kossia.ouattara@inphb.ci

A. A. Adima
e-mail: amissa.adima@inphb.ci

F. A. Adjé
e-mail: anoh.adje@inphb.ci

M.-M. Narra and S. Narra (eds.), *African Green Transition Through Innovative Pathways*, World Sustainability Series, https://doi.org/10.1007/978-3-031-87043-9_3

1 Introduction

Growing industrialization inevitably leads to a significant increase in waste. Without efficient disposal and treatment, we are witnessing an accumulation of this waste, which is a source of pollution.

Côte d'Ivoire, a country whose economy was built on agriculture and whose industry is booming, is faced with a large quantity of agro-industrial waste. This is the case for palm seed processing waste, of which Côte d'Ivoire is currently the second largest producer in Africa, with over 2.7 million tons produced in 2022 (FAOSTAT, 2022). Liquid effluents from palm oil mills represent 50–80% of processing residues (Jacquemard et al., 2012).

Various techniques have been developed to treat and manage this waste and although they offer certain advantages, they also have drawbacks such as greenhouse gas emissions, lack of valorisation of the carbon formed, etc. (Fikri Hamzah et al., 2020). In view of these facts, it is essential to find an alternative to these technologies.

Anaerobic digestion is defined as the decomposition of fermentable organic matter in the absence of oxygen. In addition to biogas, which can be converted into energy, it generates a liquid residue called digestate. In recent years, the use of anaerobic digestion has grown rapidly due to the significant and increasing deployment of renewable energies in relation to the mitigation of greenhouse gas (GHG) emissions and the need for sustainable, less harmful treatment of industrial wastewater. According to Ahmed et al. (2015), anaerobic digestion represents the efficient method for treating effluents, containing a huge organic load such as palm oil mill effluents. This degradation takes place in 4 main stages: hydrolysis, acidogenesis, acetogenesis and methanogenesis. Such a system is highly dependent on the structure of its bacterial community and several environmental factors, including pH, temperature, hydraulic retention time and C/N ratio. The process of anaerobic digestion and biogas production can therefore be inhibited by factors such as acidosis, alkalosis, H_2S accumulation, lack of micronutrients, etc.

Anaerobic digestion is an organic wastewater management technology that not only eliminates organic pollutants from liquid effluents, but also converts part of the carbon into combustible biogas, which can be valorised into heat and electricity. This technology therefore offers the prospect of energy autonomy and profitability for liquid effluent management activities.

As biogas production from anaerobic digestion is doubly beneficial, various studies are being carried out to ensure stable, high production from palm oil mill effluent. The aim of these trials is to determine the methane potential of palm oil mill effluents.

Table 1 List of main parameters to be determined

Parameters	Methods	Objective
pH	NF EN ISO 10523	Indication of acidity
TS (%)	NF EN 12,880	An overview of the total load
VS (%)	NF EN 12,879	Key indicator of substrate biodegradability
TSS (mg/L)	NF EN 872	Concentration of non-dissolved elements
COD (g/L)	NF T90-101	Evaluation of water pollutant load
BOD_5	NF EN ISO 5815–1	Evaluation of the concentration of organic pollutants
TOC	NF EN 1484	Measurement of the quantity of organic compounds contained in Water
VFA	NF T90-114	Assessment of potential physico-chemical disturbances in the digester
C/N Ratio		Assessment of substrate decomposition capacity

2 Materials and Methods

2.1 Inoculum and Substrates

The inoculum used is digested sludge from a digester in stationary condition (no significant variations in biogas production during the last residence time) with a diversified microbial flora adapted to the digestion of complex organic matter. The inoculum should have a low methane production, i.e. less than 20% of the total methane production of the inoculum + substrate, measured during the BMP test.

It was stored at room temperature (around 27°C) for 5 days before use, to deplete residual substrates from the digester.

Palm Oil Mill Effluent (POME) was used as the substrate in this experiment. The substrate was collected, transported and refrigerated at 4°C before use.

2.2 Characterization of Inoculum and Substrates

The parameters sought are defined in Table 1.

2.3 Determination of Methanogenic Potential

Biochemical Methane Potential (BMP) tests are used to determine the anaerobic degradability and methane production potential of organic substrates under stable and controlled conditions. These tests involve placing one or more substrates in contact with an inoculum in bottles or closed, single-feed reactors, and continuously measuring biomethane production until the substrate or substrates are exhausted.

The tests were carried out in 1000 mL bottles at 37 °C. To keep the flasks at a constant temperature throughout the tests, they will be placed in a water bath. Each bottle contained 491 mL of inoculum and 204 mL of POME, while the blank bottles were filled with 700 mL of sludge. To create anaerobic conditions, the gas in each bottle was replaced by nitrogen (N_2).

Methane production in each test flask is the combination of two parameters: the volume of biogas produced and the percentage of methane in this biogas. These two parameters will be determined respectively by measuring the pressure rise in the gas stream and by analyzing a gas sample collected in a specific bag. The pressure inside the bag is measured using a manometer fitted with a needle that is simply inserted into the rubber septum. The gas produced is collected in a bag with a tap, and then connected to a gas analyser to determine its composition. However, it may be that daily production is too low, in which case sampling can be carried out over several days, as the bags are well sealed.

Methanogenic potential is determined by the Eq. 1:

$$BMP = Y_{CH4-substrat} = \frac{V_{CH4-total} - VS_{inoc} * Y_{CH4-blank}}{SV_{substrate}} \tag{1}$$

$Y_{CH4\text{-}blank}$ (in NL/kgVS) is the methane yield of the blank using the volume of CH_4 measured in the blank ($V_{CH4\text{-}blank}$, in NL) and the Volatiles Solids initially introduced into the blank (VS_{blank}, in kgVS) as in Eq. 2:

$$Y_{CH4-blanc} = \frac{V_{CH4-blank}}{MV_{blank}} \tag{2}$$

$V_{CH4\text{-}total}$ is the volume of CH_4 (NL) measured in a condition, VS_{inoc} and $VS_{substrate}$ the Volatiles Solids (kgVS) of the inoculum and substrate in this condition, and $Y_{CH4\text{-}substrate}$ the methane yield of the substrate (NL/kgSV), which is the desired value.

$$V_{CH4-total} = MV_{inoc} * Y_{CH4-blanc} + MV_{substrat} * Y_{CH4-substrat} \tag{3}$$

Table 2 Characterization of POME

Parameters	POME	Units
pH	4.74 ± 0.15	-
TSS	18.33 ± 0.91	g/L
TS	$3.45\% \pm 0.02\%$	% RM
VS	$88.33\% \pm 0.2\%$	% TS
Total COD	60.72 ± 9.88	g O_2/L
Soluble COD	18.62 ± 8.48	g O_2/L
BOD_5	75.0 ± 3.1	mg O_2/L
TOC	$58.010\% \pm 0.001\%$	% MS
VFA	5003.33 ± 403	mg/L
C/N Ratio	15.33	-

3 Results and Discussion

3.1 Characterization of Palm Oil Mill Effluents

COD and VS indicate that POME can be an excellent substrate for biogas production, as it is rich in biodegradable organic matter, which is also confirmed by the VS/TS ratio of 0.88 (Vanegas et al., 2024).

The low pH and relatively high charge and volatile fatty acid content indicate a risk of acidosis in the digester, so a potential inhibition of methane production. It is therefore essential to monitor this parameter during digester tests.

Table 2 shows the average characterization results for palm oil mill effluent.

3.2 Methanogenic Potential of Palm Oil Mill Effluents

Figure 1 shows the cumulative biogas production during the BMP test of palm oil mill effluent.

Biogas production started off quickly and began to stabilize from day 13, reaching volumes close to 5,000 mL biogas. Each of the effluent samples produced respectively: 4518.04, 4887.34 and 4387.13 Nml of biogas during 29 days of digestion.

From the analysis of the biogas obtained during the test period (see Table 3), we can obtain the volume of methane.

By relating this cumulative methane production to the quantity of volatile matter used to carry out these BMP tests, we determine the methanogenic potential expressed in NmL CH_4/gVS. See Fig. 2.

The curve is divided into two main parts:

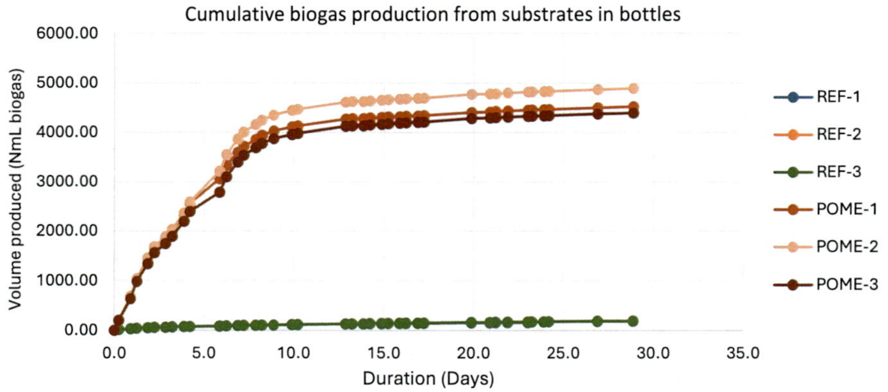

Fig. 1 Cumulative biogas production

Table 3 Analysis results of biogas samples

TRIALS		CH4 (%)	CO2 (%)	H2S (%)
Inoculum		28,30	19,50	0,02
POME—1	Day 2	33,60	32,00	0,05
	Day 6	57,80	28,80	0,02
	Day 29	58,10	26,40	0,01
POME—2	Day 2	35,00	33,20	0,05
	Day 6	57,90	28,50	0,03
	Day 29	59,90	27,10	0,02
POME—3	Day 2	32,80	33,00	0,05
	Day 6	56,30	30,50	0,03
	Day 29	59,30	26,60	0,02

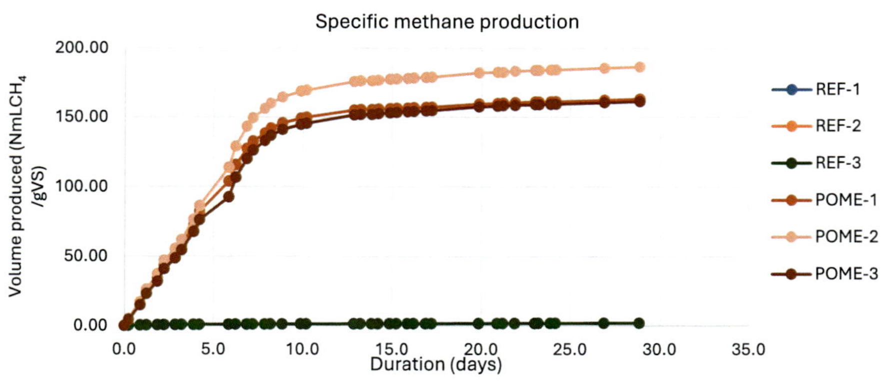

Fig. 2 Specific methane production

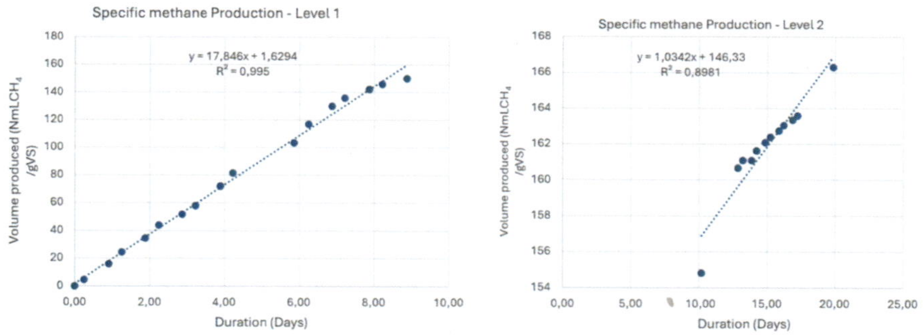

Fig. 3 Specific methane production levels

The slope or methane production rate at the start of the tests, is used to indicate the rate at which the organic matter is being degraded. In our case, the slope is subdivided into two levels:

The first stage, with a rate of 17.846 mL CH_4/gVS/Day, shows a rapid degradation of the substrate.

The second stage, with a rate of 1.034 mL CH_4/gVS/Day, indicates substrate exhaustion, meaning that the organic matter is almost completely degraded.

The plateau represents the total methanogenic potential of the substrate, i.e. the maximum quantity of methane that the substrate can produce under optimum conditions. The value obtained for the methanogenic potential of POME is 170.26 mL CH_4/gVS.

Figure 3 shows the two levels of specific methane production.

4 Conclusion

Palm oil mill effluent represents an excellent source of clean energy thanks to its high organic matter load, with a methanogenic potential of 170.26 NmL CH4/g VS. The relatively rapid organic matter degradation rate of 17.846 CH4/gVS/Day is an important factor in determining the hydraulic retention time. With a VFA content of 5000.33 mg/L, this is an important parameter to monitor during digester trials to reduce the risk of inhibiting methane production. Converting POME into biogas offers an answer to the growing need for energy. The search for optimum parameters to achieve such yields is envisaged for larger-scale trials.

Abbreviations

BMP Biochemical Methane Potential
BOD Biochemical Oxygen Demand

C/N	Carbon/Nitrogen Ratio
COD	Chemical Oxygen Demand
RM	Raw Materials
TOC	Total Organic Carbon
TS	Total Solids
TSS	Total Suspended Solids
VFA	Volatile Fatty Acids
VS	Volatile Solids

References

Ahmed, Y., Yaakob, Z., Akhtar, P., & Sopian, K. (2015). Production of biogas and performance evaluation of existing treatment processes in palm oil mill effluent (POME). *Renewable and Sustainable Energy Reviews, 42*, 1260–1278.

Angelidaki, I., Alves, M., Bolzonella, D., Borzacconi, L., & Campos; J. L., Guwy; A. J., Kalyuzhnyi, S., Jenicek, P., van Lier, J. B. (2009). Defining the biomethane potential (BMP) of solid organic wastes and energy crops: A proposed protocol for batch assays. *Water Science & Technology., 59*(5), 927–934.

FAO. (2022). FAOSTAT database, food and agriculture organization of the united nations. Available from: https://www.fao.org/faostat. (Last Accessed: Dec 15, 2024)

Fikri Hamzah, M. A., Abdul, P. M., Mahmod, S. S., Azahar, A. M., & Jahim, J. M. (2020). Performance of anaerobic digestion of acidified palm oil mill effluent under various organic loading rates and temperatures. *Water, 12*(9), 2432. https://doi.org/10.3390/w12092432

Holliger, C., Alves, M., Andrade, D., Angelidaki, I., Astals, S., Baier, U., Bougrier, C., Buffière, P., Carballa, M., de Wilde, V., Ebertseder, F., Fernández, B., Ficara, E., Fotidis, I., Frigon, J.-C., Fruteau de Laclos, H., Ghasimi, D. S. M., Hack, G., Hartel, M., … S., Torrijos, M., van Eekert, M., van Lier, J., Wedwitschka, H., Wierinck, I. (2016). Towards a standardization of biomethane potential tests. *Water Science and Technology., 74*(11), 2515–2522.

Jacquemard, J.-C., Berthaud, A., Ollivier, J., Ollivier, L., Verwilghen, A., Graille, J., & de Franqueville, H. (2012). *Le palmier à huile*. CTA, Presses agronomiques de Gembloux.

Ramón Vanegas, A. A., Vásquez, J., Molina, F., Peñuela Vásquez, M. (2024). Evaluation de la digestion anaérobie des effluents d'huilerie de palme (POME) en utilisant différentes boues comme inoculum, *Water Resources and Industry,* 31

Strategies for Improving the Quality of Low-Pressured Rice Husk Briquette: A Mini-Review

S. U. Yunusa, E. Mensah, S. Narra, A. Saleh, and K. Preko

Abstract Recently, there has been a growing interest in biomass briquette production. While this has been encouraged through the advocacy for reducing, reusing, recycling, and recovering waste as part of measures for attaining a sustainable green future and mitigating climate change, it is on the other hand a result of the energy deficit in most parts of developing countries. Despite this advancement, the processing of rice husk briquette using the low-pressured technique has been quite slow and a bit challenging due to the recalcitrant nature of rice husk resulting from its large grade sizes, low density, high silica content, and ash content. This paper combines recent literature on the strategies for improving the quality of rice husk briquettes produced through low-pressure densification to establish the current state of knowledge. These include pretreatment measures such as milling, carbonization, and torrefaction intended to improve the physical and thermal properties of the biomass and briquettes. In addition, co-densification with other biomass or the use of blends to enhance bonding and curtail ash content, and by increasing the application pressure and temperature of the machines are additional quality enhancement methods reported in this review. These measures enhance the compression process and yield dimensionally stable, durable, and thermally efficient briquettes. The paper also reports the potential of applying the measures mentioned above

S. U. Yunusa (✉)
WASCAL Graduate Research Programme On Climate Change and Land Use, Department of Civil Engineering, KNUST-Kumasi, Ghana
e-mail: suyunusa@abu.edu.ng; yunusa.s@edu.wascal.org

S. U. Yunusa · A. Saleh
Department of Agricultural and Bio-Resources Engineering Ahmadu Bello University, Zaria, Nigeria

E. Mensah
Department of Agricultural and Biosystems Engineering, KNUST-Kumasi, Ghana

S. Narra
Professorship of Material and Energy Valorization of Biogenous Residues, University of Rostock, Rostock, Germany

K. Preko
Department of Physics, KNUST-Kumasi, Ghana

M.-M. Narra and S. Narra (eds.), *African Green Transition Through Innovative Pathways*, World Sustainability Series, https://doi.org/10.1007/978-3-031-87043-9_4

to other biomasses intended for low-pressured densification. Having looked at the reviewed strategies, the paper assessed the impact of some of the methods on the quality of briquette.

Keywords Rice husk · Briquette · Energy · Pretreatment · Quality

1 Introduction

Household energy use is among the most studied energy utilization patterns globally. A majority of households in developing countries live in energy poverty (Li et al., 2023). In these parts of the world, cooking is the most energy-consuming act in the households (Dioha & Kumar, 2020). It accounts for more than 90% of the gross energy utilization (Sedighi & Salarian, 2017). However, many households find it difficult to afford or access clean energy (Yunusa et al., 2024b). Hence, they rely mainly on wood fuel energy sources like charcoal and firewood to meet their energy needs. This has contributed to deforestation and climate change. To augment the energy shortfall in this part of the world, waste biomass valorisation to energy such as fuel briquettes is encouraged. Briquette production is gaining popularity as it is not just a method of producing energy but is also considered a waste treatment technique. Unfortunately, the recalcitrant nature of most biomass has been a limiting factor in the biorefining process (Sidana & Yadav, 2022). This is more critical in physical conversion processes like briquette production. On this note, several measures such as biomass pretreatments, blends or additives, and other process adjustments like upgrading the production systems are employed to improve the quality of the biomass and the resulting fuel.

Rice husk is one of the most common feedstocks for bio-energy production. This is partly because rice husk has good potential for energy production and is abundant in most countries of the world (Jyothsna et al., 2024). In addition, rice husk has a relatively good calorific value usually between 14.7 and 16.6 MJ/kg (ISO 17225-1, 2020), conforming with the ENplus limit of 16.56 MJ/kg for fuel pellets (EPC, 2015). However, as earlier explained, due to the properties of rice husks, only a few studies have utilized it in unpretreated form for briquette production. Studies prefer rice husks in carbonized, torrefied, or blended form with another biomass to curtail the recalcitrance.

In addition to pretreatments, improving process inputs like pressure and temperature enhances the briquette's quality. Generally, briquettes are produced at low and high pressure and temperature. Although the low-pressure and temperature method of production is easier and less costly (Yank et al., 2016), processing rice husk at this pressure and temperature is usually very challenging due to its physical properties. On this basis, additives and/or binders are typically required for improved quality and durability (Yunusa et al., 2024a). Although binders enhance the briquette's quality, they sometimes influence the thermal and emission quality. On this note, the percentage of binders must be kept as low as possible. For low-pressured production, the binder must not exceed 20% (ISO 17225–1, 2020). Although some studies on low-pressure rice husk briquette production such as Yunusa et al., (2024a) and Yank et al., (2016) complied with the ISO 17225 standard on the use of additives, the quality of briquettes obtained was less than those made from other biomass under low pressure such as sawdust and corncob (Mitchual et al., 2013), and oil palm trunk and corn cob (Kpalo et al., 2020). This then verifies the assertion of Lubwama and Yiga (2018) who report that the physical properties of rice husk limit its performance when densified to briquette. On this note, pretreatment measures and other process adjustment techniques become imperative in processing rice husk into briquettes.

Against this background, this paper reviewed the literature on briquette production and quality enhancement spanning the last decade. Special emphasis was given to studies that used rice husk as a feedstock or co-feedstock in low-pressured briquette production. As shown in Fig. 1, the review stems from the outlook of Authors studies on rice husk briquette production and optimization (Yunusa et al., 2024a, 2024c), through which literature was sourced from the last 10 years (2014–2024) focusing on articles on briquette quality improvement, such as pretreatment, use of blends and additives, etc. The articles were critically reviewed to highlight the most used methods and their proven impacts by discussing the key findings.

Fig. 1 Review flowchart

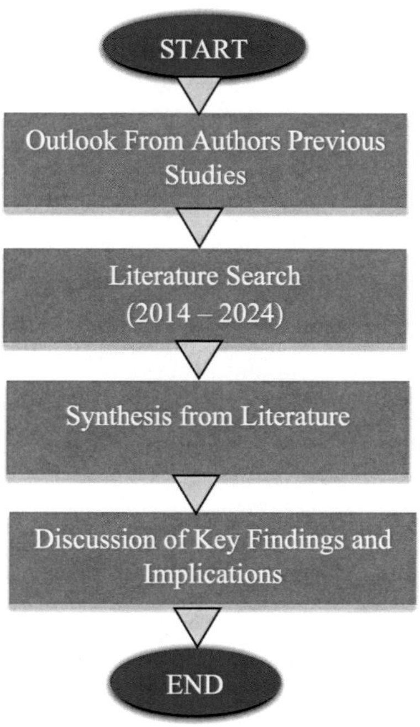

1.1 Briquette Quality Metrics and Methods of Improvement

The quality metrics of briquettes are the various parameters used to ascertain the briquette's strength, durability, and efficiency. Conventionally, briquettes are evaluated for physical/mechanical and thermal performances. The metrics used to determine the physical performance of the briquettes are density (compressed and relaxed), relaxation ratio, impact resistance, water absorption resistance, mechanical durability, and compressive strength (Yunusa et al., 2023). The metrics used for thermal performance are specific fuel consumption, burning rate, time taken to cook or boil water, and emissions from the briquette (CO_2, CO, NO, HC, etc.). These methods have been exhaustively discussed in the literature (Aggarwal & Chandel, 2022; Obi et al., 2022; Yunusa et al., 2024a). However, because there are several methods reported for evaluating each quality parameter, there is a need to have a uniform and standardized testing protocol (Obi et al., 2022). This would reduce uncertainties and biases in measurements.

Although the quality of briquettes made for industrial use is usually higher than those made for domestic application, Richards (1990) recommends a minimum impact resistance index of 50% and a compressive strength above 350 kPa for briquettes developed for industrial or domestic use. On this note, it is important to ensure the quality meets the standard. However, attaining this standard minimum has been challenging for briquettes developed from a low-pressure system. Therefore, measures such as biomass pretreatments, co-densification or use of blends/ additives, and adjustments of process conditions like pressure, and temperature are pertinent to achieve the desired quality. The framework considered in this review is summarized in Fig. 2

1.2 Pretreatment

The pretreatment process breaks the lignocellulosic chain and makes the polysaccharide elements accessible to the hydrolytic enzymes for the release of fermentable sugars (Sidana & Yadav, 2022). Through this process, the lignin in the biomass is disintegrated from the complicated structure of cellulose and hemicellulose (Mankar et al., 2021). The process enhances the surface area and activates core constituents of the biomass like lignin. Pretreatment methods are classified as physical, chemical, thermal, and biological. However, the physical and thermal pretreatment methods are the most reported for briquette production. Hence, these are the only methods considered in the current review.

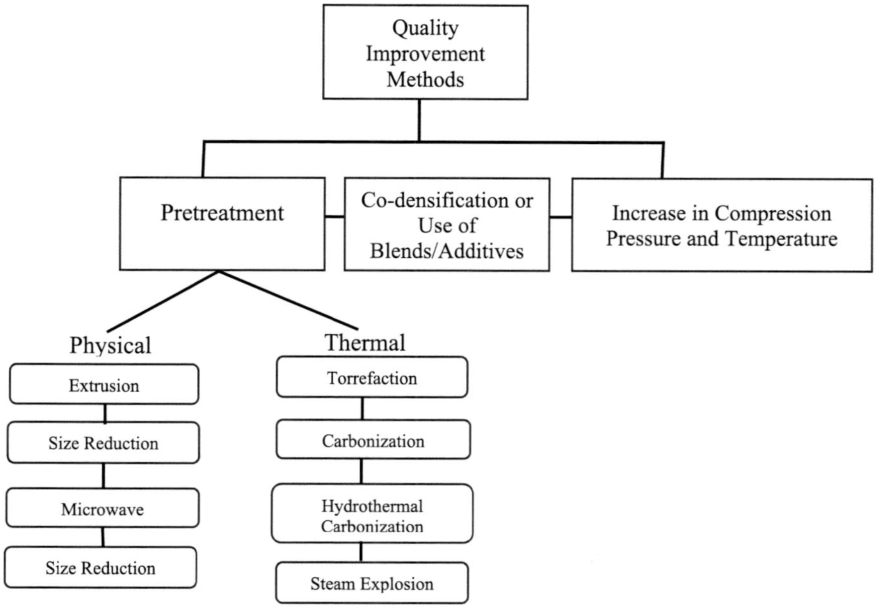

Fig. 2 Quality improvement framework

1.2.1 Physical Pretreatment

Physical pretreatment is commonly employed to transform the physical structure of the biomass. This mostly reduces the particle sizes or transforms the textural and physical form through heating and extrusion. Physical pretreatment measures include extrusion, size reduction (shredding, chipping, chopping, milling, crushing, and grinding), microwave pretreatment, and ultrasound pretreatment (Mankar et al., 2021). However, the most reported physical pretreatment method for briquette production is size reduction, especially milling. Studies that used this method on rice husks include Yunusa et al. (2024a) who milled to $\leq$ 2 mm, Arewa et al. (2016) milled to $\leq$ 1 mm, and Ndindeng et al. (2015) milled to 0.3 to 1 mm.

Extrusion: The extrusion process is a physical treatment that uses a thermo-mechanical technique to transform biomass by using the shear force generated from the rotation of the extruder screw (Mankar et al., 2021). The advantage of this process is that it requires less to moderate temperature, and does not generate microbial inhibitors (Sidana & Yadav, 2022). However, it is more suitable for herbaceous biomass as it partially degrades the lignocellulosic chain (Sidana & Yadav, 2022). Another advantage of extrusion pretreatment is that it is flexible and adaptable to other uses outside biomass pretreatment (Konan et al., 2022). On this basis, the extruders are employed in a screw press for biomass pelleting or briquette production. When using screw extruders for briquette production, binders are not required because the biomass is compressed to a very high pressure (Dinesha et al., 2019).

Studies that used extruders in briquette production include Saneewongnaayuttaya et al. (2020) who used it to produce rice husk briquette, and Yang et al. (2021) who used it to produce sawdust briquette. However, Höntsch et al. (2024), used a twin-screw extruder to comminute wheat straw for briquette production.

Size Reduction: Size reduction is the most common form of physical pretreatment. It involves reducing the sizes of biomass through chipping, chopping, crushing, milling, and grinding. The goal is usually to reduce the particle sizes and increase the surface areas. Through this, there would be an improved interparticle bonding at both core and surface layers during densification, yielding high-density briquettes. Another advantage of size reduction is that it reduces cellulose crystallinity, improves enzymatic hydrolysis, reduces cellulose polymerization, and improves mass transfer due to particle size reduction (Mankar et al., 2021). Milling is the most reported size reduction pretreatment for briquette production. Several studies have employed milling on various biomass using different screen sizes and devices. Yunusa et al. (2024a), used a hammer mill of screen size of 2 mm to mill rice husk (Fig. 3), Ndindeng et al. (2015) also milled rice husk and bran to ≥ 1 and ≤ 0.30 mm using mesh 18 and 50. Others include Unrean et al. (2018) who milled rice husk to ≤ 1 mm, and Yank et al. (2016) who milled rice husk to ≤ 2 mm.

Microwave Pretreatment: Microwave pretreatment or irradiation degrades the biomass by inducing an explosion within the particles to facilitate the disruption of recalcitrant structures (Mankar et al., 2021). The technology has been hailed as one of the best non-conventional pretreatment methods especially as it is proven essential for obtaining polysaccharide streams which include cellulose and hemicellulose (Sidana & Yadav, 2022). Jayakumar et al. (2022) employed microwave-heated alkali pretreatment on rice husks and observed an excellent impact on the quality of pellets. In some literature, it is classified as a physical pretreatment (Mankar et al., 2021), while in some it is classified as a thermal pretreatment.

Fig. 3 Rice husk **a** as received (6–8mm) **b** milled ($\leq 2mm$)

Ultrasound Pretreatment: Ultrasound pretreatment utilizes acoustic cavitation in the form of high-frequency waves to produce microbubbles which eventually collapse and release a high temperature of about 5000 °C and a pressure of 50 MPa which ruptures the glycosidic bonds in the intricate lignocellulosic chain of the biomass (Mankar et al., 2021). The method has been noted to have an excellent impact with 100% lignin removal and more than 90% reduction in the yield of sugar at the laboratory scale (Sidana & Yadav, 2022). However, this pretreatment method has not been largely employed for pellet or briquette production.

1.3 Thermal Pretreatment

Thermal pretreatment involves the application of heat to the biomass material to enhance its biorefinig efficiency. Thermal pretretament methods include torrefaction, carbonization, hydrothermal carbonization, and steam explosion.

Torrefaction: This is a thermal pretreatment method whereby the feedstock is degraded at a temperature range of 200°C to 300°C under an inert environment (Amira Atan et al., 2018). Torrefaction can be wet, dry, or ionic-liquid-assisted based on the operating condition (Olugbade & Ojo, 2020). However, dry torrefaction is the most used method for solid fuels like briquettes (Olugbade & Ojo, 2020). When employing this pretreatment method, it is important to maintain an optimum torrefaction temperature as a very high temperature may increase ash content, and reduce the density and compressive strength of briquettes (Mohd Faizal et al., 2018). Wu et al. (2018) observed that torrefaction negatively affects densification when applied on cotton stalk and wood sawdust, suggesting hydrothermal carbonization as a better option for pre-treating the stated biomass.

Carbonization: This is a thermal pretreatment technique that involves heating a biomass in an air-tight medium at a temperature range of 300 – 400°C for low-temperature carbonization or a temperature above 400°C for high-temperature carbonization (Boumediene Allaoua et al., 2018). Through this process, the biomass cell is partially degraded and the volatile content is expelled, thereby increasing the flame quality and other physicochemical properties (Nazari et al., 2020; Ngene et al., 2024). Thus, when the biomass is used in briquette production, it densifies easily due to the presence of free radicals (Lubwama et al., 2020), and also yield compact and durable briquettes with good thermal and mechanical properties (Yiga et al., 2021).

Hydrothermal Carbonization (HTC): HTC is a thermochemical process whereby a biomass feedstock is treated in an aqueous environment at a temperature of 180–300 °C and a generated pressure to produce a solid hydrochar (Pauline & Joseph, 2020). Within this subcritical temperature (< 374 °C), the cell walls of the biomass are denatured and the hemicellulose degraded and lignin transformed into sugar/syngas thereby releasing acids and inhibitors from the biomass (Sankaran et al., 2020). Studies that employed HTC include but are not limited to Wang et al. (2020) Who pretreated wood vinegar for briquette production, Wu et al. (2018) where cotton stalk and pine wood sawdust were pretreated at a laboratory scale.

Steam Explosion: This is the process where the biomass feedstock is decompressed to open up the fibres through the application of hot steam (180 to 240 °C) released at a pressure level of 1 to 3.5 MPa, thereby enhancing the accessibility of the polymers during the subsequent processes like fermentation, hydrolysis, pelleting, or briquetting (Stelte, 2013). For briquetting and pelleting, the process softens the lignin and distributes it evenly onto the raw material. This then improves bonding and results in improved density (630 to 750 kg/m^3) of pellets compared to unpretreated biomass (Stelte, 2013).

The process exposes the biomass to saturated steam using high and rapidly dropping pressure, which eventually causes the material to decompress explosively (Hoang et al., 2023). Steam explosion is more applicable in biorefineries involving cellulosic ethanol production and acid hydrolysis (Sidana & Yadav, 2022). While it can be employed on biomass intended for briquette production, there are limited applications.

1.4 Increase in Compression Pressure/temperature

The application pressure and temperature are very essential process factors in briquette production. Depending on the type of machine used, briquetting pressure is largely classified as low, medium, and high-pressure (Ibitoye et al., 2021). Machines with a pressure range of $\leq$ 5 MPa are classified as low-pressure machines, those with a pressure range between 5 and 100 MPa are classified as medium pressure, and machines with a pressure of 100 MPa and above are high-pressure machines (Eriksson & Prior, 1990). While the compression temperature is classified as low or room temperature and high-temperature compression. Briquettes produced with small-scale presses like plungers and hydraulic piston presses are mostly within the low-pressure range. Thus, yield briquettes with lower density and durability than those made with medium to high-pressured presses. While several studies, including Yunusa et al. (2024a), and Yank et al. (2016) justified the use of low-pressured machines to cost and energy consumption, it is important to note that, the pressure of these machines can be slightly improved to the lower medium range of 5 MPa to about 20 MPa, or above without significantly changing the production cost and energy. Studies that produced briquettes within this pressure range using manual presses include that of Sunnu et al. (2021) who densified biomass including rice husk at 10 to 30 MPa, Orisaleye et al. (2022) densified sawdust at 9 to 15 MPa, and Navalta et al. (2020) who densified rice bran and sugarcane bagasse at 8 to 12 MPa. The findings from these studies have proved that the quality of briquettes can be

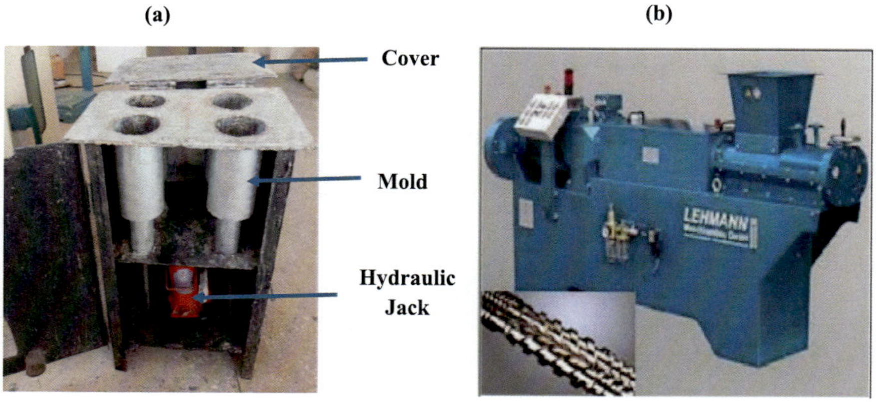

Fig. 4 **a** Hydraulic Piston Press (Yunusa et al., 2024a), **b** Screw Extruder (Narra et al., 2012)

improved using the same machine. Figure 4 shows examples of a low-pressure press (Fig. 4a) and a high-pressure press (Fig. 4b).

1.5 Co-Densification or Use of Blends/additives

In briquette production, co-densification implies blending and densifying two or more biomass feedstocks. This is broadly considered to improve the briquette's quality and performance especially if the primary biomass has high recalcitrance like ash content, silica, and emission constituents like sulphur and nitrogen. Although rice husk has a relatively good heating value (HHV) between 14.7 MJ/kg and 16.6 MJ/kg (ISO 17225–1, 2020), and fixed carbon between 10.20% and 14.50% (Mansaray & Ghaly, 1997), it is still considered a recalcitrant biomass as it has a high ash content of 13% to 23% (ISO 17225–1, 2020). In this case, feedstock like corn cob with an ash content of 3.2% and a fixed carbon of 20.7% (Okot et al., 2018), or sugarcane bagasse with ash content and fixed carbon of 3.3% and 15.8%, respectively, can be blended with rice husk to improve the thermal performance.

1.6 Effect of Blends and Selected Pretreatment Methods on the Quality Metrics of Briquettes

Having highlighted the various biomass pretreatment techniques and the potential benefits of co-densification or the use of blends and additives on rice husks, it is important to know how they impact the final quality of briquettes. Tables 1 and 2 highlight some quality improvement methods and the resultant impact on selected quality metrics (density, compressive strength, impact resistance, and heating value). In Table 1, the study of Yunusa et al. (2024a) which was carried out using purely rice husk was considered as the baseline. When compared with other studies that co-densified or blended rice husk with other biomasses, there is a significant improvement in the quality. For example, despite using an extremely low compression pressure of 0.025–0.065 MPa, Ajimotokan et al. (2019) obtained a higher compressed density of 1100–1860 kg/m^3 and relaxed density of 420–780 kg/m^3 by co-densifying rice husk and maize cob. This is significantly higher than the baseline study in which only rice husk was used.

Table 2 presents the effect of selected pretreatment methods on the quality metrics. As shown, studies that employed multiple pretreatment methods (e.g. milling and carbonization) like Ndindeng et al. (2015) and Nagarajan and Prakash (2021) had briquettes with better calorific value than those that used a single pretreatment. This is because carbonization partially degrades the biomass cell and removes the volatile content of the lignin and hemicellulose, thereby increasing the flame quality (Nazari et al., 2020). Conversely, those that employed only the physical pretreatment (milling/grinding) like Arewa et al. (2016) and Lakshika et al. (2024) had better density. This could be because milling reduces the particle sizes of the biomass and increases the surface areas. Through this, there is a corresponding increase in interparticle bonding at the core and surface layer of the briquette during compression. This further shows that the pretreatment methods affect the quality metrics differently. Therefore, a careful selection is needed based on the quality of interest.

Table 1 Effect of feedstock blends and compression pressure on the mechanical properties of briquettes

Feedstock	Co-feedstock/ Blend	Compression Pressure (MPa)	Compressed Density (kg/m^3)	Relaxed Density (kg/m^3)	Compressive strength (kN/m^2)	Impact Resistance (%)	Reference
Rice husk	-	4.5	495–691	196–306	20–410	12.5–100	(Yunusa et al., 2024a), AS
Rice husk	Rice bran	4.2	–	471.3	2.54 kN, 0.0306 N/ mm^3	–	(Yank et al., 2016)
Rice husk	Citrus Peel	0.001	350–460	–	230 – 370	–	(Magnago et al., 2020)
Rice husk	Sawdust and Corncob	0.0012	518–675	–	–	–	(Adadu et al., 2024)
Rice husk	Corncob	0.025–0.065	1100–1860	420–780	39–111	–	(Ajimotokan et al., 2019)

Note AS is Authors study

Table 2 Effect of other pretreatment measures on the mechanical properties of briquettes

Feedstocks	Type of Pretreatment	Compression Pressure (MPa)	Compressed Density (kg/m^3)	Relaxed Density (kg/m^3)	Compressive strength (kN/m^2)	Impact Resistance (%)	Heating Value (HHV) MJ/kg	Reference
Rice husk	Milling (≤ 2mm)	4.5	495–691	196–306	20–410	12.5–100	13.54–18.74	(Yunusa et al., 2024a), AS
Rice husk and bran	Milling (≤ 0.3to ≥ 1 mm), carbonization	7.74	506–605		–	–	9.66–19.23	(Ndindeng et al., 2015)
Rice husk	Milling (≤ 1mm)	8	1080.8–1159.6	552.3–632.2	–	–	–	(Arewa et al., 2016)
Rice husk, corncob, sugarcane bagasse	Carbonization, Milling (≤ 1mm),	–	530–625	–	–	95.4	26.48–27.07	(Nagarajan & Prakash, 2021)
Rice husk	Carbonization	≤ 7	–	–	–	≥ 94	9.5–16.6	(Lubwama & Yiga, 2018)
Rice husk	Grinding	≤ 5	504.2–834.0	–	–	–	15.28–15.45	(Lakshika et al., 2024)

Note AS is Authors study

2 Conclusion

This paper reports the methodologies for improving the quality of rice husk briquette. The review aligns with the need to make rice husk briquette a sustainable energy solution especially in low to middle-income countries where many households live in energy poverty. Although the review primarily targets rice husks, the measures reported are generic and could be employed on other biomasses, especially when aimed for low-pressure densification. Like many other biomasses, rice husk is highly recalcitrant, making it difficult to be valorised into fuel briquettes at low pressure. Measures such as pretreatments, blends/additives, and other process adjustments like increasing compression pressure and temperature may be employed to curb its recalcitrance and improve the quality of the briquette. Through these measures, the complex lignocellulosic chains are degraded, thereby disintegrating the lignin from the structure of cellulose and hemicellulose. Out of all the methods of quality improvements reported in this review, physical pretreatment through size reduction, thermal pretreatment through carbonization, and co-densification with other biomass are the most reported/used quality improvement methods for briquette. The study therefore suggests using these measures to improve the quality of low-pressure rice husk briquettes.

Acknowledgements The Authors are grateful to the German Federal Ministry of Education and Research (BMBF) for funding the study through the West African Science Service Centre on Climate Change and Adapted Land Use (WASCAL), under the Graduate Research Programme on Climate Change & Land Use (CCLU), College of Engineering, Kwame Nkrumah University of Science and Technology, Kumasi, Ghana. The Authors are also grateful to Ahmadu Bello University Zaria, Nigeria for supporting the lead Author with a study fellowship.

References

Adadu, C. A., Ekpo, D. D., & Okpo, E. E. (2024). Characterization of fuel pellets made from rice husk, sawdust and corncob bond with cissus populnea and manihot esculenta. *International Journal of Multidisciplinary Research and Analysis, 7*(07), 3150–3158.

Aggarwal, R. K., & Chandel, S. S. (2022). A comprehensive review of four decades of thermally efficient biomass cookstove initiatives for sustainable development in India. *International Journal of Ambient Energy.*

Ajimotokan, H. A., Ibitoye, S. E., Odusote, J. K., Adesoye, O. A., & Omoniyi, P. O. (2019). Physico-Mechanical characterisation of fuel briquettes made from blends of corncob and rice husk. *Journal of Physics: Conference Series, 1378*(2).

Amira Atan, N., Mohamed Nazari, M., & Ayuni Azizan, F. (2018). Effect of torrefaction pretreatment on physical and combustion characteristics of biomass composite briquette from rice husk and banana residue. *MATEC Web of Conferences, 150.*

Arewa, M. E., Daniel, I. C., & Kuye, A. (2016). Characterisation and comparison of rice husk briquettes with cassava peels and cassava starch as binders. *Biofuels, 7*(6), 671–675.

Boumediene Allaoua, Brahim Mebarki, Salah Tamalouzt, Nabil Benyahia, Abdelmounaim Tounzi, Amar Bousbaine, et al. (2018). Renewable energy: resources, challenges and applications. In *The Economics and Politics of China's Energy Security Transition.*

Dinesha, P., Kumar, S., & Rosen, M. A. (2019). Biomass briquettes as an alternative fuel: a comprehensive review. *Energy Technology, 7*(1801011).

Dioha, M. O., & Kumar, A. (2020). Exploring sustainable energy transitions in sub-Saharan Africa residential sector: The case of Nigeria. *Renewable and Sustainable Energy Reviews, 117*.

Eriksson, S., & Prior, M. (1990). The briquetting of agricultural wastes for fuel. In *FAO Environment and Energy* (11th ed., Vol. 11, p. 137). Food and Agriculture Organization.

European Pellet Council (EPC). (2015). Quality certification scheme for wood pellets. Part 3: Pellet Quality Requirements. In *EN plus Handbook* (Issue 3.0, pp. 1–10).

Hoang, A. T., Nguyen, X. P., Duong, X. Q., Ağbulut, Ü., Len, C., Nguyen, P. Q. P., Kchaou, M., & Chen, W. H. (2023). Steam explosion as sustainable biomass pretreatment technique for biofuel production: Characteristics and challenges. *Bioresource Technology, 385*, 129398.

Höntsch, S., Fehse, F., Schröder, H. W., Herdegen, V., & Braeuer, A. S. (2024). Influence of comminution and briquetting parameters on the agglomeration behaviour of wheat straw. *Biomass and Bioenergy, 182*, 107077.

Ibitoye, S. E., Jen, T. C., Mahamood, R. M., & Akinlabi, E. T. (2021). Densification of agro-residues for sustainable energy generation: an overview. *Bioresources and Bioprocessing, 8*(75).

International Organization for Standardization ISO 17225–1. (2020). *Solid biofuels—Fuel specifications and classes—Part 1: General requirements.*

Jayakumar, E., Chittibabu, S., Shanmugasundaram, S., & Tabil, L. G. (2022). Synergetic effect of microwave heated alkali pre-treatment on densification of rice (Oryza sativa) husk biomass grinds. *Energy Sources, Part a: Recovery, Utilization and Environmental Effects, 44*(2), 5094–5105.

Jyothsna, G., Bahurudeen, A., & Sahu, P. K. (2024). Sustainable utilisation of rice husk for cleaner energy: A circular economy between agricultural, energy and construction sectors. *Materials Today Sustainability, 25*, 100667.

Konan, D., Koffi, E., Ndao, A., Peterson, E. C., Rodrigue, D., & Adjallé, K. (2022). An overview of extrusion as a pretreatment method of lignocellulosic biomass. *Energies, 15*(9).

Kpalo, S. Y., Zainuddin, M. F., Manaf, L. A., & Roslan, A. M. (2020). Production and characterization of hybrid briquettes from corncobs and oil palm trunk bark under a low pressure densification technique. *Sustainability (Switzerland), 12*(2468).

Lakshika, H. N. M., Arachchige, U. S. P. R., & Koliyabandara, P. A. (2024). Production of rice husk derived biomass briquettes for boilers. *SLIIT International Conference on Engineering and Technology*, 286–298.

Li, H., Leng, X., Hu, J., Cao, A., & Guo, L. (2023). When cooking meets confucianism: Exploring the role of traditional culture in cooking energy poverty. *Energy Research and Social Science, 97*, 102956.

Lubwama, M., & Yiga, V. A. (2018). Characteristics of briquettes developed from rice and coffee husks for domestic cooking applications in Uganda. *Renewable Energy, 118*, 43–55.

Lubwama, M., Yiga, V. A., Muhairwe, F., & Kihedu, J. (2020). Physical and combustion properties of agricultural residue bio-char bio-composite briquettes as sustainable domestic energy sources. *Renewable Energy, 148*, 1002–1016.

Magnago, R. F., Costa, S. C., Assunção Ezirio, M. J. de, Godoy Saciloto, V. de, Cremona Parma, G. O., Gasparotto, E. S., Gonçalves, A. C., Tutida, A. Y., & Barcelos, R. L. (2020). Briquettes of citrus peel and rice husk. *Journal of Cleaner Production, 276*.

Mankar, A. R., Pandey, A., Modak, A., & Pant, K. K. (2021). Pretreatment of lignocellulosic biomass: A review on recent advances. *Bioresource Technology, 334*, 125235.

Mansaray, K. G., & Ghaly, A. E. (1997). Physical and thermochemical properties of rice husk. *Energy Sources, 19*(9), 989–1004.

Mitchual, S. J., Frimpong-mensah, K., Darkwa, N. A., & Akowuah, J. O. (2013). Briquettes from maize cobs and Ceiba pentandra at room temperature and low compacting pressure without a binder. *International Journal of Energy and Environment, 4*(38).

Mohd, H., Shamsuddin, H. S., Heiree, M., & M. H., Muhammad Ariff. (2018). Torrefaction of densified mesocarp fibre and palm kernel shell. *Renewable Energy, 122*, 419–428.

Nagarajan, J., & Prakash, L. (2021). Preparation and characterization of biomass briquettes using sugarcane bagasse, corncob and rice husk. *Materials Today: Proceedings, 47*, 4194–4198.

Narra, S., Narra, M., & Ay, P. (2012). Particle size distribution of comminuted and liberated cereal straws measured with different image analysis systems and their characteristic influence on mechanical pellets quality. *International Mineral Processing Congress(IMPC) Proceedings, February 2015.*

Navalta, C. J. L. G., Banaag, K. G. C., Raboy, V. A. O., Go, A. W., Cabatingan, L. K., & Ju, Y. H. (2020). Solid fuel from Co-briquetting of sugarcane bagasse and rice bran. *Renewable Energy, 147*, 1941–1958.

Nazari, M. M., Idroas, M. Y., & Ayuni, F. A. (2020). Carbonization effect on EFB briquettes prepared with different type of binders. *IOP Conference Series: Earth and Environmental Science, 476*(012072).

Ndindeng, S. A., Mbassi, J. E. G., Mbacham, W. F., Manful, J., Graham-Acquaah, S., Moreira, J., Dossou, J., & Futakuchi, K. (2015). Quality optimization in briquettes made from rice milling by-products. *Energy for Sustainable Development, 29*, 24–31.

Ngene, G. I., Bouesso, B., González Martínez, M., & Nzihou, A. (2024). A review on biochar briquetting: Common practices and recommendations to enhance mechanical properties and environmental performances. *Journal of Cleaner Production, 469*, 143193.

Obi, O. F., Pecenka, R., & Clifford, M. J. (2022). A review of biomass briquette binders and quality parameters. *Energies, 15*(2426).

Okot, D. K., Bilsborrow, P. E., & Phan, A. N. (2018). Effects of operating parameters on maize cob briquette quality. *Biomass and Bioenergy, 112*, 61–72.

Olugbade, T. O., & Ojo, O. T. (2020). Biomass Torrefaction for the Production of High-Grade Solid Biofuels: A Review. *Bioenergy Research, 13*(4), 999–1015.

Orisaleye, J. I., Jekayinfa, S. O., Braimoh, O. M., & Edhere, V. O. (2022). Empirical models for physical properties of abura (Mitragyna ciliata) sawdust briquettes using response surface methodology. *Cleaner Engineering and Technology, 7*(100447).

Pauline, A. L., & Joseph, K. (2020). Hydrothermal carbonization of organic wastes to carbonaceous solid fuel – A review of mechanisms and process parameters. *Fuel, 279*,

Richards, S. R. (1990). Physical Testing of Fuel Briquettes. *Fuel Processing Technology, 25*, 89–100.

Saneewongnaayuttaya, N., Panyoyai, N., Khamdaeng, T., & Wongsiriamnuay, T. (2020). Briquette production from rice husk by using screw compaction. *IOP Conference Series: Earth and Environmental Science, 463*(1).

Sankaran, R., Parra Cruz, R. A., Pakalapati, H., Show, P. L., Ling, T. C., Chen, W. H., & Tao, Y. (2020). Recent advances in the pretreatment of microalgal and lignocellulosic biomass: A comprehensive review. *Bioresource Technology, 298*, 122476.

Sedighi, M., & Salarian, H. (2017). A comprehensive review of technical aspects of biomass cookstoves. *Renewable and Sustainable Energy Reviews, 70*, 656–665.

Sidana, A., & Yadav, S. K. (2022). Recent developments in lignocellulosic biomass pretreatment with a focus on eco-friendly, non-conventional methods. *Journal of Cleaner Production, 335*, 130286.

Stelte, W. (2013). Steam explosion for biomass pre-treatment. In *Danish Technological Institute.*

Sunnu, A. K., Adu-Poku, K. A., & Ayetor, G. K. (2021). Production and characterization of charred briquettes from various agricultural waste. *Combustion Science and Technology.*

Unrean, P., Lai Fui, B. C., Rianawati, E., & Acda, M. (2018). Comparative techno-economic assessment and environmental impacts of rice husk-to-fuel conversion technologies. *Energy, 151*, 581–593.

Wang, C., Zhang, S., Wu, S., Sun, M., & Lyu, J. (2020). Multi-purpose production with valorization of wood vinegar and briquette fuels from wood sawdust by hydrothermal process. *Fuel, 282*, 118775.

Wu, S., Zhang, S., Wang, C., Mu, C., & Huang, X. (2018). High-strength charcoal briquette preparation from hydrothermal pretreated biomass wastes. *Fuel Processing Technology, 171*(July 2017), 293–300.

Yang, I., Cooke-willis, M., Song, B., & Hall, P. (2021). Densification of torrefied Pinus radiata sawdust as a solid biofuel: Effect of key variables on the durability and hydrophobicity of briquettes. *Fuel Processing Technology, 214*, 106719.

Yank, A., Ngadi, M., & Kok, R. (2016). Physical properties of rice husk and bran briquettes under low pressure densification for rural applications. *Biomass and Bioenergy, 84*, 22–30.

Yiga, V. A., Lubwama, M., Olupot, P. W., Lubwama, M., & Olupot, P. W. (2021). Effect of alkaline surface modification and carbonization on biochemical properties of rice and coffee husks for use in briquettes and fiber-reinforced plastics effect of alkaline surface modification and carbonization on biochemical properties of rice a. *Journal of Natural Fibers, 18*(4), 620–629.

Yunusa, S. U., Mensah, E., Preko, K., Narra, S., Saleh, A., & Sanfo, S. (2023). A comprehensive review on the technical aspects of biomass briquetting. *Biomass Conversion and Biorefinery*.

Yunusa, S. U., Mensah, E., Preko, K., Narra, S., Saleh, A., Dalha, I. B., & Abdulsalam, M. (2024a). Optimizing selected quality metrics of rice husk briquettes: a response surface methodology approach. *Biomass Conversion and Biorefinery, 0123456789*.

Yunusa, S. U., Mensah, E., Preko, K., Narra, S., Saleh, A., & Sanfo, S. (2024b). Assessing the nexus between household dynamics and cooking energy choice: Evidence from Kaduna state, northwestern Nigeria. *Energy Nexus, 15*, 100310.

Yunusa, S. U., Mensah, E., Preko, K., Narra, S., Saleh, A., Sanfo, S., & Sadiq, N. S. (2024c). Characterization of three lignocellulosic biomass materials as novel binders in briquette production. In *Innovations in Circular Economy and Renewable Energy in Africa* (pp. 153–166). Springer Nature Switzerland.

Carbonization of Mango Waste for Use as Biofuel

Wanignon Ferdinand Fassinou, Dokatiéné Abdoulaye Coulibaly, and Olga Yolande Ako

Abstract In face of climate change and the harmful effects of the misuse of fossil fuels, the use of new and renewable energies is essential as a solution for the salvation of humanity. In addition to the harmful effects of the misuse of fossil energy sources, there is environmental pollution by plant waste such as post-harvest waste (cocoa pods, rice husks and straws, coffee and peanut shells, etc.) but also waste from the consumption of fruits such as mango pits and skins or other foodstuffs of plant origin. This plant waste, left in nature, becomes putrescible and releases toxic gases that are harmful to both the environment and human health, whereas they can be recycled into bioenergy or biofertiliser. As for mango, a very perishable fruit, there is a lot of post-harvest loss due to diseases and pests, but there is also waste coming out of the processing units installed in the localities shown in Fig. 1. It is with a view to recovering this waste energetically that samples of mango pits from Côte d'Ivoire were collected to be carbonized in a muffle oven. The aim is to determine the experimental conditions for the manufacture of a bio-charcoal from these mango pits. From the experimental results obtained, it appears that a heating rate of 10°C/min to reach a maximum temperature of 250°C and a holding time at this temperature of two (2) hours make it possible to obtain a good-textured coal with a calorific value of 34.98 MJ/kg. These results show that mango pits can be validly transformed into biochar and used as fuel just like charcoal. Its large-scale production can therefore become a source of income for the producers of this fruit who suffer a lot of post-harvest losses in the fields and also for the processing units that could use these pits as a source of energy to power their units.

Keywords Mango · Mango pit · Experimental conditions · Higher calorific value · Côte d'Ivoire

W. F. Fassinou (✉)
Physics Department/Material Sciences, Environmental and Solar Energy Laboratory/Felix Houphouet Boigny University, Abidjan, Côte d'Ivoire
e-mail: faswaniferd@yahoo.fr

D. A. Coulibaly · O. Y. Ako
Environment and Sustainable Development Center/Institute of New Energy Research (IREN)/ Nangui Abrogoua University of Abidjan Côte d'Ivoire, Abidjan, Côte d'Ivoire

M.-M. Narra and S. Narra (eds.), *African Green Transition Through Innovative Pathways*, World Sustainability Series, https://doi.org/10.1007/978-3-031-87043-9_5

1 Introduction

Most of Ivorian electricity production is based on the use of natural gas, which emits greenhouse gases. According to the official website of the Ministry of Mines, Petroleum and Energy of Côte d'Ivoire, the target to be achieved by 2030 is 45% for renewable energies in the Ivorian energy mix (www.energie.gouv.ci, 2024). These figures clearly show the effort made by political decision-makers to integrate renewable energies into the Ivorian energy mix, including bioenergy. In the long term, the objective is to reduce greenhouse gases and also to achieve energy security for the satisfaction of national and sub-regional coverage. It is with this in mind that the following actions have been taken in Côte d'Ivoire. The foundation stone of a biomass power plant using oil palm residues with a capacity of 46 MW in the department of Aboisso was laid on July 22, 2023. This biomass power plant will be the first ever to be connected to the national electricity grid. Also, projects to build biomass power plants based on cocoa residues in the regions of Gagnoa (20 MW) and Divo (73.6 MW) and cotton stalk in the Boundiali region (25 MW) are being studied, according to the official website of the Ministry of Mines, Petroleum and Energy (www.energie.gouv.ci, 2024).

As far as the energy used in cooking is concerned throughout the country, statistics show that it is largely dominated by firewood, charcoal and bio-waste compared to butane gas and electricity, which are two sources of greenhouse gas emissions. According to the Directorate of Operations and Industries, in 2015, wood energy production was 381,820 tons, of which 376,910 tons were charred and 4,910 tons used as firewood (MMPE et al., 2016). Forest cover has decreased from nearly 16 million hectares at the beginning of the twentieth century to about 9 million hectares in 1965 then to 3 million hectares in 1991. Today, the majority of forests are made up of national parks, reserves and classified forests, some of which are severely degraded by clandestine human activity (MMPE et al. 2016). This makes Côte d'Ivoire the most deforested country in Africa. This deforestation is partly due to the use of charcoal and firewood. However, Côte d'Ivoire is a country with high and diverse agricultural resources that generate waste and residues that can be used in the country's energy mix to combat greenhouse gas emissions, deforestation and environmental pollution.

Mango is one of the most widely grown fruits and its sector produces a lot of waste and residues. According to FIRCA and the mango producer's association called INTERMANGUE, the country produces between 150,000 and 180,000 tons of mangoes per year (Coulibaly, 2022). According to the Ministry of Trade and Industry, annual mango losses are estimated at 40% of total annual production (Abidjan.net, 2024). These losses are therefore estimated at between 60,000 and 72,000 tons for a market value of more than 12 billion CFA francs at factory prices. Globally, annual mango production was estimated in 2021 at 57.01 million tons (Statista Research Department, 2023). However, post-harvest losses in mango are globally estimated at 80% of global mango production (Kansci et al., 2003). West Africa's share of this annual mango production is about 4% and the main producing countries are Côte d'Ivoire, Burkina Faso, Mali, and Senegal (PROMAK AFRIC, 2022). As far as Côte d'Ivoire is concerned, mango trees are found throughout the country, but the main production areas are in the north of the country (Fig. 1). Given the fairly large losses of mango, some studies have been directed towards the recovery of this waste. For example, Elizalde-Gonzalez et al. (2007) produced activated charcoal from mango pit for wastewater purification in Mexico. Also, in Burkina Faso, mango peels have been used to produce bioethanol (Somda, 2012) and mango pit shells have been used to produce activated carbon for gas purification (Mantchong, 2014).

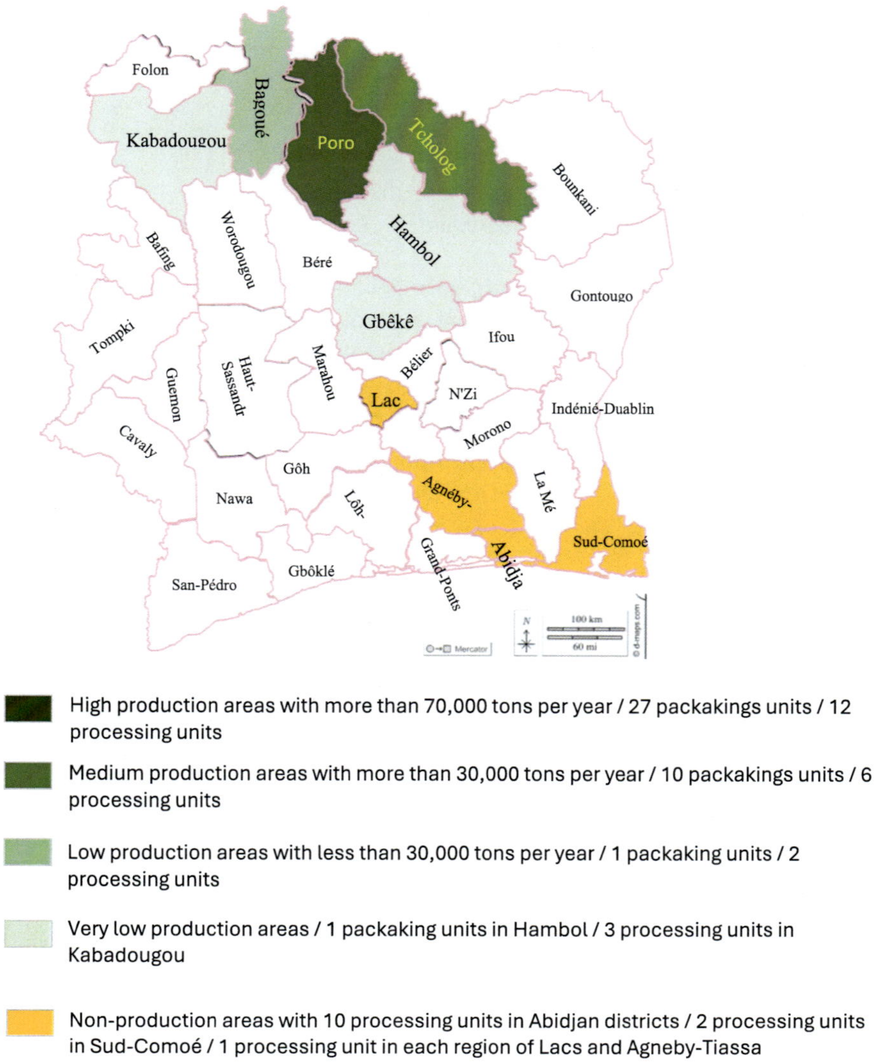

Fig. 1 Geographical distribution of mango areas, and processing units

This paper is devoted to the experimental determination of the optimal conditions for obtaining biochar from whole mango pits collected in landfills in Abidjan and the determination of its calorific value. The map in Fig. 1 shows the spatial distribution of mango production areas and processing units. It was established from data from Sékongo (2022; PROMAK AFRIC 2022 and Pugnet 2018) noted in the literature.

2 Materials and Methods

2.1 Materials

The equipment used in this work which consists of:

- mango pits collected during the 2022 and 2023 seasons (Fig. 2)
- an electronic scale << Denver Instrument >> S.602 with a maximum load of 600 g with an accuracy of 0.01 g
- a Nabertherm muffle furnace allowing temperatures between 30 °C and 3000 °C to be programmed, used for the carbonization of samples,
- a Parr-type calorimetric bomb, model 6200, used to determine the higher calorific value of the samples)
- a Memmert brand oven with a maximum temperature of 65 °C.

This material is available at the Environmental Science Laboratory (LES) of Nangui Abrogoua University except for the calorimetric bomb which is available at the University of Man. These two universities are located in Côte d'Ivoire.

Fig. 2 On the left in the figure mango pits from season 2022 and on the right, mango kernels from 2023

3 Methods

3.1 Determination of the Optimal Conditions for Obtaining Biochar

Mango pits were collected without distinction of variety at the mango unloading dock located in Adjamé in Abidjan on the outskirts of the Banco classified forest (2023 season). These pits have been washed beforehand with clear tap water and dried in the sun for a day. A sample of these pits was extracted and weighed using the Denver Instrument electronic scale before being fed into the Naberthem muffle furnace where it is heated to a fixed carbonization temperature, temperature rise rate, and residence time. Due to the unavailability of the desiccant, the sample was left to cool inside the furnace until the temperature stabilized. The mass of this sample is measured at the exit of the furnace and the processing yield has been calculated. The carbonization yield evaluates the mass of coal obtained from the initial mass of biomass used. This is the ratio of the mass of coal obtained to the initial mass of biomass used.

Several carbonization tests were performed with different samples under different heating conditions. The experimental conditions used in these different tests are:

- Carbonization temperatures: 150 °C, 200 °C, 250 °C, 300 °C and 400 °C
- Heating speed: 5 °C/min and 10 °C/min
- Duration of the temperature rise: 16 min, 22 min, 25 min, 27 min and 30 min
- Residence time: 1 h, 1 h 30 min and 2 h.

3.2 Determination of the Higher Calorific Value of Biochar

The calorimetric bomb is first calibrated using a fuel of a known calorific value which is benzoic acid (HC = 26,453 J. g −1). This calibration makes it possible to determine the water equivalent of the calorimeter. The calorific value of the fuel is systematically calculated according to the Eq. 1:

$$m_{comb} \times HCV = \left(m_e + m_{eqcal}\right) \times C_P(T_F - T_0) \quad (1)$$

m_{comb}, m_e and m_{eqcal} are respectively the masses of the fuel, water in the calorimeter and equivalent water in the calorimeter

C_P is the heat capacity of water (4180 $J.kg^{-1}K^{-1}$).

$(T_F - T_0)$ is the difference between the temperature of the water and that of the calorimeter during the experiment.

Carbonized pits were selected and crushed. Samples were extracted from this shredded material to be analysed in the presence of oxygen in the calorimetric bomb to determine its HCV.

3.3 Determination of the Humidity Level

Carbonized pits were selected and then crushed. The crushed material was sieved with a sieve of mesh sizes smaller than 2 mm. From this sieved material, a 2 g sample is extracted to be dried at 105 °C for 24 h in the oven until a constant mass is obtained according to the AFNOR XP CEN/TS 14,774–3 standard. Three tests were carried out and the moisture content calculated is the average of the three moisture levels determined from these three tests.

4 Results and Discussion

The results of all the carbonization tests carried out can be found in the appendix. Table 1 shows an excerpt from three of these carbonization tests. Test 5 has a lot of unburnt materials although it has a high carbonization efficiency (77.42%) while that of tests 4 and 3 is respectively, 40.84% and 29.75% for the same heating rate of 10°C/min and the same carbonization temperature residence time of 2 h. The carbonization temperatures of tests 3, 4 and 5 are 300 °C, 250 °C and 200 °C, respectively. Although the carbonization efficiency of test 3 (29.75%) is close to that of wood (Table 2), the choice for optimal carbonization would be that of test 4. The carbonization conditions of this test were chosen because the coal obtained is firmer and free of unburned than that of test 3 and the other tests performed.

These carbonization conditions are: the rate of temperature rise to reach 250 °C is 10 °C/min and the residence time is 2 h at a constant temperature of 250 °C. It's a compromise between tests 3 and 5. The higher calorific value of the coal in this test was determined experimentally by the Parr Model 6200 calorimeter bomb. A value of 34.98 MJ/kg was found. As for its humidity level, also determined experimentally, it is 8.49%. By way of comparison, Table 2 below gives some values of the HCV of charcoal and cashew nutshell charcoal found in the literature, as well as that of mango pits in the present study.

As a result, the carbonization temperature of the mango pit is lower while the carbonization yield is higher. The HCV of mango pits charcoal is slightly higher than that of charcoal given by the literature.

For the moisture content, it is higher than that of cashew nut shells determined by Kouamé (2022) (Table 2) and slightly higher than that of charcoal given by Kumar et al. (2017) which is 7.5% (Zinla, 2022). This relatively high moisture content favours the decrease in the lower calorific value (LCV) compared to that of cashew hull charcoal determined by Kouamé (2022). These initial results show strong potential for the use of mango pits charcoal as biofuels in heating and/or cooking processes. But additional studies of elemental, immediate, combustion smoke and ash analyses will be able to confirm these initial results.

Table 1 Excerpt from three (3) carbonization tests (full data can be seen in Annex)

Essay	Initial mass of samples m_p (g)	Speed of temperature rise (°C/min)	Temperature rise time (min)	Carbonization Temperature (°C)	Carbonization temperature duration (h)	Mass of sample after carbonization m_p (g)	Yield $\frac{m_p}{m_i} \times 100$	Pictures
1	569.25	5	25	150	1	510.94	90	
2	557.74	10	30	400	2	99.71	17.87	
3	526.18	10	27	300	2	156.52	29.75	

Table 2 Some charcoal HCV

	Pyrolysis temperature (°C)	Yield (%)	HCV (MJ/kg)	Humidity level (%)	
Mango pit charcoal	250 (laboratory)	40.84	34.98	8.49	Our current study
Charcoal (parcouri)	500 (laboratory)	>30	32	Not given	CIRAD (1992)
Charcoal	500 (traditional)	*33*	Between 27.2 and 33.5	Not given	Loupe et al. (2014)
Cashew shell charcoal	undefinied (industrial oven)	*17*	*29.24*	5.86	Kouamé (2022)

Mango pits collected from landfills (Fig. 2) without distinction of variety were carbonized in a muffle oven. The carbonization tests carried out showed that the optimal carbonization conditions would be 250 °C for the carbonization temperature reached with a temperature rise rate of 10°C/min and a residence time of 2 h. The coal obtained under these conditions has a processing efficiency equal to 40.84%, a HCV of 34.98 MJ/kg and a moisture content of 8.49%. Based on the HCV of charcoal noted in the literature, it can be said that mango pit biochar can be used as a fuel. A more relevant conclusion can be drawn after a comprehensive study of the environmental and human health impacts of the combustion of this biochar.

5 Annex

Essay	Initial mass of samples m_p (g)	Speed of temperature rise (°C/min)	Temperature rise time (min)	Carbonization Temperature (°C)	Carbonization temperature duration (h)	Mass of sample after carbonization m_p (g)	Yield $\frac{m_p}{m_i} \times 100$	Pictures
1	569.25	5	25	150	1	510.94	90	
2	557.74	10	30	400	2	99.71	17.87	
3	526.18	10	27	300	2	156.52	29.75	

(continued)

(continued)

4	571.20	10	22	250	2	233.29	40.84	
5	663.45	10	16	200	2	513.68	77.42	
6	603.20	10	22	250	1	419.60	69.56	

(continued)

(continued)

7	578.24	10	22	250	1,5	407.82	70.52	

Acknowledgements The authors thank WASCAL- Côte d'Ivoire who fully financed the participation of the presenter (Dr. FASSINOU W. F) of the communication at ICERA 2024 as well as the Nangui Abrogoua University which authorized the use of some equipment for the realization of the carbonization tests.

References

Abidjan.net. (2024). Réduction des Pertes Post- Récolte: le Ministre Souleymane Diarrassouba préside la cérémonie de signature de convention entre I2T et O5 Technologie. Published 21 march 2024 on Abidjan.net. (Last accessed: 09.01.2025)

CIRAD. (1992). Le parcouri et le saint-Martin, bois jaune de Guyane. *Bois Et Forêts des Tropiques, No, 230*, 71.

Coulibaly, B. (2022). Etude sur le financement agricole: cas de la filière mangue en Côte d'Ivoire. *Programme Hortifresh en Afrique de l'Ouest en collaboration avec l'université Wageningen*, pp. 8.

Elizalde-Gonzalez, M. P., & Hernandez-Montoya, V. (2007). Caractérisation du noyau de mangue comme matière première dans la préparation du charbon actif pour le traitement des eaux usées. *Journal De Chimie Biochimique, 36*(3), 230–238.

Kansci, G., Koubala, B. B., & Lape, I. M. (2003). Effect of ripening on the composition and suitability for jam processing of different varieties of mango (Mangifera indica). *African Journal of Biotechnology, 2*(9), 301–306.

Kouamé, K. M. (2022). Valorisation thermochimique durable des déchets de l'anacarde : Etude expérimentale, évaluation des potentialités énergétiques et contribution à la mitigation du changement climatique local. *PhD thesis at l'université Felix Houphouët Boigny d'Abidjan*, n° d'ordre 2522/2022.

Kumar, A. (2017). A review on technological options of waste to energy for effective management of municipal solid waste. *Waste Management., 69*, 407–422.

Loupe, D., & CIRAD. (2014). Carbonisation fabrication du charbon de bois. Projet Makala, 17–29. (http ://ur-bsef.cirad.fr/) (Last accessed: 09.01.2025)

Mantchong, M. T. (2014). Élaboration de charbons actifs microporeux à base de résidus de mangues en vue d'épurer le biogaz. *Mémoire de master en ingénierie de l'eau et l'environnement*, 2iE Ouagadougou.

MMPE (Ministère des mines du Pétrole et de l'Energie de Côte d'Ivoire) & CEREEC (2016). Plan d'Actions National des Energies Renouvelables (PANER) Côte d'Ivoire, Period 2016–2020/ 2030. *Politique d'énergies renouvelables de la CEDEAO (PERC*), P.10 et 48.

PROMAK AFRIC (2022). Description et analyse des chaînes de valeurs de la filière mangue. *Programme d'Appui au Développement des Filières Agricoles (PADFA),* N°2000001058

Pugnet, V. (2018). La mangue en Côte d'Ivoire. *Fiche Pays Producteur FRuiToP., 5*(255), 80–83.

Sekongo, S. (2022). Les unités de transformation de la mangue en Côte d'Ivoire. Interprofession de la mangue Intermangue, 25 p., https://jimdo-storage.global.ssl.fastly.net/file/c5239707-7e13-4b5e-ba6f-f0383b9f67fb/Étude_Unités%20de%20transformation%20de%20la%20mangue.pdf (Last accessed: 09.01.2025)

Site officiel du Ministère des mines du Pétrole et de l'Energie (MMPE) (2024). Energie et Energies renouvelables. Centrale solaire photovoltaïque de Boundiali, www.energie.gouv.ci(Last accessed: 09.01.2025)

Somda, K., M. (2012). Contribution à la dépollution de l'environnement par la valorisation de la biomasse végétale au Burkina Faso: cas de la biotransformation des résidus de mangues. *PhD thesis at l'Université de Ouagadougou*

Statista Research Department (2023). Volume de la production de fruits dans le monde en 2021, https://fr.statista.com (last accessed: 09.01.2025)

Zinla, B., T., D. (2022). Etude de la production d'électricité à partir d'une centrale thermique de 5 MW utilisant de la biomasse : cas des résidus de culture de riz, café et cacao provenant de Kodaya. *PhD thesis at l'Ecole Doctorale Polytechnique de Yamoussoukro*

Sustainable Resource Management and Circular Economy

Testing Used Tyres as Soil and Water Conservation Control Measures to Favour Circular Economy in Mali

Mahamadou Lamine Katilé, Aboubakar Bengaly, Souleymane Sidi Traoré, Jesús Rodrigo-Comino, and Hama Kassé

Abstract Microplastics are an emerging global threat in all ecosystems and have potential negative effects through multiple exposure pathways. In Mali, innovations such as the manufacture of chairs, sandals, decorative objects, etc. are being developed to reuse used tyres in order to combat environmental pollution. The special feature of this study is the experimentation with old tyres in agricultural plots to examine their effects on infiltration, runoff, moisture and soil loss. "Pneusol" works on the basis of gathering old tyres, cutting them into arcs, and positioning them in the ridges perpendicular to the direction of flows. The experimental set-up composed by control plot pair with a "pneusol" plot (30 m long and 10 m wide) has been installed on erosion crusted soil in Djindjila. Each plot has been isolated by corrugated sheets 30 cm high above the ground, embedded 20 cm deep. The results show the effectiveness of pneusol is quite significant, since it reduces the amount of water runoff by 34%, increase average moisture by 22.33% and reduce soil lost by 52.63% compared to control plot. The ability of tyres to limit runoff, improve infiltration and increase soil moisture, thereby stimulating agricultural production, has been demonstrated in this study. This result offers interesting prospects for farmers who are already experiencing the harmful effects of climate change. The effect of capturing and infiltrating runoff water, combined with the soil amendment, make "pneusol" an effective

M. L. Katilé (✉)
West African Science Service Centre on Climate Change and Adapted Land Use, Bamako, Mali
e-mail: mahamadoulaminekatile@gmail.com

A. Bengaly
Département des sciences et techniques agricoles/Institut Polytechnique Rural de Formation et de Recherche Appliquée (IPR/IFRA) de Katibougou, Bamako, Mali

S. S. Traoré
Université des Sciences Sociales et de Gestion de Bamako (USSGB), Bamako, Mali
e-mail: souleymane-sidi.traore@mesrs.ml

J. Rodrigo-Comino
Department of Regional Geographical Analysis and Physical Geography, Faculty of Philosophy and Letters University of Granada, Granada, Spain
e-mail: jesusrc@ugr.es

H. Kassé
International Crops Research Institute for the Semi-Arid Tropics (ICRISAT), Bamako, Mali

M.-M. Narra and S. Narra (eds.), *African Green Transition Through Innovative Pathways*, World Sustainability Series, https://doi.org/10.1007/978-3-031-87043-9_6

agricultural production technique on degraded land, directly from the first year of implementation. Also, the repurposing of used tyres in a circular economy makes it possible to turn a potentially problematic waste into a resource, promoting a more environmentally friendly and long-lasting model.

Keywords Agriculture · Soil and water conservation · Circular economy · Tyres · Mali

1 Introduction

It is undeniable that environmental degradation, the pressures of human activity on biodiversity and climate change are having a considerable negative impact on populations and lifestyles (IPCC, 2022). Soil conservation has become even more important to safeguard the natural resources of our world as the human population grows and the demand for land resources rises (Keesstra et al., 2016). Soils are essential but finite resource for maintaining life on Earth (Dagmar et al., 2005; Paustian et al., 1997). Preserving pure water is just as important as conserving soil (Kumar, 2024). Reducing soil erosion and conserving water are major challenges for achieving food security in sub-Saharan Africa (Akplo et al., 2022).

One of the best strategies to reduce soil erosion and associated degradation processes is the use of nature-based solutions such as soil-vegetation (based on the concept of soil health) and Landscape solutions (based on the concept of connectivity) (Kestra et al., 2018). However, in many cases, other strategies must be designed considering other available resources. More than 1.5 billion tyres are thrown away annually (Innes, 2024; Williams, 2017), generating vast areas of soil contamination, even worse when they are burnt. The smoke they give off causes serious health problems. Reusing tyres on a large scale could be a serious solution to this bulky waste. In environmental terms, tyre management will be a serious problem if they are not recycled. Tyres are solid, heavy and reflective objects because they are made of rubber. Luo et al. (2020) showed that tyre microplastics are identified as one of the most abundant types of microplastics in water and soils and may directly or indirectly cause adverse impacts on the environment and human health. Microplastics are an emerging global threat in all ecosystems and have potential negative effects through multiple exposure pathways (Kumar et al., 2020; Rodríguez-Seijo et al., 2019). In Ngaoundéré (Cameroon), research has been carried out in 2016 to evaluate the metal contamination of soils exposed to the burning of car tyres. In this study, 30 soil samples were collected in small wells and around them in three sites where the incineration activity takes place. The results show that all of heavy metals have high levels in the studied soils compared to the reference values used (Nga et al., 2016).

In recent years, the recycling of used tyres into construction and road materials has emerged as a potential innovative solution to the growing used tyre dilemma (Mohajerani et al., 2022). There has been growing interest in the use of recycled tyres in civil engineering applications, particularly in the context of retaining structures, slope stiffening, tracks with non-cohesive materials of low stability, energy absorbers, lightweight backfill on compressible soils. This aspect is clearly visible in the study on the effect of the use of rubber grains on the shear strength behaviour of sand from the river Chlef in Algeria (Kheira & Ahmed, 2019).

In 2019, 300 tyres were recycled to transform rubbish dumps in Conakry's Ratoma district. In 2020, more than 500 tyres were recycled to decorate small crossroads in Taouyah, in Conakry's Ratoma district. In 2021, production has been tripled to reach the recycling target of 1,500 tyres. The area of operation is due to be extended to other districts and communes in the city of Conakry (Initiatives Climat, 2021). In Mali, innovations such as the manufacture of chairs, sandals, decorative objects, etc. are being developed to reuse used tyres to combat pollution (Dagnoko, 2017).

Experiments or trials highlighting the importance of recycling and reusing old tyres, and how to apply the technology in farmers' fields to increase agricultural productivity is not widespread. Especially, the technology does not exist in the WOCAT database with more than 1446 Sustainable Land Management Technologies and 549 Sustainable Land Management Approaches. Some authors defined the soils where tyres are integrated into the soil profiles after long term as "Pneusoil" or "Tyresoil". It can be defined as the association of two elements, used tyre and soil (Long, 1993; Abdurahman, 1995).

To avoid a simple accumulation of this product and find a utility for the Pneusoils, we designed an experimental trial with used tyres in agricultural fields was designed to examine their effects increasing infiltration and reducing runoff and soil loss. The experimental set-up was composed by a control plot compared to a "pneusol" one (30 m long and 10 m wide) which has been evaluated on highly eroded crusted soils, located in Djindjila (Mali). Each plot has been isolated by corrugated sheets 30 cm high above the ground, embedded 20 cm deep. Its efficiency as soil erosion control measure will be tested but at the moment the logical effects of soil contamination due to micro-plastic contents, which is also a key topic nowadays (Rodrigo-Comino & Seeger, 2024), will be omitted.

2 Methodology

2.1 Study Area

The trial was conducted during the 2023 rainy season in the village of Djindjila, located in the rural commune of Meguetan which experiences a Sudanese-Sahelian climate (Coulibaly, 2022). The zone is characterized by three distinct seasons: (i) the hot dry season lasts from February to June, marked by winds throughout the day and night. The period is characterized by relatively low humidity and high evapotranspiration; (ii) the cool dry season occurs from November to February, with relatively low wind and temperature; and, (iii) the wet, warm season from June to October, being the most favourable period for vegetation growth. Throughout the year, the temperature in varies between 17 and 45 °C, while the average annual rainfall ranges from 700 to 900 mm.

Djindjila is characterized by hydromorphic soils (IUSS-WRB, 2022). These have an average depth of about 125 cm at foot slopes and are considered heavy and challenging to work by farmers. Approximately, 41% of Méguétan farmers cultivate this soil and the main crops grown in these soils are sorghum, maize and millet (Coulibaly, 2022). They are also affected by streams or temporary channels, which can be transformed into rills and ephemeral gullies after strong rainfall events. These soils also experience gleyfication in very deep horizons (140 cm) located in the temporary flood zone of the swamps. Tropical ferruginous soils can be found on spreading pediments with a depth of 140 cm. They account for a significant portion of the zone's land and are utilized for cultivating sorghum, cotton, millet, and maize. We can also find soils with high gravel content and crust blocks which are less prevalent in the area. They have a depth of less than 50 cm and are not suitable for mechanization using animal traction. The main crops grown in these soils include cowpeas, sorghum, millet, and groundnuts. Finally, leptosoils located in slightly sloping areas have a shallow crust of about 25 cm (International Union of Soil Sciences, 2022).

Soil management is primarily focused on both food and cash crop, although there are still some underexploited plains. Various food crops cultivated in the region include *Zea mays* (maize), *Pennisetum glaucum* (millet), *Sorghum bicolor* (sorghum), *Oryza sativa* (rice), *Vigna unguculata* (cowpea) and *Vigna subterranea* (wandzou). Cash crops, on the other hand, consist of *Gossypium spp.* (cotton), *Rachis hypogaea* (groundnut) and *Sesamum indicum* (sesame). Also, some fruit trees can be found such as *Vitellaria paradoxa* (shea), *Parkia biglobosa* (néré), *Adansonia digitata* (baobab), *Coccoloba uvifera* (grape) as well as plantations such as *Magifera indica* (mango), orange trees, *Psidium guajava* (guava), *Citrus limon* (lemon) and *Anacardium occidentale* (cashew). However, agricultural sector in the region faces several challenges including: (i) non-degradation of plastic waste in nature; (ii) loss of land due to erosion and soil poverty; and, (iii) the scarcity and irregular distribution of rainfall. To address these challenges and enhance agricultural productivity, farmers use a combination of agricultural inputs. These inputs include chemical fertilizers like diammonium phosphate, urea, cereal complex, and cotton complex, as well as organic fertilizers like manure and other organic materials (Coulibaly, 2022). By utilizing a mix of these inputs, farmers aim to improve soil fertility and enhance crop yields in the area.

2.2 Experimental Setup

The experimental set-up (Fig. 1) has been installed on erosion crusted soil in Djindjila. It was composed of a complete randomized block with two repetitions A and B. Each block consisted of two plots (30 m long and 10 m wide) each receiving a specific treatment: a control plot pair with a "pneusol" plot. Each plot has been isolated by

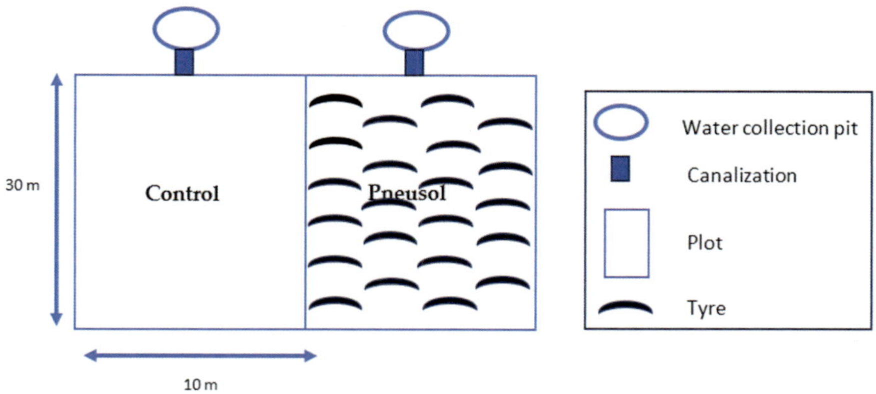

Fig. 1 Illustration of experimental setup

Fig. 2 Image of the tyres in the plot

corrugated sheets 30 cm high above the ground, embedded 20 cm deep. The principle behind "pneusol" is to collect used tyres, cut them into arcs and place them in the ridges perpendicular to the direction of flows (see Figs. 1, 2).

2.3 *Estimating Runoff and Erosion*

First, a Time-Domain Reflectometry (TDR) probe at 10 days intervals was sampled to measure soil moisture in both managed and control plot. Runoff and soil loss were measured using soil erosion plots after building two experimental areas. Water runoff, noted Lr, expressed in millimetres, corresponding to the quantity of rain that did not infiltrate the soil was then measured. Its value has been calculated from the volume of water runoff measured from the barrels (Eq. 1) (Le Bissonnais et al., 1996).

$$\text{Lr} = \frac{\text{V} \times 1000}{\text{S}} \tag{1}$$

The runoff coefficient (Cr) is expressed as a percentage, and represents the proportion of precipitation (P in mm) that did not infiltrate the soil. It was determined using Eq. 2 designed by Le Bissonnais (Le Bissonnais et al., 1996).

$$\text{Cr} = \frac{\text{Lr}}{\text{Rainfall}} \times 100 \tag{2}$$

For the erosion measurement, after each rainfall event resulting in runoff, the water collected in the drum has been thoroughly mixed and a 1 L sample has been taken. These samples are then sent to the laboratory for oven drying and weighing of the sediment. These data are scaled to determine sediment erosion and nutrient content per hectare.

The total soil loss has been calculated using Eq. 3 (Kouelo, 2016):

$$\text{PT} = \text{Tf} + \text{Cs}\left(\frac{\text{Vr}}{\text{Qa}}\right) \times 100 \tag{3}$$

Tf = sediments deposited at the bottom of the tanks; Vr: volume of runoff; Cs: Sediment load and Qa: amount of aliquot taken.

2.4 *Crop Growth Parameter*

This operation was carried out by selecting 10 plants in each experimental unit on days 30, 45 and 60 after plantation. Observations were made on each plot in accordance with the experimental protocol throughout the trial and focused on: (i) measurement of plant height (from the crown to the terminal bud); (ii) diameter at the crown (using a calliper); (iii) number of leaves by counting the number of leaves on each plant; (iv) weighting of 1000 seeds to assess the physiological maturity. After harvesting, the panicles were dried and dehulled. The seeds were weighed to determine the weight of 1000 seeds.

2.5 *Statistical Analysis*

A three-way ANOVA using PROC MIXED procedure was conducted. Tillage system and fertilizer input rates were taken as a fixed effect, while pneusol was considered as a random effect. Significant fixed effects were further dissected by extracting means and performing Tukey's Honestly Significant Difference pairwise comparisons with an alpha of 0.05. The normality and homogeneity of the data for each variable were tested by the Shapiro–Wilk test (Shapiro & Wilk, 1965) and by Bartlett's test (Bartlett, 1937), respectively. All statistical analyses were conducted in SAS9.4 (SASInstitute, 2013).

3 Results and Discussion

3.1 Effect of Treatment on Soil Moisture

For the management of water resources and agricultural productivity, soil moisture is essential (Gupta et al., 2024). Figure 3 shows the effect of treatment on soil moisture over time for both the control plot and the Pneusol-treated plot. The soil moisture levels are plotted across different dates, highlighting the differences between the two treatments. At initial condition, the control plot starts with a soil moisture level of 15.33%, while the Pneusol-treated plot has a higher value at 16.8%. This indicates that the pneusol treatment potentially retains more moisture than the control initially. Both plots reach their peak soil moisture levels on September 8, 2023. The control plot has 19.21%, while the Pneusol plot reaches 22.33%. The consistent higher value for the pneusol treatment indicates that it is more efficient at retaining moisture. The results showed that moisture levels were always higher in the managed plots than the controls one. Previous studies show that contour ridging, slope ridging, and no-tillage (Akplo et al., 2022) mulch which is widely recognized (Araya et al. 2015; Toom et al., 2019) and crop residues as mulch at the soil surface (Balesdent et al., 2000) maintain soil moisture.

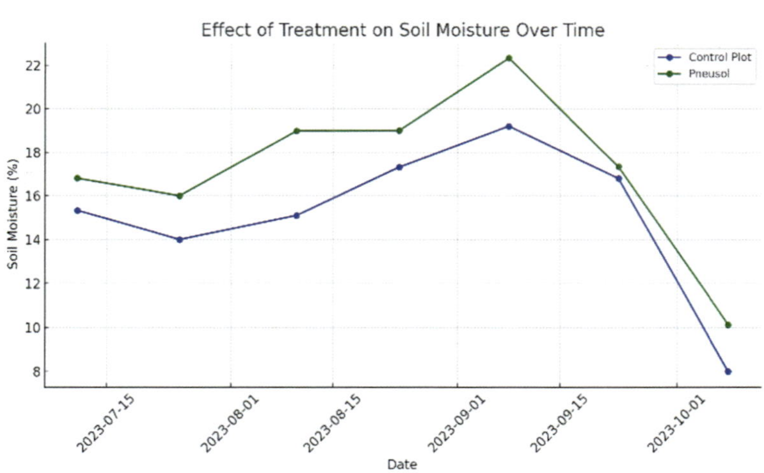

Fig. 3 Effect of treatment on soil moisture over time

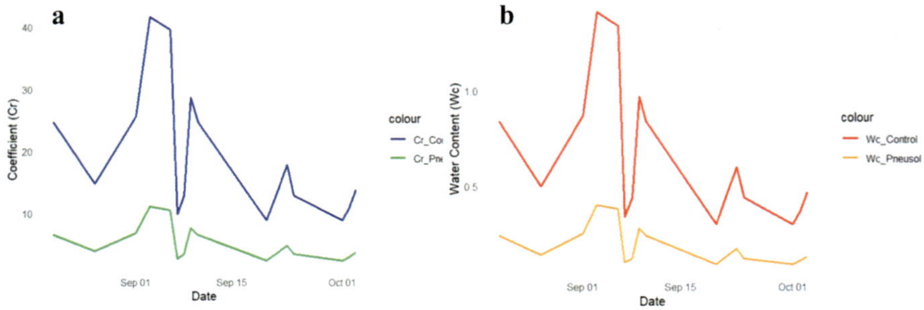

Fig. 4 Runoff coefficient (**a**) water collected (**b**)

3.2 Effect of Treatment on Runoff Coefficient

The ability of tyres to reduce the runoff coefficient (Fig. 4a) and the amount of water collected (Fig. 4b) after rainfall has been observed. The control plot had the highest runoff coefficients, which led to the migration of soil particles.

The results of Zoure et al. (2017) in Burkina Faso show that agricultural practices reduce runoff by almost 30% (± 5%). Pneusol provides a better runoff rate, but the scientific literature does not contain any contradictory publications. This finding agrees with the result of (Abu-Zreig et al., 2011)., which indicated that plots with 5% and 15% rock fragment coverage had runoff depth reductions of 17% and 30%, respectively, as compared to the control. In their overview of soil and water conservation, (Gachene et al., 2019) also confirm that crop management practices (Nyawade et al., 2018b), conservation tillage, soil management practices (Otieno et al., 2018; Rodrigues et al., 2018), contour farming and Contour Bands (Thierfelder & Wall, 2009), grass strips and vegetative buffers, diversion bitches, infiltration ditches, waterways, terraces, water retention pits, structural measures of soil and water conservation slow down runoff. A study on the effect of tillage and mulching on soil water erosion in Linsinlin watershed, centre of Benin indicated that these factors were significant on decreasing runoff and soil loss (Akplo et al., 2017).

3.3 Effect of Treatment on Soil and Nutrient Losses

Pneusol consistently shows reduced nutrient loss across all parameters (nitrogen, phosphorus, and potassium) as well as soil loss compared to the control (see Fig. 5). It is effective at minimizing nutrient runoff, which could lead to better soil fertility retention. The results concerning soil loss corroborate those of Ouédraogo et al. (2019) in Burkina Faso, Bwandamuka et al. (2021) in R D Congo, Dicko et al., 2022 at Katibougou research station (Mali), and Akplo et al. (2022) in central Benin. All these studies have demonstrated that soil and water conservation managed plots have a much greater tendency to reduce soil and nutrients loss than control plots. Using

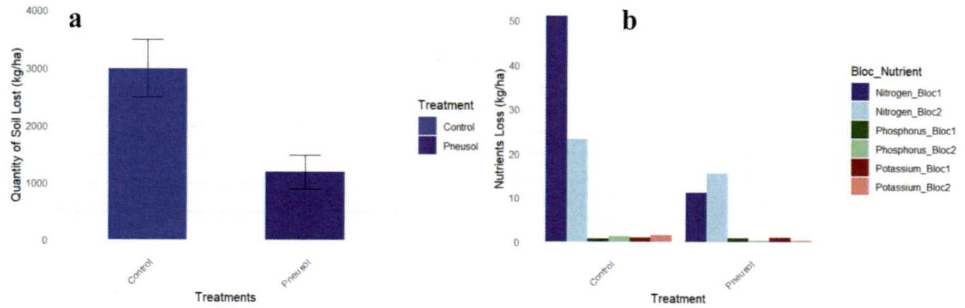

Fig. 5 Soil (**a**) and nutrient (**b**) loss under pneusol and control plot

Soil and Water Conservation (SWC) measures on farmland greatly enhanced the majority of the soil's physicochemical characteristics when compared to farms that did not use SWC measures. SWC measures have been developed to reduce soil loss (Leykoun et al., 2023). In southern Ethiopia, the study on assessing soil erosion and farmers' decision of reducing erosion for sustainable soil and water conservation showed that farmers were using traditional techniques to reduce erosion and that the government had introduced conservation measures such as soil and stone bunds (Djillo et al., 2024).

3.4 Effect of Treatment on Crop Growth Parameters

The need to effectively maximize the use of arable land through the adoption of sustainable agricultural practices will grow as agriculture attempts to fulfil the expanding demands of a growing human population for enough food and textiles (Lizotte & Locke, 2018).

For this reason, effective soil management practices need to be implemented on agricultural land. Regarding soil and nutrient loss, the results showed a significant effect of the use of used tyres. During the experiment, not only number of leaves but also plant height and the diameter have progressively increased (see Fig. 6). According to Pieters et al. (2022), plants are complex organisms that are subject to a variety of environmental factors that dynamically affect their physiology and phenotypic. An intriguing substitute for conditioning their surroundings is the use of used tyres as a Soil and Water Conservation (SWC) strategy, particularly in the Sahelian context where temperatures are rising more frequently.

The average grain and straw yields attained during the experiment has been determine in this study. We note that the yields from pneusol plots (2,816 kg/ha) are higher than those from the control plot (2,035 kg/ha). Similarly, the biomass obtained on the managed plot was 5,608 kg compared with 4,988 kg/ha for the control plot. The results of the study conducted in drought agro-ecological zones in Rwanda showed that all water conservation techniques tested had a highly significant (P <

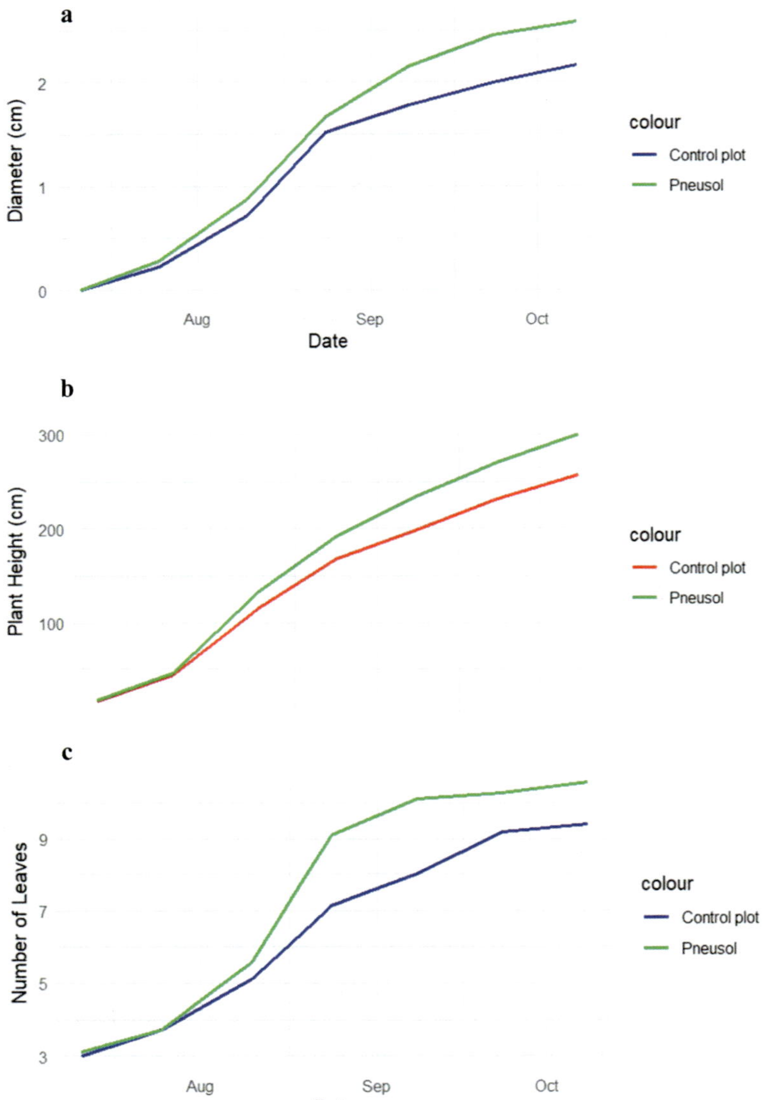

Fig. 6 Effect of treatment on Diameter (**a**), Plant height (**b**), Number of leaves (**c**)

0.001) impact on maize grain yield and yield components, including plant height, cob diameter and length, number of leaves, 100 grain weight, and yield per plant (Uwizeyimana et al., 2018).

4 Challenges

Almost 280 million tyres reach the end of their useful life every year around the world. This figure underlines the importance of finding effective solutions for managing this waste. Although around 30 million tyres are reprocessed or reused, a significant proportion (around 250 million) require appropriate management, either through safe disposal or recycling. Used tyres are not just waste, they are also a resource. They can be transformed into a multitude of useful products, from sports surfaces to building materials and everyday objects. The main stages in tyre recycling, are shredding the rubber; removing the steel and textile fibres; cryogenic shredding; and pyrolysis. In terms of challenges and prospects, it is important to further develop tyre recycling and reuse. Recycling this waste offers a number of advantages, including a reduction in the environmental footprint, the creation of new resources and economic development. Recycling and reusing tyres are essential solutions for managing this waste in a sustainable way and recovering a resource that would otherwise be lost. The use of used tyres as a water and soil conservation technique is a technological innovation that could improve tyre reuse processes.

5 Conclusion

This study demonstrates the impressive benefits of using used tyres as a water and soil conservation technique. The technique's ability to limit runoff by 34% and add moisture to the soil by 22.33%, thereby boosting agricultural production by 2,816 kg/ha through the retention of soil particles and nutrients, was highlighted. This result offers interesting prospects for farmers who are already suffering the adverse effects of climate change, particularly in the West African context, which is prone to erosion and unpredictable pockets of drought. Tyres are economical and therefore competitive with traditional materials and are easy to install. Old tyres are a waste product that is well spread throughout the country and, overall, easy to find, collect, transport and use. Reusing them promotes the circular economy by turning a potential source of pollution into a beneficial resource for both agriculture and the environment. Despite the achievement of research objectives highlighting the use of used tyres in water and soil conservation, it would be interesting to investigate the potential toxic leachate from the rubber material.

Acknowledgements The authors would like to express their gratitude to the German Federal Ministry of Education which financed this study through the West African Science Service Centre on Climate Change and Adapted Land use (WASCAL) within the framework of the PhD scholarship in Climate Change and Agriculture. Sincere thanks and gratitude to the International Conference on Circular Economy, Renewable Energies and Green Hydrogen in Africa (ICERA); and EGEMAP lab (Environment, Geography and Mapping) at UGR, IPR/IFRA and USTTB.

References

Abbas Mohajerani, Halenur Kurmus, David Conti, Lucinda Cash, Adrian Semcesen, Mohammed Abdurahman, Md Tareq Rahman, (2022). Environmental impacts and leachate analysis of waste rubber incorporated in construction and road materials: *A review, Science of The Total Environment, Volume 835*, 155269, ISSN 0048-9697, https://doi.org/10.1016/j.scitotenv.2022.155269

Abdurahman, M. (1995). Modélisation et simulation des ouvrages en Pneusol, *Doctoral dissertation*, Nantes.

Abu-Zreig, M. M., Tamimi, A., & Alazba, A. A. (2011). Soil erosion control and moisture conservation of arid lands with stone cover. *Arid Land Research and Management, 25*(3), 294–307. https://doi.org/10.1080/15324982.2011.565859

Akplo, T. M., Alladassi, F. K., Zoundji, M. C. C., Avakoudjo, J., Houngnandan, P., & Dagbénonbakin, D. G. (2022). Impact of conservation tillage on runoff, soil loss, and soil properties on acrisols and ferralsols in central Benin. *Canadian Journal of Soil Science., 102*(03), 659–671.

Akplo, T. M., Alladassi, F. K., Houngnandan, P., Benmansour, M., Rabesiranana, N., Mabit, L., & Alohoutade, F. M. (2017). Effect of tillage and mulching on soil water erosion in Linsinlin watershed, Centre of Benin. https://doi.org/10.18006/2017.5(4).515.524

Balesdent, J., Chenu, C., & Balabane, M. (2000). Relationship of soil organic matter dynamics to physical protection and tillage. *Soil Tillage Research, 53*, 215–230. https://doi.org/10.1016/S0167-1987(99)00107-5

Bartlett, M. S. (1937). Properties of sufficiency and statistical tests. *Proceedings of the Royal Statistical Society, Series. A, 160*, 268–282. https://doi.org/10.1098/rspa.1937.0109

Bwandamuka, F. M. K., Mukalay, J. B., Ndandula, F. K., Masangu, F. K., Kasongo, M. M., & Makanda, G. M. (2021). Évaluation du risque d'érosion sur quelques sols de la plaine de Lubumbashi, RD Congo. *International Journal of Biological and Chemical Sciences., 15*(5), 2095–2117.

Coulibaly M., (2022). Pratiques de gestion des espaces agricoles et leurs impacts sur les ressources naturelles dans la zone soudano-sahélienne: cas de la commune de Méguetan. *IPR/IFRA* de Katibougou.

Dagnoko M. (2017). Mali: une solution innovante pour les pneus usés. https://www.youtube.com/watch?v=aV8bmS7nKlM (Last accessed: 09.01.2025)

Dicko, G., Bengaly, A., & Ballo, S. (2022). Effets des pratiques agricoles sur l'érosion hydrique en zone soudanienne du Mali (station de l'IPR/IFRA de Katibougou). *International Journal of Biological and Chemical Sciences, 16*(1), 345–352.

Djillo, S. C., Wolka, K., & Tofu, D. A. (2024). Assessing soil erosion and farmers' decision of reducing erosion for sustainable soil and water conservation in Burji woreda, southern Ethiopia. *Science and Reports, 14*, 8638. https://doi.org/10.1038/s41598-024-59076-6

Gupta, S. K., Singh, S. K., Kanga, S., Kumar, P., Meraj, G., Sahariah, D., Debnath, J., Chand, K., Sjan, B., & Singh, S. (2024). Unearthing India's soil moisture anomalies: Impact on agriculture and water resource strategies. *Theoretical and Applied Climatology, 155*, 7575–7590. https://doi.org/10.1007/s00704-024-05088-1

Handiso, M. A., Hemacho, A. H., Bongido, B. L. et al. (2023). Effect of conservation tillage on soil and water conservation, yield and yield components of Maize (*Zea mays* L.) in South Ari District, Southern Ethiopia. *Discov Agric* 1, 10 https://doi.org/10.1007/s44279-023-00008-9

Initiatives Climat. (2021). Pneus usagés: un problème environnemental négligé. *Association Initiatives Climat*. 37.

International Union of Soil Sciences—World Reference Base for Soil Resources (IUSS-WRB). (2022). International soil classification system for naming soils and creating legends for soil maps. (4th edition). *International Union of Soil Sciences (IUSS).*

Intergovernmental Panel on Climate Change, (2022). Climate change: a threat to human wellbeing and health of the planet. Taking action now can secure our future, Working Goup II, https://www.ipcc.ch/2022/02/28/pr-wgii-ar6/.

Innes, J. R. (2024). *One-and-a-half billion tyres wasted annually–there's a better way to recycle them.* University of Bradford.

IUSS-WRB. (2022). *International soil classification system for naming soils and creating legends for soil maps.* 4th ed. International Union of Soil Sciences (IUSS) (ISBN 979-8-9862451-1-9 4th edition).

Keesstra S., Nunes J., Novara A., Finger D., Avelar D., Kalantari Z., Cerda A. (2018). The superior effect of nature-based solutions in land management for enhancing ecosystem services, *Science of The Total Environment, 610–611*, Pages 997–1009, ISSN 0048-9697, https://doi.org/10.1016/j.scitotenv.2017.08.077.

Keesstra, S. D., Bouma, J., Wallinga, J., Tittonell, P., Smith, P., Cerdà, A., Montanarella, L., Quinton, J., Pachepsky, Y., & Van Der Putten, W. H. (2016). Forum paper: The significance of soils and soil science towards realization of the UN sustainable development goals (SDGs). *Soil Discussions,* 1–28.

Kheira, N., Ahmed, A. (2019). A direct shear investigation on the determination of the shearing resistance of reinforced soil with waste rubber. In: Kallel, A., et al. Recent advances in geo-environmental engineering, geomechanics and geotechnics, and geohazards. CAJG 2018. *Advances in Science, Technology & Innovation. Springer, Cham.* https://doi.org/10.1007/978-3-030-01665-4_68.

Kouelo, A. F. (2016). *Effects of farming practices on soil degradation in the three watersheds of southern Benin [Effets des pratiques culturales sur la dégradation du sol au niveau des trois bassins versants du sud Bénin].* University of Abomey-Calavi, Benin.

Kumar, P. (2024). Soil Conservation for Global Sustainability. In: Kulkarni, S., Haghi, A. K. (eds.) *Global Sustainability. World Sustainability Series.* Springer, Cham. https://doi.org/10.1007/978-3-031-57456-6_6

Kumar, M., Xiong, X., He, M., Tsang, D. C., Gupta, J., & Khan, E. (2020). Microplastics as pollutants in agricultural soils. *Environmental Pollution., 265*, 114980.

Le Bissonnais, Y., Benkhadra, H., Gallien, E., Eimberck, M., Fox, D., Martin, P., Douyer, C., Ligneau, L., Ouvry, J. F. (1996). Runoff and interrill erosion on silty loamy soils: analysis of scale transfer from m^2 to small cultivated catchments. *Géomorphologie: Relief, Processus, Environnement, 3* pp. 51–64.

Leykun, S., Teklay, A., Gurebiyaw, K., Dile, Y. T., Bayabil, H. K., & Ashenafi, M. (2023). Impacts of soil and water conservation measures on soil physicochemical properties in the Jibgedel Watershed. *Ethiopia. Environ Monit Assess, 195*, 447. https://doi.org/10.1007/s10661-023-110 59-2

Lizotte, R. E., & Locke, M. A., Jr. (2018). Assessment of runoff water quality for an integrated best management practice system in an agricultural watershed. *Journal of Soil Water Conservation, 73*(3), 247–256.

Long, N. T. (1993). Le Pneusol: Recherches-réalisations-perspectives. Institut National Des Science Appliquées de Lyon, Lyon, France, Ese de Doctorat.

Luo, Z., Zhou, X., Su, Y., Wang, H., Yu, R., & Zhou, S. (2020). Environmental occurrence, fate, impact, and potential solution of tire microplastics: Similarities and differences with tire wear particles. *Science of the Total Environment., 2021*(795), 148902.

Nebo, G. I., Manyevere, A., Araya, T., & van Tol, J. (2020). Short-term impact of conservation agriculture on soil strength and saturated hydraulic conductivity in the South African semiarid areas. *Agriculture, 10*(9), 414. https://doi.org/10.3390/agriculture10090414

Nga, L. E., Sadjo, S. M., & Zame, P. Z. (2016). Evaluation de la contamination métallique des sols exposés à l'incinération des pneus d'automobiles dans la ville de Ngaoundéré (Cameroun). *Journal Materials Environmental Science, 7*(12), 4633–4645.

Nguyen, T. L. (1993). Le Pneusol: Recherches—Réalisations—Perpectives. Institut national des sciences appliquees de lyon. Génie civil : Sols—Matériaux—Structures et Physique du Bâtiment. Laboratoire Central des Ponts et Chaussées (LCPC). p 42

Nyawade, S., Karanja, N., Gachene, C., Parker, M., & Schulte-Geldermann, E. (2018). Susceptibility of soil organic matter fractions to soil erosion under potato-legume intercropping systems in Central Kenya. *Journal of Soil and Water Conservation, 73*, 568–577.

Otieno, H.M., Chemining'wa, G. N., Zingore, S. (2018). Effect of farmyard manure, lime and inorganic fertilizer applications on soil pH, nutrients uptake, growth and nodulation of soybean in acid soils of Western Kenya. *Journal of Agricultural Science 10*(4):199–208

Paustian, K. A. O. J. H., Andren, O., Janzen, H. H., Lal, R., Smith, P., Tian, G., & Woomer, P. L. (1997). Agricultural soils as a sink to mitigate CO_2 emissions. *Soil Use and Management, 13*, 230–244. https://doi.org/10.1111/j.1475-2743.1997.tb00594.x

Pieters, O., De Swaef, T., Stock, M. et al. (2022). Leveraging plant physiological dynamics using physical reservoir computing. *Sci Rep* 12, 12594 (2022). https://doi.org/10.1038/s41598-022-16874-0

Rodrigo-Comino, J., Seeger, M. (2024). Exploring the unintended contribution of soil erosion research to microplastic contamination, *Results in Earth Sciences*, *2*, 100043, ISSN 2211-7148, https://doi.org/10.1016/j.rines.2024.100043

Rodrigues, M. Â., Ladeira, L. C., & Arrobas, M. (2018). Azotobacter-enriched organic manures to increase nitrogen fixation and crop productivity. *European Journal of Agronomy, 93*, 88–94.

Rodríguez-Seijo, A., Pereira, R. (2019). Microplastics in agricultural soils: Are they a real environmental hazard? In *Bioremediation of agricultural soils*, (pp. 45–60). CRC Press.

SAS Institute (2013). Base SAS 9.4 procedures guide. SAS Institute, Cary, NC.

Shapiro, S. S., & Wilk, M. B. (1965). An analysis of variance test for normality (complete samples). *Biometrika, 52*(3–4), 591–611. https://doi.org/10.1093/biomet/52.3-4.591

Schröter, D., Cramer, W., Leemans, R., Prentice, C., Araujo, M. B., Arnell, N. W., Boneua, A., Brugman, H., Carter, T. R., (….) Zierl, B. (2005). Ecosystem Service Supply and Vulnerability to Global Change in Europe. *Science 310*, 1333–1337(2005). https://doi.org/10.1126/science.1115233

Thierfelder, C., & Wall, P. C. (2009). Effects of conservation agriculture techniques on infiltration and soil water content in Zambia and Zimbabwe. *Soil Tillage Res, 105*, 217–227.

Toom, M., Tamm, S., Talgre, L., Tamm, I., Tamm, Ü., Narits, L., et al. (2019). The effect of cover crops on the yield of spring barley in Estonia. *Agriculture, 9*, 172. https://doi.org/10.3390/agriculture9080172

United Nation Convention to Combat Desertification (2024). Global database on sustainable land management. https://qcat.wocat.net/en/wocat/

Uwizeyimana D., Mureithi S.M., Karuku G., Kironchi G. (2018). Effect of water conservation measures on soil moisture and maize yield under drought prone agro-ecological zones in Rwanda, *International Soil and Water Conservation Research*, *6*(3), pp 214–221, ISSN 2095-6339, https://doi.org/10.1016/j.iswcr.2018.03.002

Williams J. (2017). Car tyres in a circular economy: What can the world do with 1.5 billion waste tyres? *The Earthbound Report*

Zouré, C.O., Karambiri, H., Dial, N., Queloz, P., Koita, M., Fowé, T. (2017). Impacts des pratiques culturales sur le fonctionnement des hydrosystèmes dans le bassin versant de Tougou au Burkina Faso. In 2–4

The Case for Integrated Tyre Recycling Systems in West Africa: Environmental and Health Impacts of Tyres as Fuel for Meat Preparation

Mona-Maria Narra, Djangbadjoa Gbiete, Damgou Mani Kognine, and Michael Nelles

Abstract The unsustainable practice of using end-of-life tyres (ELTs) as fuel for meat preparation in West Africa poses significant environmental and public health risks. This study examined heavy metal contamination in tyre-grilled meats from Lomé, Togo, analysing samples of beef and goat meat for cadmium (Cd), lead (Pb), chromium (Cr), iron (Fe), and mercury (Hg) using atomic absorption spectrometry. Results revealed concerning levels of heavy metals exceeding European Union Maximum Residue Limits (MRLs), with unwashed beef skin containing particularly high concentrations (Cd: 0.124 mg/kg, Pb: 0.715 mg/kg, Hg: 1792.475 μg/kg). Comparative analysis with similar studies across West Africa demonstrated consistently elevated heavy metal concentrations in tyre-singed meats. The findings highlight the urgent need for structured tyre recycling systems and alternative meat preparation methods to protect public health and environmental safety. Implementation of tyre buyback programmes and raising awareness about health risks are recommended as crucial steps toward sustainable ELT management in the region.

Keywords End-of-Life tyres · Heavy metal contamination · Food safety · West Africa

M.-M. Narra (✉) · D. Gbiete · M. Nelles
Professorship of Waste and Resource Management, Faculty of Agricultural and Environmental Sciences, University of Rostock, Rostock, Germany
e-mail: mona-maria.narra@uni-rostock.de

D. M. Kognine
Laboratoire Sur L'Energie Solaire (L.E.S), Département de Physique, Faculté Des Sciences (FDS), Université de Lomé (UL) 01BP1515, Lomé, Togo

M.-M. Narra and S. Narra (eds.), *African Green Transition Through Innovative Pathways*, World Sustainability Series, https://doi.org/10.1007/978-3-031-87043-9_7

1 Introduction

The management of end-of-life tyres (ELTs) poses significant environmental and health challenges globally, especially in developing countries with limited waste management systems. In Togo, like in many West African countries, tyres are often not properly discarded as no system for recycling is in place. As Narra et al. (2024) highlighted, approx. 10% of all ELTs are used in an artisanal or construction way in Togo. A quarter of all tyres are simply dumped and the majority (65%) of ELTs are burned. Burning tyres majorly includes the singeing of animal carcasses for food preparation.

While the practice of burning tyres is cost-effective and readily accessible, it poses critical environmental and public health concerns. Tyres are composed of complex mixtures of synthetic rubbers, heavy metals, and other hazardous chemicals that release toxic compounds, such as polycyclic aromatic hydrocarbons (PAHs) and dioxins, when burned (Jeong, 2022). These emissions contribute to air pollution and pose risks to both human health and ecological systems (Jimoda et al., 2018).

Despite these risks, the informal use of tyres as fuel persists, largely due to a lack of structured recycling systems and public awareness about safer alternatives. Currently, Togo lacks comprehensive policies and infrastructure to address the proper treatment of ELTs, leading to unsustainable practices that undermine efforts to achieve environmental sustainability. The absence of a formal recycling framework not only perpetuates hazardous practices but also misses opportunities for economic development through material recovery and resource optimization.

This article examines the use of tyres by butchers for food preparation in Togo as a case study to highlight the urgent need for structured tyre recycling systems. By analysing the environmental and health implications of current practices the necessity of transitioning to sustainable tyre management will be highlighted. It also advocates for the implementation of policies and technologies that align with global best practices for waste management and environmental protection.

1.1 Toxic Emissions from Tyre Combustion

Waste tyres are primarily composed of synthetic rubber, carbon black, and various chemical additives designed to enhance performance and durability. When these tyres are burned, particularly under uncontrolled conditions such as open-air grilling, they release a wide range of hazardous chemicals into the environment. These chemicals include polycyclic aromatic hydrocarbons (PAHs), volatile organic compounds (VOCs), and heavy metals (Jimoda et al., 2018).

Polycyclic Aromatic Hydrocarbons (PAH) are a class of organic compounds that are formed during the incomplete combustion of organic matter, including rubber. PAHs are well-established carcinogens and mutagens (Sombiri et al., 2024). When tyres are burned for grilling meat, PAHs can adhere to the surface of the meat being cooked, leading to direct ingestion by consumers. PAH exposure is associated with increased risks of cancer, particularly lung, skin, and gastrointestinal cancers (Sombiri et al., 2024). Additionally, PAHs can accumulate in fatty tissues, leading to chronic health issues over time.

Tyre combustion also releases Volatile Organic Compounds (VOCs), including benzene, toluene, and xylene (Mentes et al., 2022). These compounds are known for their harmful effects on the respiratory system. Benzene is a potent carcinogen that has been linked to leukaemia and other forms of cancer (Wang et al., 2022). Toluene and xylene, on the other hand, can cause neurological damage, liver and kidney dysfunction, and respiratory irritation upon prolonged exposure (Saeedi et al., 2024). During incineration of ELTs, VOCs can be inhaled by individuals in proximity to the fire, increasing the risk of respiratory ailments.

Waste tyres contain significant quantities of heavy metals, including lead, cadmium, zinc, and mercury (Jeong, 2022). When tyres are burned, these metals can be released into the environment, contaminating the air and food. Cadmium and lead exposure from tyre combustion can result in neurotoxicity, kidney damage, and developmental issues in children (Okonkwo et al., 2018). In particular, lead is a well-known neurotoxin that can impair cognitive development, especially in young children, while cadmium exposure is linked to lung cancer and bone disease (Jomova et al., 2024).

2 Materials and Methods

2.1 Sample Collection

In order to determine the direct health impacts of tyre grilled meats, beef and goat meat has been procured in Gbossimé abattoir in Lomé, Togo in February 2024. The meat has been procured as normal food for consumption and was transported in small plastic bags on a motorcycle to the cooling area. It was then cut into smaller sizes and frozen in sealed deep-freezer bags in a laboratory freezer at −15 °C until the analysis.

The meat was observably procured as meat grilled on tyre fire. The procured meat was:

Beef skin, washed after grilling.

Beef skin, not washed after grilling.

Goat meat, washed after grilling.

Goat skin, washed after grilling.

The meats were then analysed for heavy metal contamination in a laboratory of the Université de Lomé in Togo. Unfortunately, it was not possible to analyses the meats for PHAs and VOCs as the necessary laboratory equipment for these analyses were not readily available in Togo. It has been elaborated to transport the samples abroad for further analyses, but this was not possible due to import rules of meats.

2.2 Sample Preparation

The meat samples were prepared for analyses through mineralization and solubilization of the samples.

The solubilization method used is mineralization by nitric acid attack. This method is carried out in a closed environment at 95 °C. For 1g of dried sample, weighed on a BEL L303i electric balance (Pmax = 310g, Accuracy = 0.001), 4 ml of nitric acid (68%) were added for mineralization. Beforehand, 10 ml of 40% hydrogen peroxide (H_2O_2) is added to each sample and left to act for 24 h before acid attack. On a hot plate, these preparations are gradually heated to a temperature of around 95°C. Mineralization for mercury (Hg) analysis was carried out cold, i.e. without heating. This process minimizes the volatile effect of Hg. After nitric acid etching, the samples are filtered using filter paper. The filtrate obtained contains the chemical element to be determined.

Internal controls were carried out during handling. During the analyses, duplicates were randomly inserted into the batches, and a blank was prepared simultaneously using the same reagents as for the samples of interest and under the same experimental conditions. This blank was used to zero the instrument and is measured together with the standard solutions at regular intervals during analysis.

2.3 Determination of Heavy Metals in the Samples

Determination of the trace metal content of the samples was carried out using the SAA atomic absorption spectrophotometer (iCE 3000 SERIES THERMO FISCHER, Fig. 1) in relation to the calibration curve, using a series of standard solutions of well-known concentrations (see Table 1).

Fig. 1 Atomic absorption spectrometer (Brand iCE 3000 SERIES THERMO FISCHER)

Table 1 Technique, specific equipment and reference standard for the analysed elements

Analytic parameter	Technique	Specific instrument used	Reference standards
Cadmium (Cd)	Flame atomic absorption spectrometry	SAA iCE 3000 SERIES THERMO FISCHER	NF EN ISO 5961
Lead (Pb)	Flame atomic absorption spectrometry	SAA iCE 3000 SERIES THERMO FISCHER	FD T 90–112
Iron (Fe)	Flame atomic absorption spectrometry	SAA iCE 3000 SERIES THERMO FISCHER	FD T 90–112
Chromium (Cr)	Flame atomic absorption spectrometry	SAA iCE 3000 SERIES THERMO FISCHER	FD T 90–112
Mercury (Hg)	Cold steam hydride	SAA iCE 3000 SERIES THERMO FISCHER + VP100	NF EN 1483

The standard solutions are used to calibrate the measuring instrument. The concentrations of the standard solutions are chosen taking into account the foreseeable concentrations of the solutions studied. These must lie within the ranges of the standard solutions. These concentrations were read under the following experimental conditions (Table 2).

Mercury is specifically analysed using an atomic absorption spectrophotometer (iCE 3000 SERIES THERMO FISCHER brand) coupled to a VP100 hydride generator. Due to the unique properties of Hg, it was possible to perform a vapor analysis of the element at room temperature. Taken in its most elementary sense, the VP100 accessory reduced Mercury to an elemental vapor and uses the Atomic Absorption technique to determine its quantity.

The spectrometer displayed the concentration of the element studied in the samples directly to the computer. These concentrations were expressed in mg/l (Cd, Pb, Cr and Fe) or µg/l (Hg). However, the actual concentrations (C) of the elements measured in the samples are expressed in mg/Kg or µg/Kg dry weight respectively. These concentrations are calculated according to the Eq. 1:

$$C\left(\frac{mg}{kg}\right) or \left(\frac{\mu m}{kg}\right) = \frac{(spectrometer * filtrationvolume)}{masssamples} \tag{1}$$

Table 2 Summary of experimental conditions for elemental analysis

Elements analyzed	Wavelength (nm)	Slot width	Type of flame
Cadmium (Cd)	228,8	0,5	Air-Acetylene
Lead (Pb)	217,0	0,5	Air-Acetylene
Chromium (Cr)	357,9	0,5	Air-Acetylene
Iron (Fe)	248,3	0,2	Air-Acetylene
Mercury (Hg)	253,7	0,5	No heating

Heavy metal analysis was carried out for Fe, Cd, Pb, Cr, and Hg. Analysis for Arsenic (As) was unfortunately not possible as important materials for analysis were not able to be secured in a long time in Togo. For each sample four separate analyses were carried out.

3 Results and Discussion

The samples were analysed and the average results of the heavy metal analyses for Fe, Cd, Pb, Cr, and Hg are displayed in Table 3.

In order to compare the results of the samples with safe limits, no specific national standards or law directly regulating the levels of heavy metals in meats were found for Togo. However, food safety regulations in many countries and also the Economic Community of West African States (ECOWAS) region are influenced by international guidelines, such as those established by the Codex Alimentarius, which set maximum residue limits (MRLs) for contaminants in food, including heavy metals like lead, cadmium, and arsenic. These standards aim to protect consumer health by defining safe limits for hazardous substances in various foods.

The Codex Alimentarius provides guidelines for maximum levels of heavy metals in food, including meats, as part of its General Standard for Contaminants and Toxins in Food and Feed (Codex Stan 193–1995). However, the specific limits for all heavy metals (like Fe, Cd, Pb, Cr, and Hg) in beef and goat meat are not explicitly detailed in the publicly available Codex documentation. Limits are often tailored for specific contaminants based on food type and regional regulations.

The European Union (EU) regulates heavy metals in food under Regulation (EC) No. 1881/2006, which establishes maximum levels for contaminants like lead, cadmium, mercury, and others in various food categories. These levels are regularly updated based on scientific advice from the European Food Safety Authority (EFSA).

According to European legislation, specifically Regulation (EC) No 1881/2006 and its subsequent amendments, the maximum allowed levels of heavy metals in beef and goat meat are transcribed to Table 4. This value is part of the broader European framework aimed at limiting contaminants in food to ensure consumer safety. This regulation covers a wide range of foodstuffs, including meat products, and sets specific thresholds for various heavy metals and other hazardous substances (Regulation (EC) No 1881/2006).

Table 3 Average results of heavy metal analysis for the sample meats grilled on ELTs

Sample	Characteristics	Fe (mg/kg)	Cd (mg/kg)	Pb (mg/kg)	Cr (mg/kg)	Hg (μg/kg)
Beef skin	Washed	142,453	0,103	0,269	0,462	634,392
Beef skin	Not washed	387,745	0,124	0,715	1,265	1792,475
Goat meat	Washed	235,753	0,078	Not detected	1,687	612,239
Goat skin	Washed	32,919	0,102	0,102	0,823	2018,384

Table 4 MRLs in the EU for heavy in beef and goat meat

Heavy metal	Max. level in beef and goat meet (EU)	Notes
Cadmium (Cd)	0,05 mg/kg	Applies to muscle meat. Higher levels allowed in offal like liver and kidney
Lead (Pb)	0,1 mg/kg	Consistent for most meats according to EU standards
Chromium (Cr)	Not specifically regulated	Chromium limits are not directly addressed in EU regulations for food
Mercury (Hg)	0,03–0,05 mg/kg	Focus is primarily on methylmercury in muscle meats
Iron (Fe)	Not regulated	Iron is considered an essential nutrient, and limits are not typically set

The regulations focus primarily on toxic heavy metals (Cd, Pb, and Hg), which pose significant health risks at elevated levels. Chromium and iron are less commonly regulated in meat due to their variable toxicities and biological roles.

A comparison of data on heavy metal concentration in tyres singed meats (goats as well as cows) from literature results from West Africa shows that heavy metal levels are nearly always exceeding the EU standards for safe foods. The data for Cd compares three studies (Mensah et al., 2019), (Obiri-Danso et al., 2008) and (Yaya et al., 2024) with this study and the EU MRL. Except for Yaya et al. (2024) all results report higher than recommended Cd levels in the respective meats (Fig. 2).

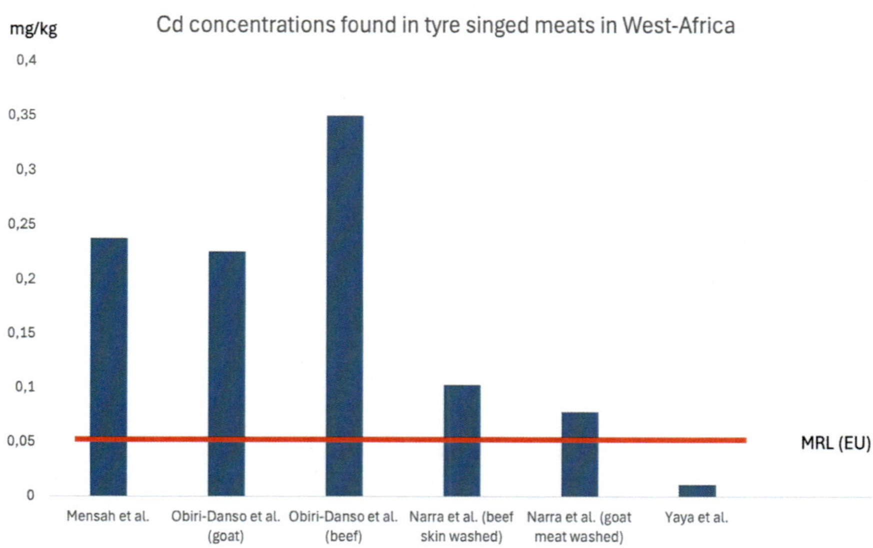

Fig. 2 Comparison of Cd concentration found in tyre singed meats in West Africa and their results in comparison the the EU MRL levels

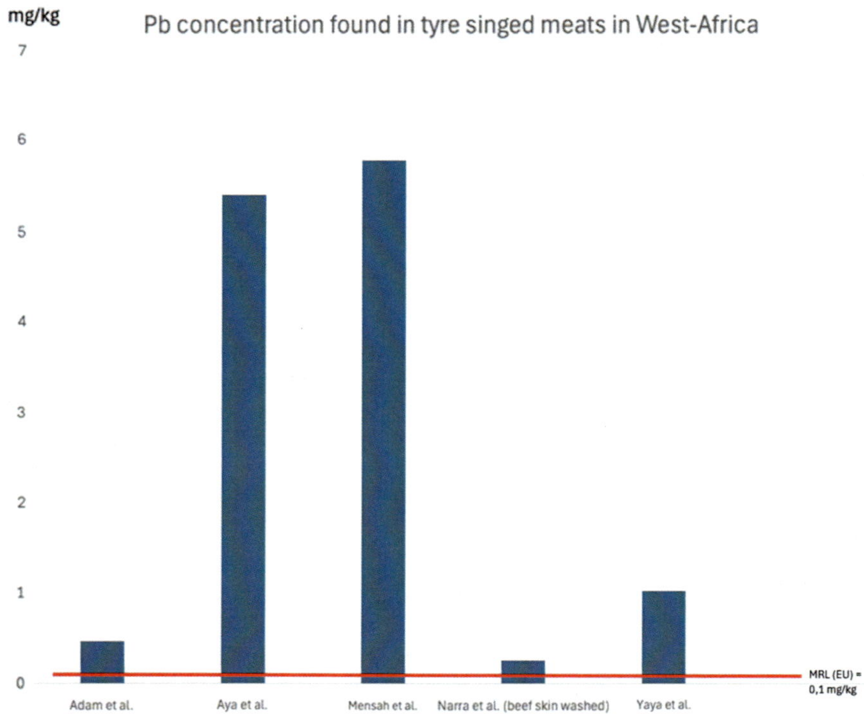

Fig. 3 Comparison of Pb concentration found in tyre singed meats in West Africa and their results in comparison the the EU MRL levels

The data for Pb compares four studies (Adam et al., 2013), (Aya & Nwite, 2016) (Mensah et al., 2019) and (Yaya et al., 2024) with this study and the EU MRL. All results report higher than recommended Pb levels in the respective meats (Fig. 3), with some reaching alarming Pb levels.

The data for Hg compares one study (Yaya et al., 2024) with this study and the EU MRL. All results report higher than recommended Hg levels in the respective meats (Fig. 4).

3.1 Health Risks to Consumers and Workers

Consumers of grilled meat prepared using tyre combustion are at direct risk of ingesting these harmful chemicals. Heavy metals can adhere and accumulate in meats as is highlighted by Figs. 3–5 and accumulate in the human body even after the meat is consumed (Morais et al., 2012). This means that individuals who regularly consume grilled meat cooked over tyre fires may have chronic exposure to these carcinogenic substances.

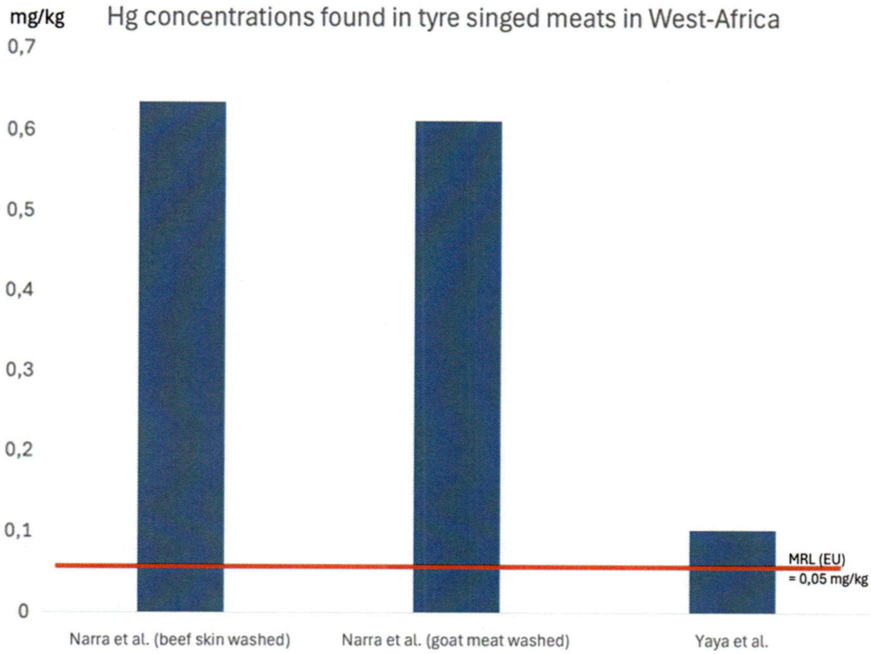

Fig. 4 Comparison of Hg concentration found in tyre singed meats in West Africa and their results in comparison the the EU MRL levels

In the human body, cadmium has adverse effects on several organs, the nervous system and among others, has negative effects on the reproductive and immunology systems (Morais, et al., 2012). Bioaccumulated lead is known to cause developmental problems such as reduced mental abilities, behavioural disorder, restricted growth, and impaired hearing as well as interfering with the vitamin metabolism (Nkwunonwo et al., 2020). Mercury (especially methylmercury) can cross the blood brain barrier and is one of the most toxic heavy metals (Zhang et al., 2014). It is known that the uptake of mercury through food passes placental barriers and causes cognitive impairment in children who are exposed to it in the womb or during breastfeeding (Booth & Zeller, 2005).

The health risks associated with tyre combustion are not limited to taken up toxic substances through food intake, but also include inhalation or environmental contamination.

In addition to consumers, workers involved in the grilling process are exposed to extremely high levels of these toxic substances. Butchers, street vendors, and individuals preparing meat for commercial sale are often in close proximity to tyre fires for extended periods. Okonkwo et al. (2018) concluded that occupational exposure to tyre combustion products significantly increases the risk of respiratory diseases, lung cancer, and cardiovascular diseases.

3.2 Environmental Impact and Policy Implications

The environmental impact of burning waste tyres is another major concern. The release of pollutants from tyre combustion contributes to air, soil, and water contamination. Tyre combustion releases particulate matter (PM), which can be carried by wind and deposited in nearby soil and water sources (Yeh & Yeh, 2023). These particles may contain toxic chemicals and heavy metals that can accumulate in the ecosystem, posing risks to wildlife and potentially entering the food chain.

Air Pollution: Airborne particles from tyre fires can travel long distances, spreading pollutants over a wide area. The release of black carbon and particulate matter from tyre fires contributes to poor air quality and respiratory diseases in humans (Han et al., 2024). In urban areas where tyre burning for grilling is common, these pollutants exacerbate existing air quality issues, increasing the prevalence of asthma, bronchitis, and other respiratory conditions among the population (Han et al., 2024).

Water and Soil Contamination: The deposition of tyre combustion byproducts into the soil and water is another environmental concern. It has been found that heavy metals from tyre fires can leach into the soil, affecting soil fertility and contaminating groundwater supplies (Raudonytė-Svirbutavičienė et al., 2022). This contamination can have long-term effects on agricultural productivity and water safety, particularly in areas reliant on groundwater for drinking and irrigation.

Despite the well-documented hazards associated with tyre combustion, the practice of using ELTs as fuel for singing or grilling meats remains common in many parts of West Africa due to the lack of effective regulation and enforcement. The ease of availability, low cost and high combustibility of ELTs are the reason for using these materials as fuel (Amfo-Out et al., 2014).

The lack of public awareness about the health risks associated with tyre combustion also contributes to the persistence of this practice. A survey (Hill, 2015) conducted in Ghana found that many individuals involved in the grilling process were unaware of the dangers posed by tyre smoke and did not take protective measures to reduce their exposure.

4 Conclusion

The pressing environmental and public health challenges associated with the unsustainable use of end-of-life tyres (ELTs) as fuel in West Africa necessitate immediate and innovative solutions. Central to addressing these challenges is the implementation of structured tyre recycling systems that not only mitigate environmental pollution but also improve public health outcomes. By attributing a monetary value to used tyres through recycling initiatives, it becomes possible to dissuade the use of ELTs in harmful practices and create a more sustainable framework for their management.

One of the primary drivers of unsustainable ELT practices in West Africa is their perceived lack of value. Tyres are often readily available and used indiscriminately as cheap alternatives to conventional fuels (Amfo-Otu et al., 2014), (Hill, 2015). By creating a system where tyres have a defined monetary worth, communities can begin to see these materials as valuable resources rather than waste. Establishing tyre buyback programs as part of structured recycling initiatives can serve as a powerful incentive for proper disposal. These programs could involve purchasing ELTs from local collectors, thereby fostering a circular economy where materials are continuously reused and repurposed.

Furthermore, recycling initiatives can generate employment opportunities and spur economic development in local communities. Sorting, transporting, and processing tyres for reuse or recycling can create jobs while reducing environmental harm. With proper support, these systems can become self-sustaining, reducing dependency on harmful practices and alleviating the public health burden linked to tyre burning (Baker, 2018).

By eliminating the economic incentives for unsustainable practices, public health improvements can be realized on a broader scale. Cleaner environments lead to healthier populations, lower healthcare costs, and increased productivity (Huang et al., 2024). As awareness grows about the link between tyre burning and health risks, communities are likely to support recycling initiatives, especially when coupled with financial benefits.

The transition to structured tyre recycling in West Africa is more than an environmental necessity; it is a pathway to sustainable development. By integrating monetary incentives for ELT collection and recycling, the economic drivers of harmful practices can be dismantled.

Reduced pollution, improved public health, and economic growth are achievable outcomes when tyres are viewed as valuable resources rather than waste. By prioritizing tyre recycling, West Africa can lead the way in demonstrating how innovative, inclusive solutions can address pressing environmental and social challenges, paving the way for a healthier and more sustainable future.

References

Adam, I., Okyere, D., & Teye, M. (2013). Assessment of heavy metal residues in hides of goats singed with tyres, and the effect of boiling on the heavy metal concentration in the hides. *Journal of Veterinary Advances, 3*(5), 165–169. https://doi.org/10.5455/jva.20130531104440

Amfo-Otu, R., Agyenim, J. B., & Adzraku, S. (2014). Meat contamination through singeing with scrap tyres in Akropong-Akuapem abbatoir. *Ghana, Applied Research Journal, 1*(1), 12–19.

Aya, F. C., & Nwite, J. N. (2016). Implication of roasting goats with tyre on human health and the environment in Abakaliki, Ebony State, Nigeria. *Journal of Pollution Effects & Control, Cont, 4*, 153. https://doi.org/10.4172/2375-4397.10000153

Baker, S. E. (2018). Exploring extended producer responsibility: Waste tyres as a resource for job creation in South Africa? ProQuest dissertations & theses global: the sciences and engineering collection. (2140330599). Available online: https://www.proquest.com/dissertations-theses/exploring-extended-producer-responsibility-waste/docview/2140330599/se-2, Last accessed: 10.12.2024.

Booth, S., & Zeller, D. (2005). Mercury, food webs, and marine mammals: implications of diet and climate change for human health. *Environmental Health Perspectives, 113*(5), 521–526. https://doi.org/10.1289/ehp.7603

Codex Stan 193–1995, General Standard for Contaminants and Toxins in Food and Feed, Codex Alimentarus, available online: https://www.fao.org/fao-who-codexalimentarius/sh-proxy/fr/?lnk=1&url=http%253A%252F%252Fworkspace.fao.org%252Fsites%252Fcodex%252FStandards%252FCXS%2B193-1995%252FCXS_193e.pdf, Last accessed: 27.11.2024.

Han, C., Jang, M., Yoon, J., Lee, B., Kim, J., Jang, H., & Benmarhnia, T. (2024). Estimating the acute health effects of smoke exposure from an urban factory fire accident: a case study of a tire factory fire in Korea. *Environmental Health Perspectives, 132*(8), 087008. https://doi.org/10.1289/EHP14115

Hill, T. (2015). Public health implications associated with the practice of utilizing tires to singe meat in three major cities of ghana: a concurrent mixed methods study. Electronic Theses and Dissertations. 1350

Huang, W., Xu, H., Wu, J., Ren, M., Ke, Y., Qiao, J. (2024). Toward cleaner air and better health: Current state, challenges, and priorities. *Science, 385*:6707: Air pollution. 386–390. https://doi.org/10.1126/science.adp7832

Jimoda, L. A., Sulaymon, I. D., Alade, A., & Adebayo, G. A. (2018). Assessment of environmental impact of open burning of scrap tyres on ambient air quality. *International Journal of Environmental Science and Technology, 15*, 1323–1330. https://doi.org/10.1007/s13762-017-1498-5

Jeong, H. (2022). Toxic metal concentrations and Cu–Zn–Pb isotopic compositions in tires. *Journal of Analytical Science and Technology, 13*, 2. https://doi.org/10.1186/s40543-021-00312-3

Jomova, K., Alomar, S. Y., Nepovimova, E., Kuca, K., & Valko, M. (2024). Heavy metals: Toxicity and human health effects. *Archives of Toxicology*. https://doi.org/10.1007/s00204-024-03903-2

Mentes, D., Tóth, C. E., Nagy, G., Muránszky, G., & Póliska, C. (2022). Investigation of gaseous and solid pollutants emitted from waste tire combustion at different temperatures. *Waste Management, 149*, 302–312. https://doi.org/10.1016/j.wasman.2022.06.027

Mensah, N. J., Antwi-Akomeah, S., Akanlu, S., Sixtus, B. M. B., & Godfred, S. E. (2019). Residual levels of heavy metal contaminants in cattle hides singed with scrap tyre and firewood fuel sources: a comparative study in the Wa Municipality of Ghana. *American Journal of Environmental Science & Technology, 3*, 11–21.

Morais, S., Garcia e Costa, F., Pereira, M. d. L. (2012) Heavy metals and human health. In: Oosthuizen, J. (Edit): Environmental Health, Emerging Issues and Practice, InTEchOpen. ISBN: 978–953–307–854–0.

Narra, M.-M., Gbiete, D., Agboka, K., Narra, S., & Nelles, M. (2024). Tracing the EoL-tyre management chain in Togo with focus of implementing a tyre recycling plant. *Sustainability, 16*(21), 9193. https://doi.org/10.3390/su16219193

Nkwunonwo, U. C., Odika, P. O., & Onyia, N. I. (2020). A review of the health implications of heavy metals in food chain in Nigeria. *The Scientific World Journal, 2020*(1), 6594109.

Obiri-Danso, K., Hogarh, J. N., & Antwi-Agyei, P. (2008). Assessment of contamination of singed hides from cattle and goats by heavy metals in Ghana. *African Journal of Environmental Science and Technology, 2*(8), 217–221.

Okonkwo, F. O., Njan, A. A., Ejike, C. E. C. C., Nwodo, U. U., & Onwurah, I. N. E. (2018). Health implications of occupational exposure of butchers to emissions from burning tyres. *Annals of Global Health, 84*(3), 387–396. https://doi.org/10.29024/aogh.2321

Raudonytė-Svirbutavičienė, E., Stakėnienė, R., Jokšas, K., Valiulis, D., Byčenkienė, S., Žarkow, A., 2022. Distribution of polycyclic aromatic hydrocarbons and heavy metals in soil following a large tire fire incident: A case study. *Chemosphere. 286* (1). https://doi.org/10.1016/j.chemosphere.2021.131556.

Saeedi, M., Malekmohammadi, B., Tajalli, S. (2024). Interaction of benzene, toluene, ethylbenzene, and xylene with human's body: Insights into characteristics, sources and health risks, *Journal of Hazardous Materials Advances, 16*, 100459, https://doi.org/10.1016/j.hazadv.2024.100459.

Sombiri, S., Balhara, N., Attri, D., Kharb, I., & Giri, A. (2024). An overview on occurrence of polycyclic aromatic hydrocarbons in food chain with special emphasis on human health ailments. *Discover Environment, 2*, 87. https://doi.org/10.1007/s44274-024-00121-6

Wang, J., Li, X., Wang, B., Xiong, J., Li, Y., Guo, Y., Zhu, T., Xu, W. (2022). Emission characteristics of volatile organic compounds during a typical top-charging coking process, *Environmental Pollution, 308*, 119648, https://doi.org/10.1016/j.envpol.2022.119648.

Yaya, C., Soro, D. B., Hampoh, A. H., N'guettia, K. R., Sanogo, S., Aboua, K. N., Meite, L., Traore, K. S., Dembele, A. 2024. Contaminant intake from fuels used when grilling meats and stripping beef hides. *Journal of Materials and Environmental Science, 15*(9), pp 1258–1267

Yeh, C.-K. and Yeh, R.-H. (2023). Combustion of scrap waste tires of a cogeneration plant. *Journal of Marine Sciences and Technology. 31*(1). https://doi.org/10.51400/2709-6998.2680.

Zhang, H., Feng, X., Chan, H. M., & Larssen, T. (2014). New insights into traditional health risk assessments of mercury exposure: implications of selenium. *Environmental Science & Technology, 48*(2), 873–1360. https://doi.org/10.1021/es4051082

Sustainable Energy Systems for Growing Cities

Sustainable Solutions for Growing Cities in Sub-Saharan Africa: Trends in Urban Energy Demand

Ghafi Kondi Akara, Dambré Koungbanane, Mouhamadou Bamba, and Cecil Ouma

Abstract Sub-Saharan Africa (SSA) is undergoing rapid urbanization, driven by population growth and rural-to-urban migration. This trend has led to a surging demand for energy services, often surpassing the capacity of existing infrastructure. The resulting energy gap exacerbates inequalities in access and contributes to negative environmental impacts. Adopting sustainable energy models is critical to achieving inclusive, low-carbon development. African energy systems are increasingly pivoting toward disruptive renewable energy sources and microgrid-based solutions. Interest in the circular economy has also gained momentum, spearheaded by institutions like the African Development Bank and national finance ministries, alongside modern circular economy principles. Africa is emerging as a pioneer in exploring innovative aspects of the circular economy, particularly in the realm of green hydrogen. This study examines the interplay between rapid urbanization, energy consumption, and CO_2 emissions in SSA cities, proposing sustainable solutions rooted in the circular economy and energy transition. Using historical data (2000–2021) and employing statistical and spatial analysis methods, the research projects trends up to 2050. Key indicators include energy consumption, urbanization rates, GDP per capita, and CO_2 emissions. Projections indicate that by 2050, SSA's urban population could reach 1.07 billion, with energy demand rising to 80,480 GWh, placing immense pressure on resources. CO_2 emissions are expected to increase to approximately 20 million metric tons per capita. While these forecasts underscore significant challenges, they also reveal opportunities: adherence to circular economy principles could mitigate urban challenges and reduce greenhouse gas emissions. These findings are particularly relevant to governments, investors, urban planners,

G. Kondi Akara (✉) · M. Bamba
African Institute for Mathematical Sciences Research and Innovation Centre (AIMS RIC), Kigali, Rwanda
e-mail: gkondi@aimsric.org

D. Koungbanane
Laboratory for Biogeographic Research and Environmental Studies, Department of Geography, University of Lome, Lome, Togo

C. Ouma
HySA Infrastructure Center of Competence, North West University, Faculty of Engineering, Potchefstroom, South Africa

M.-M. Narra and S. Narra (eds.), *African Green Transition Through Innovative Pathways*, World Sustainability Series, https://doi.org/10.1007/978-3-031-87043-9_8

policymakers, and businesses seeking to address future energy demands within the framework of a circular economy.

Keywords Sub-Saharan Africa · Urban explosion · Circular economy · Energy demand · Sustainability · CO_2 emissions

1 Introduction and Context

In Sub-Saharan Africa (SSA), rapid population growth is driving accelerated urbanization with rural–urban migration playing a key role. However, within SSA, rural regions are now on the verge of becoming predominantly urban. This can largely be attributed to spill-over effects resulting from congestion and increased competition for resources in legacy cities. Smaller towns and peri-urban areas have been forced to absorb the population influx from legacy cities hence the emergence of new settlement patterns. The United Nations Population Fund (2007) estimates that the urban population in developing countries will grow exponentially, exceeding rural populations by 850 million by 2050. This shift in settlement patterns is likely to significantly impact the demand for urban services in general and energy services in particular. Presently within SSA cities, energy needs far exceed the capacity to provide services and this challenge needs to be addressed if SSA energy systems must meet future city demands.

Furthermore, SSA urban areas are undergoing rapid transformations, accompanied by an increase in energy consumption per unit of economic activity (Borchers et al., 2018). This development, which is both energy-intensive and harmful to the environment and public health, perpetuates inequalities and vulnerability. For a sustainable future, these cities must adopt a low-carbon, inclusive, and resilient development model, requiring a revision of energy management. The instability of modern energy supply and its importance for economic well-being and poverty reduction are well-established. After centuries of colonialism, Africa faces a crucial choice: follow old industrial models or explore new paths of economic growth. Cities are now viewed as laboratories for alternative development strategies, based on urban metabolism and sustainable energy practices (Calitz, 2019).

Indeed, Sub-Saharan African cities are not only experiencing population growth but also spatial expansion. This demographic and spatial growth poses significant challenges for meeting energy service needs. As a result, providing continuous electricity service to the entire urban population is one of the most poorly delivered services in these cities, a reflection of poorly managed urbanization processes over several decades (Nyassogbo, 2010).

Consequently, urbanization in Africa presents both challenges and opportunities. African cities attract numerous migrants seeking economic opportunities or fleeing rural insecurity. This, often informal, migration complicates urban management, which lacks the resources to accommodate this rapid growth (Gwaleba, 2018). Despite these constraints, the economic potential of African cities is promising, driven by dynamic entrepreneurs, informal markets, and innovative initiatives, such as those observed in Addis Ababa, Kigali, and Nairobi (Borchers et al., 2018). However, the lack of adequate data hinders long-term urban planning. Access to such data is essential for developing sustainable and inclusive development strategies.

Many African countries have recently demonstrated significant commitment to promoting the circular economy. Regarding the energy transition, Africa still has significant potential, encouraging the adoption of renewable energies and micro-grids. Interest in the circular economy is reified by actions from the African Development Bank, finance ministries, and the increasing media attention given to the features of these new economies. Africa is thus a pioneering ground for global innovations, particularly in green hydrogen.

The need to meet the growing energy demands of cities in a sustainable way has led to the emergence of the energy transition and the circular economy. Indeed, issues related to the energy transition are at the heart of scientific debates and attract the attention of researchers, activists, and governmental and non-governmental organizations. They highlight the concept of sustainability in the energy sector (Roques & Roux, 2018). The energy transition is one facet of environmental transition aligned with the pursuit of sustainable development goals, aiming to ensure greater access to reliable, sustainable, modern, and affordable energy services for all social groups (Rafitoson, 2017). As such, it serves as a cornerstone in the context of environmental, political, and socio-economic crises.

Governments and the private sector are showing increasing interest in energy consumption in SSA cities, which remain largely dependent on fossil fuels despite efforts to adopt sustainable practices. Urban energy consumption, influenced by factors such as demography, urbanization, and GDP, is complex. Interest in energy management intensified after the 2007/2008 crisis and the spike in fossil fuel prices, spurring numerous studies on the subject (Borchers et al., 2018). Today, cities are seen as key players in the global energy transition (Grimaldi, 2018).

Many countries around the world have included the energy transition in their political agendas to ensure more carbon-efficient, effective, and sustainable economies and societies. The challenge is to implement deep, large-scale transformations of energy consumption and production systems, thereby reducing energy poverty and greenhouse gas emissions. The energy transition brings sustainable development to life by promoting renewable energies over non-renewable fossil fuels.

Moreover, SSA represents 17.51% of the global population but consumes only 6.1% of the world's energy. This energy consumption accounts for only 3.1% of global electricity production, or more than 26,823 terawatt hours, and about 3% of primary energy (World Bank, 2022). This shows that electricity supply remains insufficient, as two out of three people lack access to modern energy services, with an average electricity access rate of around 20% (World Bank, 2022). Although SSA is a lower emitter of greenhouse gases, it is the first to suffer from the harmful effects of climate change. This has compelled these countries to commit to reducing greenhouse gas emissions in line with the Paris Agreement while also reducing energy poverty in their regions.

Energy, essential for urban development, is often neglected in urban planning models (Borchers et al., 2018). In Africa, the urban growth model differs from others due to various factors, including the prevalence of informal settlements. Currently, about 37% of Africa's population lives in urban areas, but this figure will increase significantly by 2050 in the context of high population growth, climate change, and biodiversity loss (UN-Habitat, 2020).

A 2023 report by the Africa Cities Network and the Global Initiative highlights that while African cities house a young and growing population, over half of them face sustainability challenges that could lead to irreversible damage. Although they often replicate the approaches of Northern cities, it is crucial that they develop local solutions adapted to their context. Examples from Cape Town, Addis Ababa, Nairobi, and Accra show how the circular economy can address essential needs in energy, water, food, and transport (UN-Habitat, 2023, ICLEI-ACEN, 2023). The African Urban Agenda recommends maximizing local opportunities while avoiding capture by external interests.

Thus, the dynamics of urbanization present significant challenges but also invaluable opportunities for the sustainability and resilience of SSA cities. Rapid urbanization in Africa poses major challenges, notably the need to provide basic services and energy in areas often labelled as mega-slums. This growth increases energy consumption, necessitating a re-evaluation of urban processes and energy supply to ensure sustainability. African cities must shift from a linear growth model to circular development, rethinking energy planning and resource recovery (Remøy et al., 2019).

The concept of the circular economy enables cities to optimize resource flows, integrating renewable energy production such as photovoltaics and biogas to support sustainable growth (Bîrgovan et al., 2022). The interdependence between urbanization, energy consumption, and the circular economy shapes their development.

In this context, the implementation of effective strategies to reduce greenhouse gas (GHG) emissions in African cities is essential, particularly regarding the New Urban Agenda (NUA), the Sustainable Development Goals (SDGs), and Nationally Determined Contributions (NDCs). Although studies exist on urban development and energy use, the complex relationships between urbanization, energy consumption, and GHG emissions remain partially understood.

The primary objective of this study is to analyse the relationship between urban explosion in Sub-Saharan Africa and energy use for sustainable development. Specifically, this work explores (1) the current and future trends of key factors influencing urban expansion in SSA, (2) current trends and future models of energy consumption and production in SSA cities, and (3) the interrelationship between urbanization, energy consumption and production, and CO_2 emissions within the context of the circular economy. The results of this research are crucial for governments, investors, urban planners, and businesses to better anticipate and manage the evolution of energy demand within a circular economy framework.

2 Study Area

2.1 *This work focuses on the countries of Sub-Saharan Africa.*

Sub-Saharan Africa, often referred to as "the Sub-Sahara," encompasses the region of the African continent located south of the Sahara Desert (Fig. 1). It is ecologically distinct from the northern countries due to the extreme climate of the Sahara, the world's largest hot desert.

Sub-Saharan Africa consists of forty-eight countries extended across four geographical sub-regions: West Africa, East Africa, Central Africa, and Southern Africa. It covers an area of 24,328,266 km^2 (World Bank, 2021) and has a population of 1.24 billion people (World Bank, 2023a, b).

Significant socio-economic transformations are observable in the cities of Sub-Saharan African countries. This suggests that the urban explosion is impacting the socio-economic life of the population. The living standards of urban populations influence energy production and consumption patterns. This overall movement, characterized by migration, rural exodus, and other factors in Sub-Saharan Africa, is marked by both rapid urbanization and "exurbanization". Thus, the urban explosion in this part of Africa presents not only opportunities but also enormous challenges,

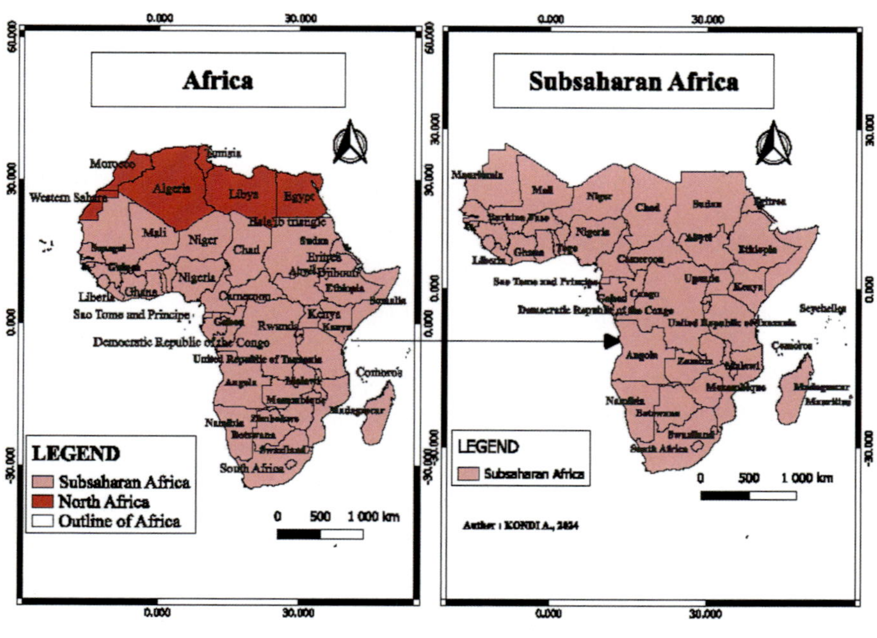

Fig. 1 Geographical division maps of Africa versus Sub-Saharan Africa

particularly regarding energy use for sustainable urban development (Saghir & Santoro, 2018; UN-Habitat, 2015).

In order to contribute to reducing CO_2 emissions in the atmosphere, in line with the Paris Agreement, and to increase resilience to extreme weather events and natural disasters, Sub-Saharan African countries must accelerate the transition to a low-carbon economy, protect the natural environment, and create green, decent, and dignified jobs. The widespread adoption of circular principles could create new business opportunities, generate jobs, and contribute significantly to many of the 17 Sustainable Development Goals (SDGs).

3 Methodological Framework and Data Sources

3.1 Approach to Data Collection

Table 1 provides an overview of the data types utilized for this study, including energy production, electricity consumption, and urbanization metrics, alongside the coverage period for each dataset (2000–2021) and their respective sources. Notably, energy data encompasses gross production from various sources such as fuel oil, natural gas, hydro, wind, and solar, as well as electricity consumption across sectors like industry, households, and agriculture. Urbanization data focuses on the extent of urban areas. To ensure completeness and reliability, a data imputation technique was applied to address and fill gaps in the datasets before initiating the analysis. This table serves as the foundation for the comprehensive examination of trends in energy and urbanization.

3.2 Methodological Insights

3.2.1 Techniques for Analysing Temporal Data

The temporal analysis method takes into account data on energy (consumption and production), urbanization (urban area, urbanization rate, and population size), GDP per capita, and CO_2 emissions. All data underwent systematic quality control to ensure reliability and to avoid the use of outlier data. After this step, missing data were filled using the linear regression method. Linear regression (Cornillon et al., 2011; Kibala, 2018) is suitable for datasets that have a linear relationship between data points. The forecast of the predicted value is calculated based on the existing values when the correlation is significant, or alternatively, using the values from years adjacent to the missing data.

The equation for this model is written as Eq. (1):

Table 1 Summary of data sources for energy and urbanization metrics (2000–2021)

Data		Types	Period	Sources
Energy	Electricity Production and Imports	Gross production (from Fuel Oil, Gas/Diesel Oil, Natural gas, Hydro, Wind, Solar Pv and Other sources)	From 2000 To 2021	AFREC Website www.au-afrec.org Specifically: https://au-afrec.org/data-statistics-energy-browser-0
	Electricity Consumption	Final consumption (from Industry, Households, Commercial and public services, Agriculture, forestry and fishing, Non-specified)		
Urbanization	Surfaces	Urban areas		World Development Indicators de la Banque mondiale Website: https://donnees.banquemondial.org
	Urbanization rate	Urbanization rate in %		
	Population size	Urban population		
Economic	GDP per capita			
Greenhouse Gas	CO_2 emissions			

$$a + bx \text{ or } a = \bar{y} - b\bar{x} \text{ and } b = \frac{\sum(x - \bar{x})(y - \bar{y})}{\sum(x - \bar{x})^2} \tag{1}$$

where X and Y are constants, with x and y representing the sample means for the independent variable (x-known) and the dependent variable (y-known). x represents the observation whose value is being predicted, y-known represents the matrix or range of dependent data, and x-known represents the matrix or range of independent data.

Next, descriptive statistical methods (mean, standard deviation, etc.) were used. To evaluate the trend in the data series, the coefficient of determination (R^2) was calculated. A higher R^2 indicates a better fit. A positive coefficient means that an increase in the independent variable leads to an increase in the dependent variable. The R^2 value ranges between 0 and 1. When R^2 is 0, it means that the model does not fit the data at all, indicating no linear relationship between the variables. When R^2 is 1, the fit is perfect, meaning that all observations perfectly match the regression line. If R^2 falls between 0.7 and 0.9, the fit is good and suggests that the model likely represents the data well. If R^2 is between 0.5 and 0.7, the fit is acceptable but indicates room for improvement. A value below 0.5 means that the model does not describe the underlying data accurately, and in this case, the model must be adjusted to obtain relevant results.

3.2.2 Techniques for Spatial Analysis

GIS (Geographic Information Systems) played a crucial role in this work. For spatial analysis, the IDW (Inverse Distance Weighted) interpolation method was used. IDW is a technique (Blanco et al., 2018; Bartier and Keller, 2016; Chen & Liu, 2012) that predicts the value of an attribute at positions where no sample is available, based on the spatial distance between that position and other positions where samples have been collected. Observations that are spatially closer to the target observation receive a higher weight in the prediction, while more distant observations have a relatively lower influence. The weight assigned to each sample is also controlled by a power parameter, often referred to as pp. As pp increases, the weight assigned to distant samples decreases. This factor can be understood as a smoothing parameter. If the smoothing effect is too strong, it can obscure interesting variability or give the impression that the predicted variable is more homogeneous than it really is. Conversely, when pp is low, both nearby and distant samples exert a strong influence on the predictions, potentially leading to overly smoothed results that do not account for local phenomena. On the other hand, if pp is too high, only very local processes will be considered, which can give the interpolated map a granular, non-smooth appearance.

The advantage of the IDW technique is that it takes more observations into account for predictions, which can improve their accuracy. However, it is essential to properly adjust the pp parameter to ensure that the interpolated map is both relevant and reliable. The equation used to express this is as Eq.(2):

$$u(X) = \frac{\sum_{k=0}^{N} \times w_k(X)^p uk}{\sum_{k=0}^{N} \times w_k(X)^p} \text{ or } w_k(X) = \frac{1}{d(X, X_k)} \tag{2}$$

is a simple weighting function, as defined by Shepard (1968), where x is the point to be interpolated, Xk is a (known) interpolation point, Uk is the value of the function u at the point Xk, d is a given distance from the interpolation point Xk to the point to be interpolated X, N is the total number of known points used in the interpolation, and p is a real positive number, called the power parameter.

Thus, through the different calculation and spatial analysis methods offered by the QGIS 3.38.1 software, operations were carried out with urbanization data (areas, urbanization rate and population size), energy data (consumption and production), GDP per capita and CO_2 emissions, and maps were produced.

4 Results and Interpretations

The analysis of the results of this study focuses on the characteristics of historical and future trends.

4.1 *Urban Explosion in Sub-Saharan Africa: Trends and Implications*

This study aims to analyse the spatio-temporal evolution of the factors contributing to urban explosion and their impact on energy use within the context of the circular economy from 2000 to 2021.

Dynamics of Urban Growth: Analysing Urban Areas, Rates, Population, and GDP.

Figure 2 illustrates the annual trends in urban area expansion, urbanization rates, and GDP per capita in Sub-Saharan African countries from 2000 to 2021.

The analysis of Fig. 2 reveals that urbanization in Sub-Saharan African countries has been evolving, with an annual increase in urban areas, indicating a high rate of urbanization during the 2000–2021 period. On average, urban areas expanded by 8,430.40 km^2 per year. The urbanization rate increased from 33.59% in 2000 to 52.26% in 2021, representing an average annual growth rate of 2.61%. During the same period, the population grew from 210,824,543 inhabitants in 2000 to 494,081,113 in 2021. The growth in both urban areas and population is highly significant, with R^2 values of 0.94 and 0.98, respectively.

Alongside the increase in urban areas and the urbanization rate, GDP per capita increased from 2000 to 2019 before experiencing a slight decline between 2020 and 2021 due to the COVID-19 crisis. GDP per capita fluctuated between $4,440.24 and $5,889.64. However, the rise in GDP per capita is not as significant as the increase in urban areas and population, as the R^2 value is 0.77.

This spatial, demographic, and economic shift has profound implications on the urban fabric, housing demand, infrastructure, and services, as well as on the economy and environment. It presents major challenges in terms of urban planning, access to basic services, and waste management. Still, it also offers opportunities for economic development, job creation, and the improvement of living conditions.

Exploring the Spatial Changes in Urban Areas and Their Effects on Economic Growth.

Figure 3 illustrates the spatial and temporal trends in urban area expansion, urbanization rates, urban population growth, and GDP per capita across Sub-Saharan African countries.

The examination of Fig. 3 reveals an uneven distribution in the growth of urban areas, urbanization rates, population, and GDP per capita. Urban development plays a key role in shaping energy use in Sub-Saharan Africa. Some countries with high urbanization—characterized by large urban areas, high urbanization rates, population size, and GDP per capita—exhibit significant spatial disparities. The spatial distribution indicates that countries such as Nigeria, South Africa, the Democratic Republic of Congo, Ghana, Ethiopia, Tanzania, Sudan, and Côte d'Ivoire lead in spatial expansion, urbanization rate, population, and GDP per capita. In contrast, countries like São Tomé and Príncipe, Seychelles, Cape Verde, and Comoros exhibit much lower values. This heterogeneity in spatial distribution of urban areas, population, urbanization rate, and GDP can be attributed to various internal and external factors. However, the overall size of the country significantly influences these parameters.

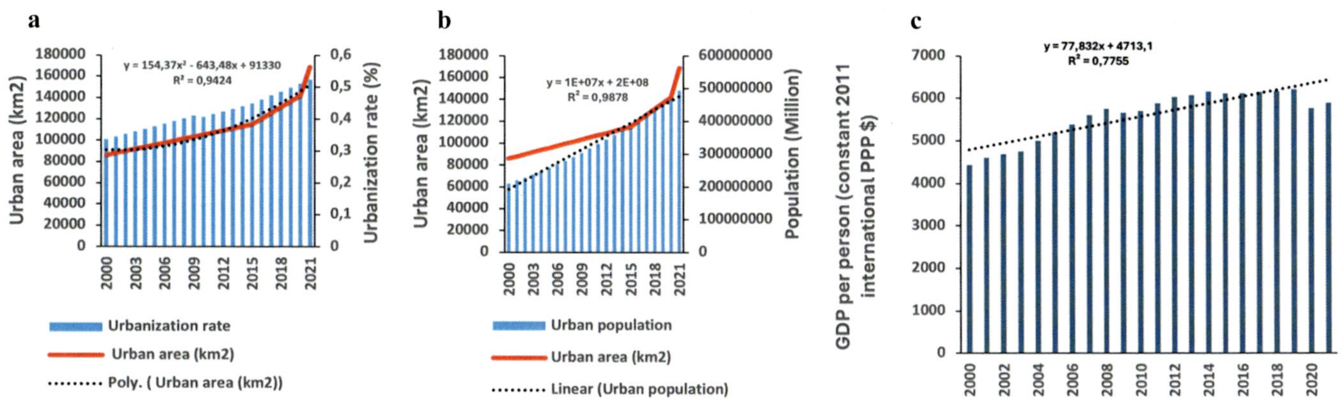

Fig. 2 Trends in urbanisation **a**, population growth **b**, and GDP per capita **c** over time

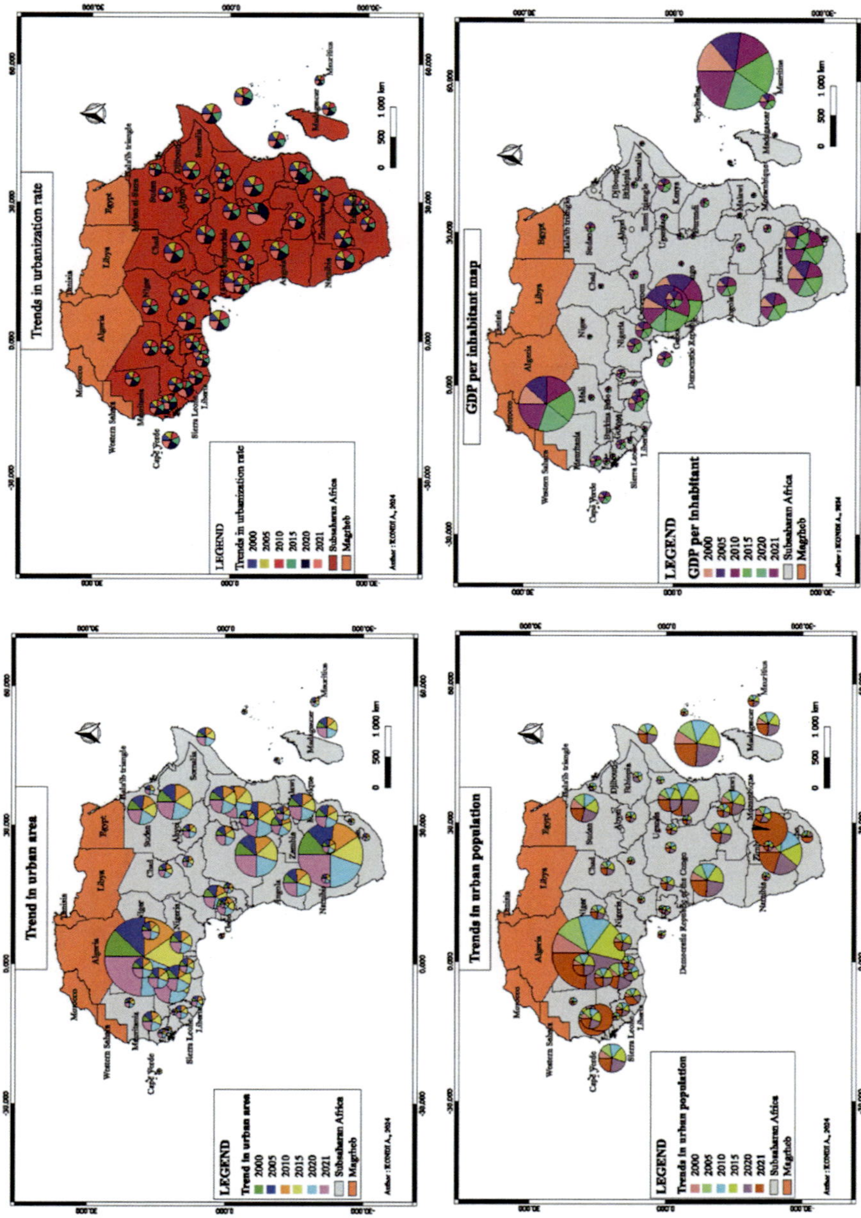

Fig. 3 Spatial and temporal trends in urban areas, urbanization rates, urban population, and GDP per capita across Sub-Africa GDP

4.2 *Evolution of Energy Use Patterns*

4.2.1 Trends in Energy Consumption Across Sub-Saharan Africa

Figure 4 illustrates the trends in energy consumption across various sectors, including industry, households, commercial services, and agriculture, as well as the overall final energy consumption from 2000 to 2021.

The analysis of Fig. 4 reveals a significant upward trend in final energy consumption across sectors in Sub-Saharan African countries from 2000 to 2021, with an R^2 value of 0.95 indicating a strong linear relationship. Final electricity consumption increased from 239,758.18 GWh in 2001 to 459,582.16 GWh in 2021, with the sharpest growth observed after 2010, as evidenced by the positive slope and high regression fit.

Among energy-consuming sectors, industry remains the largest consumer, rising from 127,677 GWh to 224,196 GWh, followed by households (49,582.74 GWh to 138,468.73 GWh), the commercial and services sector (28,904.24 GWh to 69,163.03 GWh), and agriculture (5,854.40 GWh to 12,720.90 GWh).

Agriculture remains the lowest energy-consuming sector, primarily dependent on traditional energy sources such as biomass, human labour, and animal power. The limited penetration of modern energy sources in this sector is largely due to inadequate energy infrastructure and low energy demand.

The growth of the industrial sector plays a pivotal role in driving rising energy demand, with its consumption rate outpacing other sectors. This reflects the sector's crucial role in economic production and development, where energy acts as a catalyst for industrial output and economic activities.

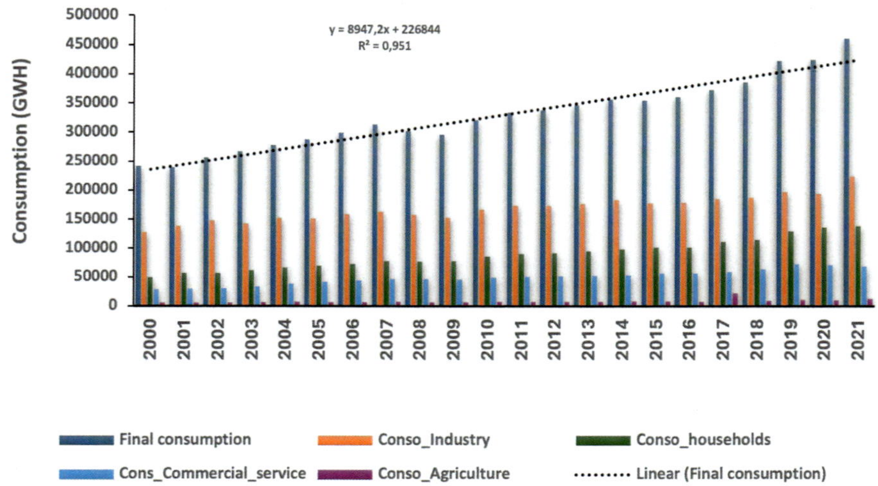

Fig. 4 Trends in energy consumption by sector from 2000 to 2021in Sub-Saharan Africa

However, energy consumption trends across Sub-Saharan Africa exhibit regional disparities, with some regions—particularly those with higher levels of industrialization and urbanization—experiencing more significant increases in energy demand. These variations reflect differences in energy access, infrastructure development, and economic drivers, highlighting the need for targeted strategies to reduce energy inequalities across regions.

While total energy consumption has risen, further analysis is required to assess per capita energy consumption. A relatively slow improvement in energy consumption per capita could indicate persistent gaps in energy access for a significant portion of the population, despite overall growth in demand. This underscores the need for equitable energy distribution policies to ensure inclusive growth.

The observed upward trend in energy consumption, particularly in the industrial sector, suggests a positive correlation between energy demand and economic activities. The industrial sector's dominant share reinforces its role as a driver of economic growth, supporting productivity, job creation, and expansion across key sectors.

4.2.2 Understanding the Landscape of Energy Production

Figure 5 depicts the annual evolution of energy production and its sources in Sub-Saharan Africa from 2000 to 2021.

Figure 5 reveals a significant upward trend in energy production in Sub-Saharan Africa, with an R^2 value of 0.9511, highlighting the consistency and strength of this increase. Gross production of electricity consistently increased, rising from approximately 310,239 GWh in 2000 to over 600,000 GWh in 2021.

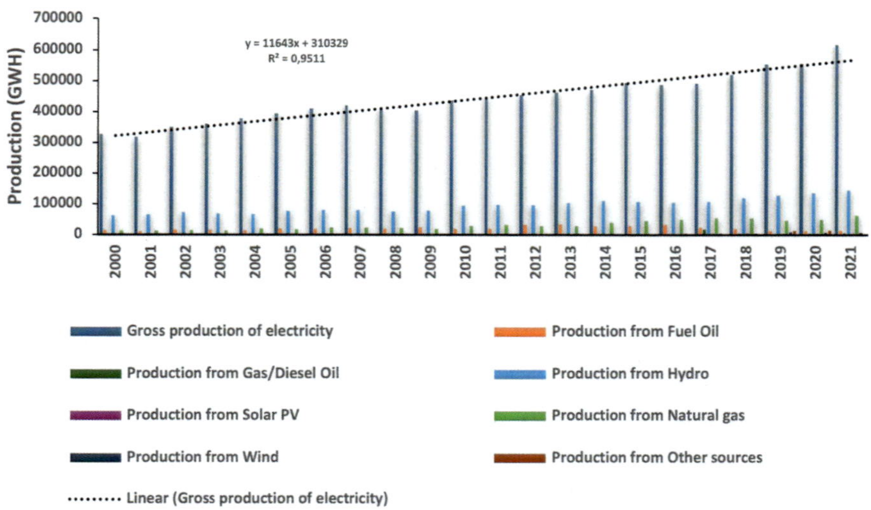

Fig. 5 Trends in gross electricity production by source in Sub-Saharan Africa over the period 2000–2021

Hydropower emerges as the dominant energy source, with production rising from 62,403.82 GWh in 2000 to 144,173.98 GWh in 2021. It is followed by natural gas, which increased substantially from 12,147 GWh to 62,175.74 GWh, and fuel oil, showing more modest growth from 11,405.24 GWh to 13,512.06 GWh.

Other sources, including diesel oil, solar PV, wind, and unspecified sources, contribute only marginally to the total electricity production. While renewables such as solar PV and wind show gradual progress, their overall contribution remains limited, reflecting the slow adoption of renewable energy technologies.

This limited penetration of renewable energy highlights the absence of robust renewable energy policies to align with the region's urban expansion and growing energy demand. Addressing this gap is critical, as it represents a major obstacle to achieving sustainable development goals (SDGs), particularly in promoting clean and sustainable energy solutions to meet the needs of a growing population and economy.

4.3 Assessing the Impact of Urban Growth on Energy Consumption

4.3.1 Correlation Between Urban Areas and Energy Consumption

The relationship between urban expansion and energy use is analysed through the correlation between urban areas and energy consumption (Fig. 6).

The analysis illustrates a strong correlation between urban area expansion, energy consumption, and energy production. Urban areas increased significantly from 87,924.91 km^2 in 2000 to 168,608.08 km^2 in 2021, reflecting rapid urbanization. During the same period, energy consumption rose from 239,758.18 GWh to

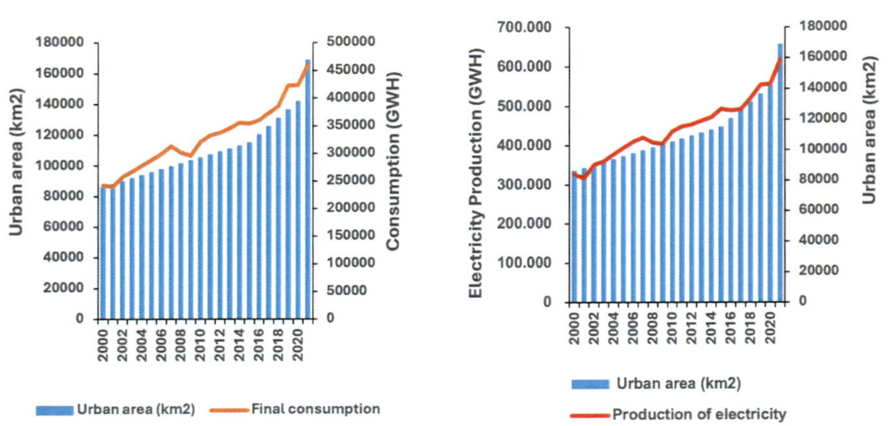

Fig. 6 Relationship between urbanization, energy consumption, and electricity production trends (2000–2021)

459,582.16 GWh, while electricity production grew from 317,236 GWh to 618,090 GWh.

The upward trend across all variables highlights the growing interdependence between urbanization and energy demand. Urban expansion drives higher energy consumption as energy-intensive sectors—such as industry, households, and commercial services—spread into newly urbanized areas. This increased demand necessitates higher energy production to support urban growth.

However, the differing growth rates also emphasize ongoing energy poverty in the region, where energy infrastructure struggles to keep pace with urbanization. Addressing this gap is critical to ensuring equitable energy access and meeting the needs of rapidly expanding urban areas.

Overall, the analysis establishes a clear relationship between urbanization, energy demand, and energy production, underscoring the need for integrated energy and urban development policies to support sustainable growth.

4.3.2 Demographic Factors Shaping Energy Consumption and Production

Figure 7 shows the demographic evolution in Sub-Saharan Africa and its impact on energy use.

The analysis of Fig. 7 demonstrates that population growth in Sub-Saharan Africa influences energy consumption and production. The population grew from 210,824,543 to 475,423,476, while energy consumption increased from 239,758.18 GWh to 459,582.16 GWh between 2000 and 2021. During this period, energy production rose from 317,236 GWh to 618,090 GWh, indicating that as the urban population grows, energy consumption follows the same trend, driving production services to increase supply. Thus, urbanization is a key factor influencing electricity production and consumption in Sub-Saharan Africa. The urban population growth drives energy demand, establishing a strong relationship between population growth and increased energy demand during 2000–2021. Urbanization, through population growth and spatial expansion, plays a critical role in energy use in Sub-Saharan African countries.

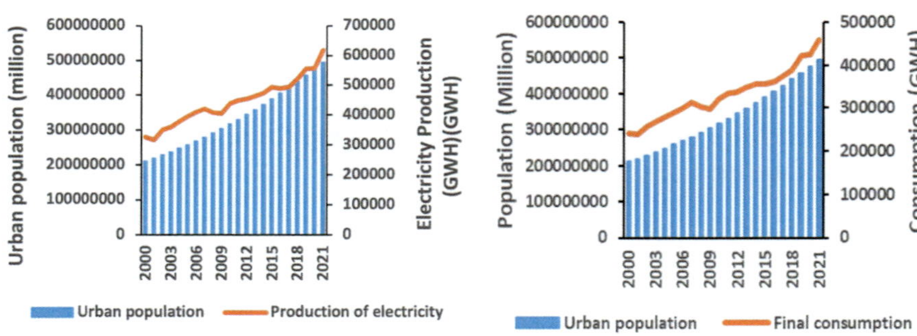

Fig. 7 Correlation of population growth with energy use and production

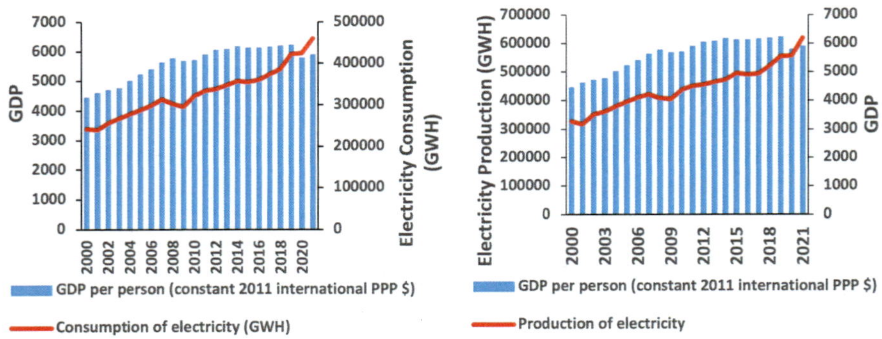

Fig. 8 Impact of GDP growth on energy use and production

4.3.3 Exploring GDP Per Capita Trends Impacts on Energy Consumption and Production

The analysis of the relationship between GDPs per capita and its impact on energy consumption and production (Fig. 8) illustrates the importance of GDP in energy use.

Economic growth is a key indicator of energy consumption and production. The analysis of Fig. 8 shows that GDP per capita increased from 2000 to 2019 before slightly declining in 2020–2021 due to the COVID-19 crisis. GDP per capita ranged from $4,440.24 to $5,889.64. Like consumption, energy production followed the same upward trend, indicating that increased energy consumption and production are signs of GDP growth in Sub-Saharan African countries. The upward trend in GDP per capita is accompanied by an increase in both energy consumption and production in Sub-Saharan African countries. This is reflected in the linear relationship observed between the growth in GDP per capita and energy usage, indicating a cause-and-effect link between the two. As the population's standard of living rises, so do their energy needs, which explains the parallel growth in GDP and energy demand.

4.4 Exploring Urbanization's Impact on Energy Use Within a Circular Economy Framework

4.4.1 Examining CO_2 Emissions in Relation to Urbanization Rates and Population Size

Figure 9 presents a correlation between the urbanization rate, population size, and CO_2 emissions.

The analysis of Fig. 9 shows that the increase in the urbanization rate and population size contributes to a rise in CO_2 emissions into the atmosphere. This correlation is strong, as evidenced by an R^2 value of 0.96. The urbanization rate increased from

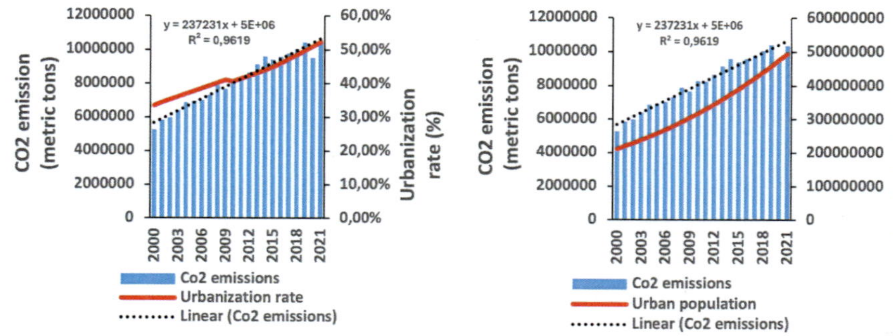

Fig. 9 Correlation between urbanization rate, population Size, and CO_2 emissions

33.59% to 52.26%, while CO_2 emissions grew from 5,242,520.21 metric tons to 10,391,291.76 metric tons. This indicates that the goal of sustainable development, within the framework of the circular economy, is at an impasse. Demographic growth plays a key role in driving CO_2 emissions, as it remains one of the leading factors behind humanity's growing impact on the natural environment. This is largely driven by energy consumption and production needed to meet the ever-increasing demands, aligned with technological advancements. In general, population growth continues to be a major driver of rising CO_2 emissions, primarily due to the use of fossil fuels. This is particularly evident in Sub-Saharan African countries, where increasing CO_2 emissions are linked to urban growth.

4.4.2 Exploring the Dynamics of Energy Consumption, Production, and CO_2 Emission

Figure 10 shows the relationship between energy consumption, production, and CO_2 emissions.

The analysis of Fig. 10, which shows the year-on-year evolution of CO_2 emissions in relation to energy usage, indicates a rise in both energy consumption and production. CO_2 emissions followed a similar trend during this period, with the

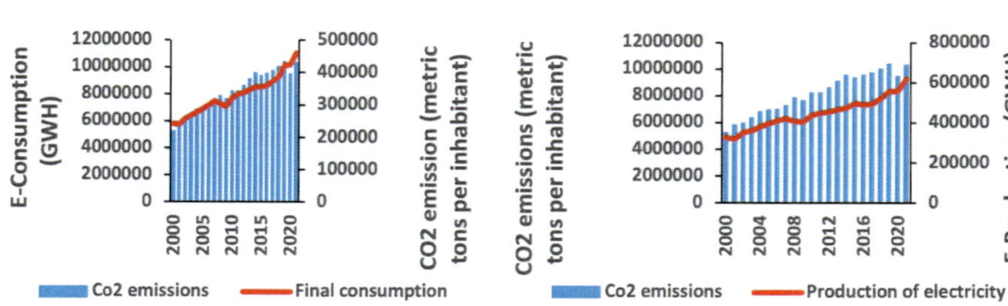

Fig. 10 Illustration of the interactions of energy use, production, and CO_2 emissions

highest emissions observed between 2004 and 2019. These emissions ranged from 5,242,520.22 to 10,391,291.8 metric tons per capita.

During the same time, energy consumption increased from 239,758.18 GWh to 459,582.16 GWh, while total energy production rose from 317,236 GWh to 618,090 GWh.

As energy consumption and production increase, CO_2 emissions also rise, leading to environmental pollution and challenging the principles of the circular economy, ultimately impeding sustainable development. There is a strong correlation between CO_2 emissions, energy consumption, and production. Energy consumption in Sub-Saharan African countries remains largely dominated by fossil fuels, such as coal and oil, which are the primary contributors to the increase in CO_2 emissions from 2000 to 2021.

The direct source of CO_2 emissions is energy consumption, and the amount of energy consumed has a significant impact on CO_2 emissions. This demonstrates the relationship between energy consumption and sustainable economic growth.

4.5 First Partial Conclusion

The study of urban growth dynamics in Sub-Saharan African countries from 2000 to 2021 reveals a significant increase in urban areas, urbanization rates, population size, and GDP per capita. Urban areas expanded by an average of 8,430.40 km^2 per year, while the urbanization rate increased from 33.59% to 52.26%. The population grew from 210,824,543 to 494,081,113 inhabitants, and GDP per capita rose from \$4,440.24 to \$5,889.64.

Simultaneously, energy usage underwent substantial changes, with energy consumption increasing from 239,758.18 GWh to 459,582.16 GWh, and energy production rising from 317,235.64 GWh to 618,090.27 GWh. This shift contributed to the rise in CO_2 emissions, which grew from 5,242,520.21 metric tons to 10,391,291.76 metric tons, undermining the principles of the circular economy and, consequently, sustainable development.

For controlled and sustainable urbanization, Sub-Saharan African countries must adhere to circular economy principles and transition to renewable energy sources.

4.6 *Future Horizons: Anticipating Key Trends and Projections*

4.6.1 City Dynamics: Analysing Urbanization, Population Growth, and Economic Indicators

Figure 11 illustrates the projected trends for urban areas, urbanization rates, population size, and GDP per capita in Sub-Saharan African countries by 2030, 2040, and 2050.

Based on the projections, the urban population is expected to grow by 2030, 2040, and 2050, reaching 675,303,039 inhabitants, 873,020,809 inhabitants, and 1,070,738,579 inhabitants, respectively. This population increase will lead to an expansion of urban areas, from 409,704.44 km^2 in 2030 to 677,589.29 km^2 in 2040 and 945,474.13 km^2 in 2050. Consequently, there will be a sharp rise in the urbanization rate, with figures projected to reach 63.34% in 2030, 75.66% in 2040, and 87.98% in 2050. From 2021 to 2030, urban areas will expand by an average of 853.55 km^2 per year, increasing to 1,411.64 km^2 annually between 2030 and 2040, and 1,712.26 km^2 from 2040 to 2050.

Similarly, an analysis of GDP per capita forecasts indicates growth through 2030, 2040, and 2050. Using 2020 as the reference year with a GDP per capita of $5,770.09,

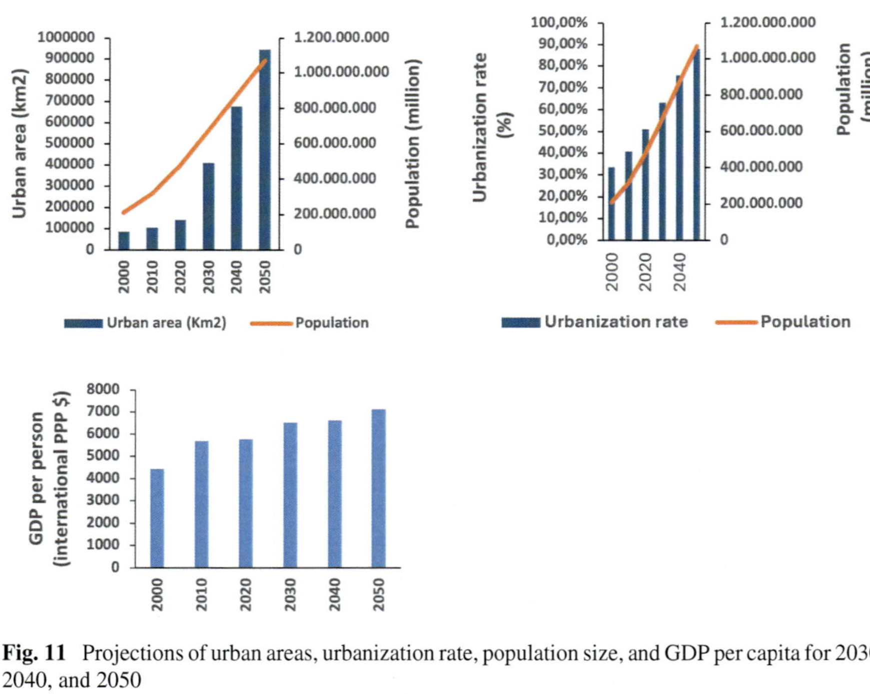

Fig. 11 Projections of urban areas, urbanization rate, population size, and GDP per capita for 2030, 2040, and 2050

this figure is projected to rise to $6,516.68 by 2030, $6,627 by 2040, and $7,123.23 by 2050.

This increase in the urban population presents an opportunity for job creation and economic growth. It is important to emphasize that economic growth has long been seen as a significant factor driving energy consumption. The relationship between these two variables is evident here through the trends in energy consumption and GDP fluctuations, forming the foundation of energy forecasts.

It is also important to highlight disparities in the projections for urban areas, population size, urbanization rate, and GDP per capita. These disparities are illustrated spatially in Fig. 12.

Figure 12 illustrates the uneven spatial distribution of urban areas in Sub-Saharan African countries projected for 2030, 2040, and 2050. These countries will not experience uniform growth in urban areas during this period. Nigeria is expected to see significant urban area expansion (90,898.30 km^2 in 2030; 150,633.25 km^2 in 2040; and 210,368.19 km^2 in 2050), followed by South Africa (90,898.30 km^2 in 2030; 150,633.25 km^2 in 2040; and 210,368.19 km^2 in 2050), the Democratic Republic of Congo (27,958.73 km^2 in 2030; 46,800.89 km^2 in 2040; and 65,643.06 km^2 in 2050), Ghana (17,293.73 km^2 in 2030; 27,961.18 km^2 in 2040; and 38,628.62 km^2 in 2050), and Tanzania (17,240.77 km^2 in 2030; 28,111.59 km^2 in 2040; and 38,982.41 km^2 in 2050). Conversely, some countries, such as São Tomé and Príncipe (11.95 km^2 in 2030; 13.51 km^2 in 2040; and 15.06 km^2 in 2050) and Seychelles (25.14 km^2 in 2030; 35.85 km^2 in 2040; and 46.56 km^2 in 2050), will exhibit smaller urban expansions.

Similarly, from 2030 to 2050, urbanization rates in Sub-Saharan African countries are projected to increase, with notable spatial disparities. Countries like Burundi (192.55% in 2030; 307.23% in 2040; and 421.91% in 2050), Equatorial Guinea (112.28% in 2030; 144.52% in 2040; and 176.76% in 2050), Malawi (96.75% in 2030; 107.61% in 2040; and 118.47% in 2050), and Sierra Leone (90.66% in 2030; 136.16% in 2040; and 181.66% in 2050) will experience high urbanization rates. Conversely, countries such as Mauritius (19.44% in 2030; 20.76% in 2040; and 22.08% in 2050), Lesotho (24.2% in 2030; 25.9% in 2040; and 27.6% in 2050), Eritrea (34.3% in 2030; 39.48% in 2040; and 44.66% in 2050), and Togo (34.3% in 2030; 39.48% in 2040; and 44.66% in 2050) are expected to have lower urbanization rates during the same period.

Regarding the projected urban population size, the spatial analysis reveals that certain countries will have much larger urban populations than others. Nigeria (153,144,850 inhabitants in 2030; 198,369,110 in 2040; and 243,593,370 in 2050) and South Africa (46,113,806 in 2030; 52,595,456 in 2040; and 59,077,106 in 2050) are projected to experience significant urban population growth. Meanwhile, countries like Seychelles (73,572 inhabitants in 2030; 78,032 in 2040; and 82,492 in 2050) and São Tomé and Príncipe (211,356 inhabitants in 2030; 260,236 in 2040; and 309,116 in 2050) will have smaller urban populations.

The spatial analysis of GDP per capita projections shows that countries such as Mauritania (USD 38,891.85 in 2030; USD 53,350.77 in 2040; and USD 53,352.22 in 2050), Somalia (USD 1,474.95 in 2030; USD 1,491.19 in 2040; and USD 1,507.43

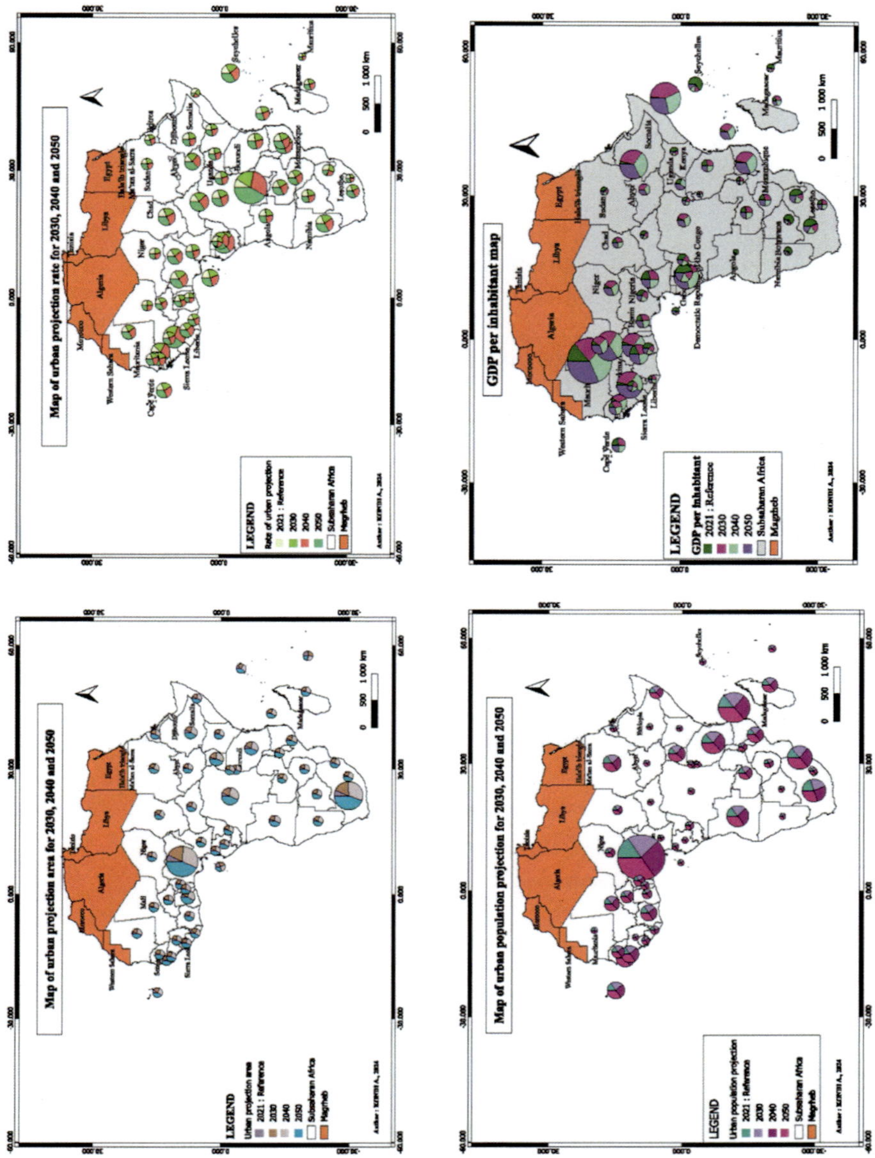

Fig. 12 Spatial distribution of urban areas, urbanization rates, population size, and GDP projections for 2030, 2040, and 2050

in 2050), Ethiopia (USD 20,150.39 in 2030; USD 20,309.12 in 2040; and USD 21,104.40 in 2050), and Burkina Faso (USD 18,683.20 in 2030; USD 19,770.26 in 2040; and USD 21,461.19 in 2050) will have higher GDPs in 2030, 2040, and 2050. Conversely, Uganda (USD 2,528.34 in 2030; USD 2,631.82 in 2040; and USD 3,093.01 in 2050) and Angola (USD 10,739.65 in 2030; USD 11,287.41 in 2040; and USD 11,835.17 in 2050) are projected to have relatively lower GDPs during this period.

It is important to note that this spatial disparity in urban areas, urbanization rates, population sizes, and GDP per capita in Sub-Saharan African countries is influenced by numerous local and external factors.

4.6.2 Energy Consumption Insights with a Look into Final Energy Consumption Behaviours

Figure 13 presents the projected final energy consumption and the various sectors driving energy consumption in Sub-Saharan African countries.

Based on trends in energy demand across various sectors, projections have been made to estimate energy demand up to 2030, 2040, and 2050. Four key sectors of energy consumption are considered in this research for future analysis: industry, household consumption, commerce and services, and agriculture. The figure shows that industry will continue to be the largest energy consumer, with a projected demand of 407,547.52 GWh in 2030, 621,457.30 GWh in 2040, and 804,808.54 GWh in 2050. It is followed by household energy demand, estimated at 155,489.39 GWh

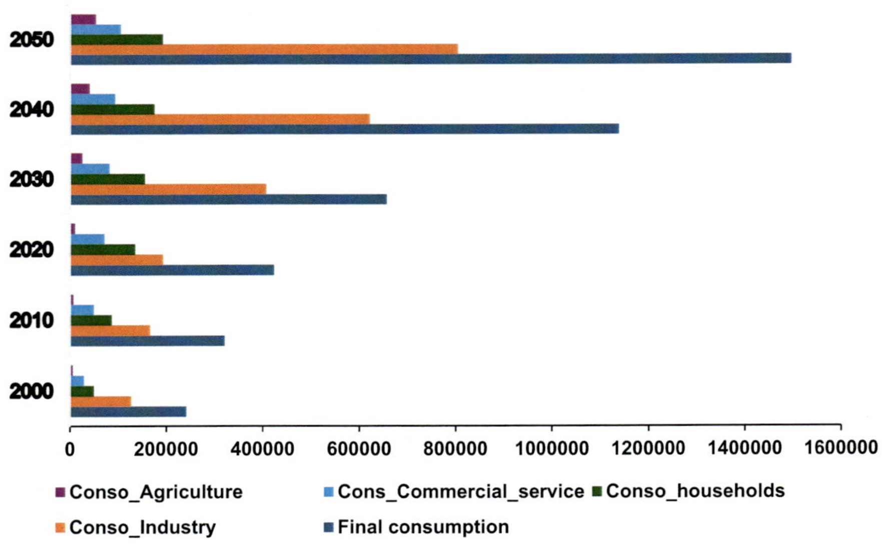

Fig. 13 Projected energy consumption by sector from 2000 to 2050 in 10-Year Intervals

in 2030, 175,346.84 GWh in 2040, and 192,367.51 GWh in 2050. The third-largest energy-consuming sector will be commerce and services, with 81,586.07 GWh in 2030, 93,175.38 GWh in 2040, and 104,764.69 GWh in 2050. The agriculture sector will demand the least energy compared to the other sectors, with 25,554.81 GWh in 2030, 40,527.67 GWh in 2040, and 53,361.55 GWh in 2050. The low level of mechanization in agriculture in Sub-Saharan African countries is a key reason why it will consume relatively little energy in the future.

The projected growth in energy consumption will significantly impact sustainable urban development. This highlights the need for Sub-Saharan African countries to explore alternative energy sources to meet the growing population's energy needs. Furthermore, these energy demand projections reveal disparities among countries regarding final energy consumption, as illustrated in Fig. 14.

The spatial analysis of projected energy consumption in Sub-Saharan African countries shows that, based on 2021 data, there will be an unequal distribution of energy demand. Some countries will experience significantly higher energy demands compared to others. South Africa (396,068.56 GWh in 2030; 602,946.11 GWh in 2040; and 809,823.67 GWh in 2050), Nigeria (75,235.49 GWh in 2030; 90,911.94 GWh in 2040; and 106,588.39 GWh in 2050), Ghana (31,914.85 GWh in 2030; 47,299.13 GWh in 2040; and 62,683.41 GWh in 2050), and Mozambique (24,381.89 GWh in 2030; 36,572.83 GWh in 2040; and 48,763.77 GWh in 2050) are projected to have the highest energy demands by 2030, 2040, and 2050.

In contrast, countries like Comoros (48.87 GWh in 2030; 61.29 GWh in 2040; and 73.71 GWh in 2050), São Tomé and Príncipe (79.4 GWh in 2030; 107.07 GWh

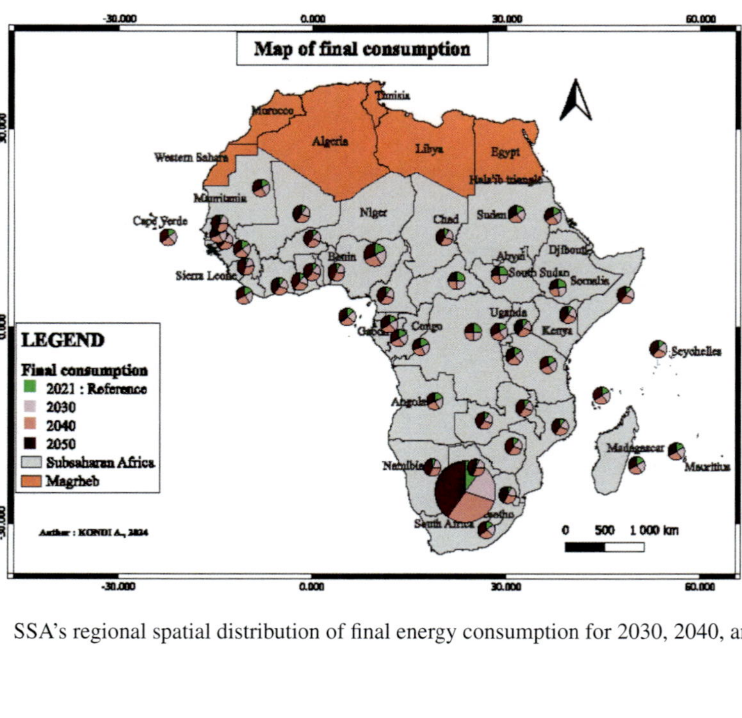

Fig. 14 SSA's regional spatial distribution of final energy consumption for 2030, 2040, and 2050

in 2040; and 134.75 GWh in 2050), the Central African Republic (162.78 GWh in 2030, 226.18 GWh in 2040, and 289.58 GWh in 2050), and The Gambia (237.98 GWh in 2030; 286.243 GWh in 2040; and 34.51 GWh in 2050) will have much lower energy demands compared to other countries.

It is important to note that several factors contribute to this unequal spatial distribution of energy demand projected for 2030, 2040, and 2050.

4.6.3 Near, Medium, and Long-Term Forecasting of Energy Generation Trends

To inform decision-making regarding energy use in Sub-Saharan African countries, a long-term outlook is essential. Figure 15 presents the projected energy production for the years 2030, 2040, and 2050.

According to the analysis of Fig. 15, all energy production sources are expected to increase in output by 2030, 2040, and 2050. Hydroelectric power will remain the leading source, followed by natural gas, diesel oil, and fuel oil. Other energy sources will contribute only marginally to the overall production, meaning that renewable energy sources (such as solar and wind) will have limited usage by these future milestones. The increase in energy production will primarily rely on conventional sources, particularly fossil fuels, with only a limited contribution from renewables.

If the same sources and levels of production persist, achieving sustainable development will be challenging. While energy production will indeed drive economic growth, it will also lead to an increase in CO_2 emissions, exacerbating environmental concerns.

Furthermore, it is noted that by 2030, 2040, and 2050, countries will not produce energy at the same levels. Some nations will be major producers, while others will

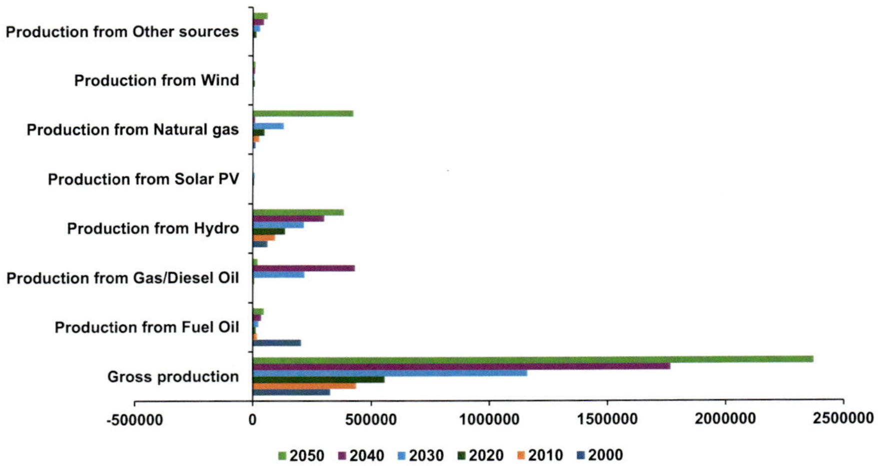

Fig. 15 Energy production projections by source from 2000 to 2050 with 10-Year Intervals in SSA

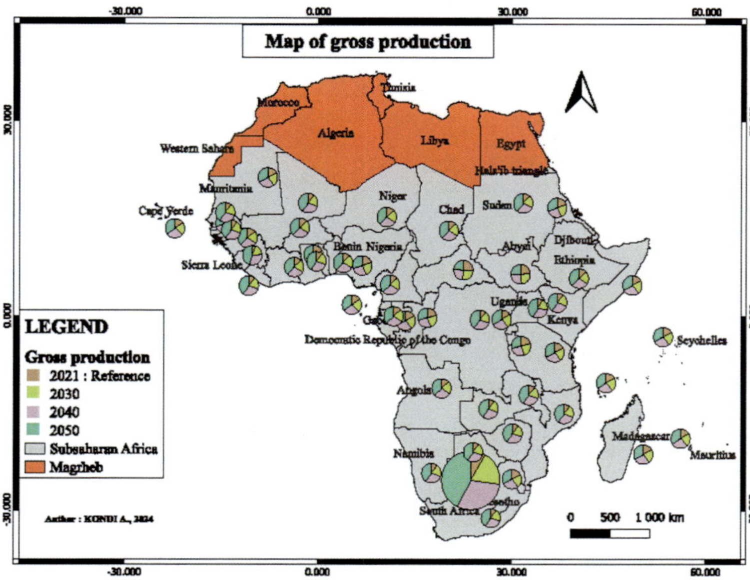

Fig. 16 Future energy production allocation across SSA's regions 2030, 2040, and 2050

have relatively low production, resulting in disparities in energy output across Sub-Saharan Africa. This disparity in energy production is illustrated in Fig. 16.

Figure 16 highlights disparities in energy production across Sub-Saharan African countries for the years 2030, 2040, and 2050. It shows that South Africa (662,230.31 GWh in 2030; 1,075,111.61 GWh in 2040; and 1,487,992.92 GWh in 2050), Nigeria (84,049.6 GWh in 2030; 96,109.28 GWh in 2040; and 108,168.96 GWh in 2050), and Mozambique (54,269.9 GWh in 2030; 81,404.86 GWh in 2040; and 108,539.81 GWh in 2050) are expected to produce significantly more energy than other countries.

Conversely, countries like Comoros (111.82 GWh in 2030; 138.58 GWh in 2040; and 169.17 GWh in 2050) and the Central African Republic (148.97 GWh in 2030; 189.59 GWh in 2040; and 233.75 GWh in 2050) will have much lower energy production. This projected energy production aligns with the projected energy consumption trends.

4.6.4 Scenario Analysis for Future Energy Consumption and Production

Figure 17 illustrates the intersection between urban areas, energy consumption, and energy production on one hand, and between urban population size, energy consumption, and production on the other, in terms of forecast scenarios.

Figure 17 illustrates that energy consumption is projected to rise from 656,839.50 GWh in 2030, corresponding to an urban area of 409,704.44 km^2 and a population

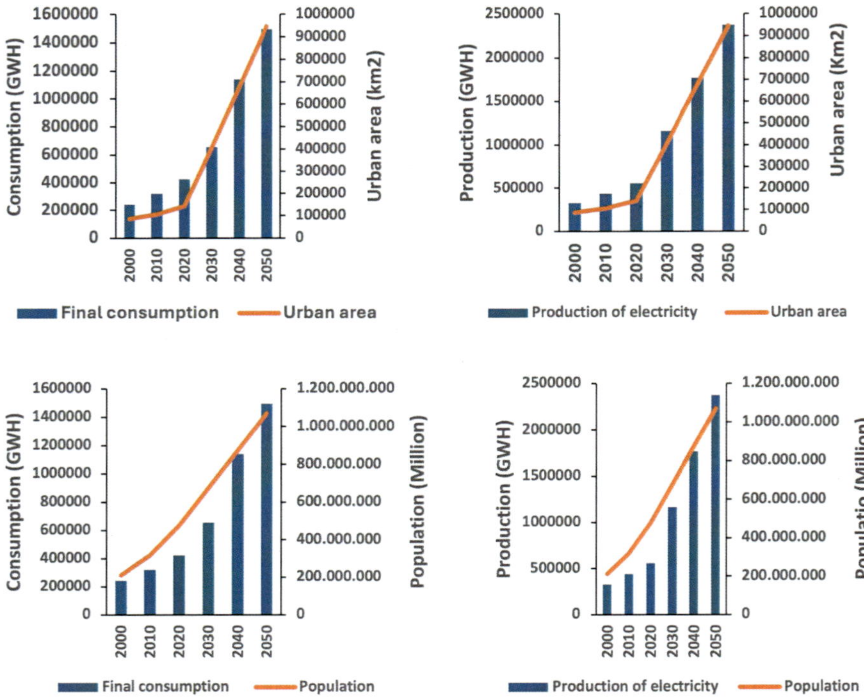

Fig. 17 Nexus of urban areas, population dynamics, and energy metrics projections by 2030, 2040, and 2050

of 675,303,039, to 1,138,557.08 GWh in 2040, with an area of 677,589.29 km^2 and a population of 873,020,809. By 2050, consumption is expected to reach 1,495,912.30 GWh, alongside an area of 945,474.13 km^2 and a population of 1,070,738,579.

In parallel with these increases in energy consumption, energy production is also projected to rise significantly, reaching 1,162,808.75 GWh in 2030, 1,767,788.20 GWh in 2040, and 2,372,614.41 GWh in 2050.

Moreover, the projected expansion of urban areas and growth in population by 2030, 2040, and 2050 will lead to higher energy consumption and production. These projections underscore the need for strategic planning to meet future energy demands.

Indeed, population growth will have a direct impact on potential economic growth and support increased energy consumption. Across all projected time horizons, the expansion of urban areas and the rise in population size will significantly influence energy demand, which is expected to grow substantially.

4.6.5 Assessing CO_2 Emissions Through the Lens of Circular Economy and Urban Energy Trends

Figure 18 presents the projections for urban area, population size, energy consumption, energy production, and CO_2 emissions for 2030, 2040, and 2050.

A prospective analysis of CO_2 emissions, based on urban and energy trends, indicates that if the rate of CO_2 emissions observed from 2000 to 2021 continues at the same pace, Sub-Saharan African countries are projected to emit 14,512,741.91 metric tons per capita of CO_2 by 2030, 16,728,309.96 metric tons per capita by 2040, and 20,486,945.83 metric tons per capita by 2050. This signifies a rising trend in CO_2 emissions over the future decades.

In other words, the projected increase in urban areas and population will lead to higher energy consumption and production, resulting in an escalation of CO_2 emissions into the atmosphere throughout the 2030s, 2040s, and 2050s. The more energy is consumed and produced, the greater the emissions.

The prospective analysis of urban areas, population size, energy consumption, and CO_2 emissions highlights key long-term trends and the potential consequences of energy choices, particularly regarding greenhouse gas (GHG) emissions and consumption patterns, on long-term global energy demand.

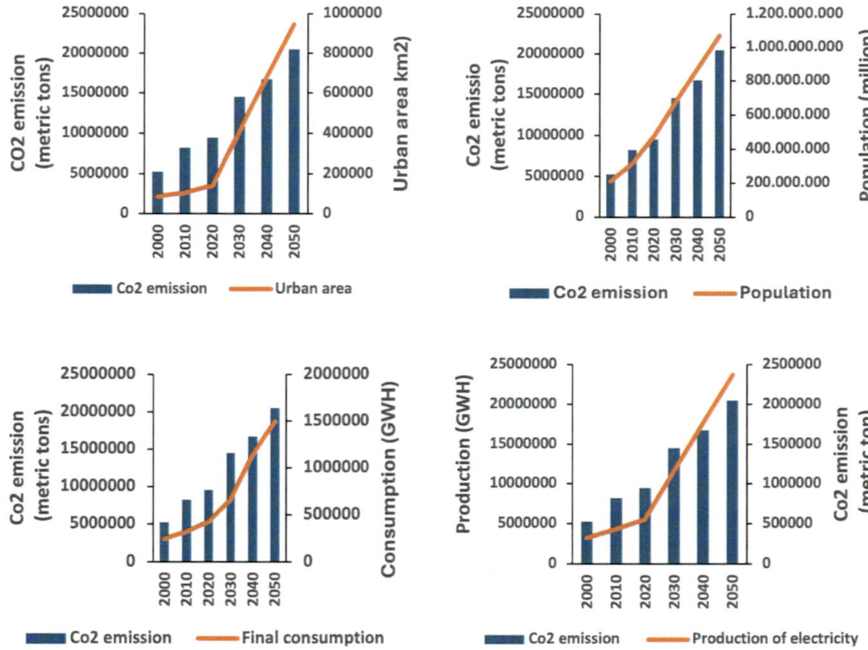

Fig. 18 Projections of urban growth, population, energy consumption, production, and CO_2 emissions by 2030, 2040, and 2050

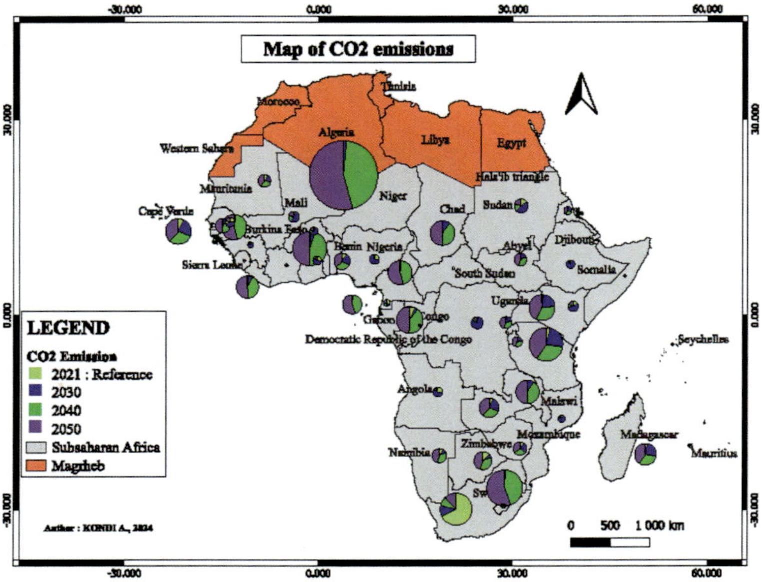

Fig. 19 Spatial analysis of projected CO_2 emissions for the next three decades (2030, 2040, and 2050)

From a spatial perspective, the prospective analysis shows an unequal distribution of CO_2 emissions across Sub-Saharan Africa by 2030, 2040, and 2050 (Fig. 19).

Figure 19 illustrates the spatial distribution of CO_2 emissions, based on 2021 levels, showing which countries in Sub-Saharan Africa are projected to emit more CO_2 by 2030, 2040, and 2050. Nigeria is expected to lead in emissions, with 153,144,850 metric tons per capita in 2030, 198,369,110 metric tons per capita in 2040, and 243,593,370 metric tons per capita in 2050. The Democratic Republic of Congo will follow, with projected emissions of 63,526,645 metric tons per capita in 2030, there Will be an increase of 84,968,185 metric tons per capita in 2040, and 106,409,725 metric tons per capita in 2050.

South Africa is also among the major contributors, with emissions projected at 46,113,806 metric tons per capita in 2030, 52,595,456 metric tons per capita in 2040, and 59,077,106 metric tons per capita in 2050. Ethiopia will contribute 38,685,093 metric tons per capita in 2030, with an increase of 3,073,592 metric tons per capita in 2040, and 3,576,892 metric tons per capita in 2050. Conversely, countries like Seychelles, with 73,572 metric tons per capita in 2030, 78,032 metric tons per capita in 2040, and 82,492 metric tons per capita in 2050, as well as São Tomé and Príncipe and Eswatini, are projected to have significantly lower CO_2 emissions.

4.7 Second Partial Conclusion

The prospective study of urban expansion and energy use in Sub-Saharan Africa reveals that urban areas, population size, urbanization rates, and GDP per capita will increase through 2030 and continue into 2050. Urban areas are expected to cover 409,704.44 km^2 in 2030, 677,589.29 km^2 in 2040, and 945,474.13 km^2 in 2050, with urbanization rates of 63.34%, 75.66%, and 87.98%, respectively. The population is projected to reach 675,303,039 people in 2030, 873,020,809 people in 2040, and 1,070,738,579 people in 2050. GDP per capita is expected to rise, reaching $6,516.68 in 2030, $6,627 in 2040, and $7,123.23 in 2050.

This economic dynamism, combined with rapid population growth and urbanization, will place significant pressure on energy demand, which is expected to continue growing. Energy consumption is projected to reach 407,547.52 GWh in 2030, 621,457.30 GWh in 2040, and 804,808.54 GWh in 2050. Energy production will rise to 1,162,808.75 GWh in 2030, 1,767,788.20 GWh in 2040, and 2,372,614.41 GWh in 2050. Simultaneously, the urban population will increase, and the growing economy, reflected in rising GDP per capita, will drive higher electricity demand and energy production using the same sources, increasing CO_2 emissions.

CO_2 emissions are projected to reach 14,512,741.91 metric tons per capita in 2030, 16,728,309.96 metric tons per capita in 2040, and 20,486,945.83 metric tons per capita in 2050. This trajectory contradicts the principles of a circular economy and will contribute to global warming, jeopardizing the sustainable development of Sub-Saharan African countries.

5 Insights and Reflections Discussion

The rapid urban expansion and energy consumption in Sub-Saharan African countries is a critical issue that involves multiple stakeholders, including researchers, policy-makers, NGOs, and civil society. Many studies have focused on recent urban growth and energy use in the region. This study demonstrates that Sub-Saharan African cities have experienced significant spatial growth in recent decades, marked by an increase in urban areas, population size, and urbanization rates. Urban areas expand by an average of 8,430.40 km^2 annually, with an average annual urbanization rate of 2.61%. Over the same period, the population grew from 210,824,543 in 2000 to 494,081,113 in 2021, aligning with the findings of (Nkulu et al.,2022). They reported a 16% population increase in the commune of Lubumbashi over the past decade, with the 2018 population representing 14% of the city's total. Similarly, Zoma et al. (2022) highlighted the population growth and urban expansion challenges in Ouagadougou.

Urban growth in Sub-Saharan Africa presents significant challenges for the population.

Rapid urbanization has driven major shifts in energy use, particularly in energy consumption and production, as revealed in this study. (Roland,2001) noted that the swift and recent nature of urbanization in Sub-Saharan Africa has set records, resulting in increased demand for basic services, especially electricity. These findings are consistent with (Goudie, 2015), who examined urban sprawl's impact on electricity distribution in Ngaoundéré, Cameroon. He emphasized that the spatial disparities in electricity access stem from unplanned urbanization and the financial capacity of residents to connect to the power grid. Similar observations were made by (Anjali et al., 2012) and the African Development Bank Group in 2022, who found that while access to electricity has improved since the 1990s, the number of people without access has increased due to population growth.

Dziwonou in2001 highlighted that the extraordinary urbanization in Sub-Saharan Africa presents enormous challenges for cities. Despite some progress in urban planning, it has not kept pace with urban growth. This imbalance is also reflected in the mismatch between urban expansion and energy consumption and production, as observed in studies by (Dziwonou, 2001; Nyassogbo, 2010; Dandonougbon and Hetcheli, 2013; Fagbedji et al., 2017) on Lomé. These concerns align with those of (Niandou, 2017), who studied urbanization and energy poverty in West Africa's major cities. Zoma et al., 2022 also noted that rapid and uncontrolled spatial expansion has strained infrastructure and urban services, causing energy access problems.

This research shows a correlation between GDP growth per capita and rising energy demand and supply. As urban populations in Sub-Saharan African countries grow rapidly, their economies have maintained strong GDP growth, as noted by (Burhama 2023). The impact of urbanization on GDP is driven by two factors: first, urbanization increases the share of the population living in more productive urban areas, which boosts national GDP and per capita income; second, cities become more productive due to the influx of additional workers, as urban productivity levels are generally higher than those in rural areas. The growth in population, improved living conditions, and economic expansion are driving higher energy needs, consistent with the findings of (Waite et al., 2017). However, despite the benefits of urbanization, it has not led to a sustainable transformation in cities' energy use, as also concluded by (Oumar, 2018).

Urbanization remains a significant concern, even in less densely populated urban centres, as population growth often outpaces economic capacity, particularly in energy use. Like urban growth, energy use in Sub-Saharan Africa has increased, with rising trends in both energy consumption and production. This aligns with the World Energy Outlook (WEO, 2014), which reported significant global energy consumption growth across various regions. According to WEO, global final energy consumption in 2012 was approximately 9 billion tons of oil equivalent (toe). This required nearly 13.4 Mtoe of primary energy production, which includes energy used in the transformation of different initial energy resources to reach this 9 Mtoe. These findings also align with (Kondi et al. 2023), who studied recent trends in urban electricity consumption for cooling in West and Central Africa. Their results showed that baseline consumption and temperature sensitivities increased significantly over time, making electricity demand patterns highly non-stationary. Additionally, a mismatch exists between energy use and urban dynamics, leading to challenges for sustainable urban development that balances spatial growth with energy consumption.

The results of this study reveal that Sub-Saharan African countries have faced rapid urbanization from 2000 to 2021, a trend expected to continue through 2030, 2040, and 2050, affecting energy use in terms of both demand and supply. Similar results were previously reported by the International Francophonie Agency in 2001 and WEO in 2014, which highlighted that current and future urbanization will drive growing demand for energy products and services. The study predicts that urban population growth, expansion of urban areas, and rising living standards, as reflected in GDP per capita, will continue to increase energy demand by 2030, 2040, and 2050 in Sub-Saharan Africa. The rapid global demographic shifts observed since the 1950s are accelerating even faster in Africa. During the 2020s, according to UN demographic projections (United Nations, 2019a), the demographic curves of China (1.439 billion in 2020), India (1.380 billion in 2020), and Africa (1.340 billion in 2020) will intersect.

According to the UN's World Urbanization Prospects (WUP) (United Nations, 2019b), the urban population is expected to grow even more rapidly. This predictable increase in population and living standards will result in greater energy needs, aligning with the findings of the United Nations Population Fund in 2007, Grosset and Van in 2016), and Burhama in 2023. Sub-Saharan Africa's development is further hindered by energy inequalities, as non-renewable energy consumption and production remain significant. Sub-Saharan African countries must better manage their natural resources by promoting local consumption and improving the use of revenue from exports, in line with circular economy principles, as previously noted by (Pierre et al., 2009). Africa should also promote the development of renewable energy, which remains largely untapped.

The current economic development and rapid urbanization of Sub-Saharan African countries contribute to mass energy production and consumption, compromising sustainable urban development and deviating from circular economy principles. This explains the upward trend in energy consumption and production in these countries, as observed in studies by (Guerrero et al., 2013) and (Sotamenou et al., 2010). The urban expansion in Sub-Saharan Africa, reflected in increasing urban areas, urbanization rates, and population size, is putting pressure on resources and the environment. Continuous spatial expansion due to urbanization also exerts growing pressure on environmental management and sustainability, as Harris in 1996 noted. Urbanization and economic development significantly affect the nature and volume of solid waste, which in turn influences CO_2 emissions. This observation was echoed by (Wang et al., 2017).

Prospective analysis of CO_2 emissions based on urban and energy trends shows an upward trajectory for future emissions, similar to findings by (Lamb et al., 2021). In rapidly industrializing regions, fossil fuel-based energy systems have continued to expand, slowing their growth only recently. High demand for materials, land, energy services, and mobility has led to increased emissions in the industrial sector. To balance urban growth with energy use and achieve sustainable development goals, Sub-Saharan African countries must embrace renewable energy in line with circular economy principles, facilitating an energy transition. Additionally, new policies or strategies are required to accelerate economic and social development while protecting the environment. While energy is a driving force in the short and long term for economic and social development, access to this critical resource remains limited. Questions remain about energy production and consumption to safeguard humanity from their negative effects. These findings echo those of (Oumar, 2018), who recommended new energy and economic growth policies.

6 Concluding and Actionable Recommendations

6.1 Conclusions

The study on urban explosion and energy usage in Sub-Saharan Africa highlights a significant increase in urban areas, urbanization rates, population size, and GDP per capita from 2000 to 2021. Urban areas have expanded by an average of 8,430.40 km^2 per year, with urbanization rates rising from 33.59% to 52.26%. The population increased from 210,824,543 to 494,081,113 people, while the GDP per capita grew from \$4,440.24 to \$5,889.64. Alongside this growth, energy consumption saw substantial changes, rising from 239,758.18 GWh to 459,582.16 GWh, while energy production increased from 317,235.64 GWh to 618,090.27 GWh. This surge in energy usage also led to a rise in CO_2 emissions, from 5,242,520.21 metric tons to 10,391,291.76 metric tons.

The prospective analysis of urban explosion and energy usage predicts continued growth in urban areas, urbanization rates, and population sizes. Urban areas are expected to expand to 409,704.44 km^2 by 2030, 677,589.29 km^2 by 2040, and 945,474.13 km^2 by 2050. Urbanization rates will reach 63.34%, 75.66%, and 87.98%, respectively, while the population will rise to 675,303,039 by 2030, 873,020,809 by 2040, and 1,070,738,857 by 2050. The GDP per capita will grow to \$6,516.68 in 2030, \$6,627.00 in 2040, and \$7,123.23 in 2050. Energy consumption and production will also follow this upward trend, with energy consumption reaching 407,547.52 GWh in 2030, 621,457.30 GWh in 2040, and 804,808.54 GWh in 2050, while energy production will rise to 1,162,808.75 GWh in 2030, 1,767,788.20 GWh in 2040, and 2,372,614.41 GWh in 2050. These urban and energy dynamics will contribute to further increases in CO2 emissions, which are projected to reach 14,512,741.91 metric tons by 2030, 16,728,309.96 metric tons by 2040, and 20,486,945.83 metric tons by 2050.

The urban explosion in Sub-Saharan Africa will thus pose significant challenges in terms of energy and the circular economy. The first major challenge linked to urban growth is energy usage. Energy consumption and production must keep pace with urban growth to alleviate energy poverty while ensuring reliable and sustainable access to energy for the rapidly growing urban population. Achieving this balance is key to fostering sustainable urban development.

A substantial reduction in energy consumption is crucial for shaping sustainable cities, requiring changes in policy, regulation, and land use planning. To promote a circular economy, policies should incorporate fiscal instruments and investments in public infrastructure to improve energy efficiency. Energy infrastructure must be flexible to accommodate diverse energy flows, and projects should be managed by local authorities, with public–private partnerships ensuring equitable distribution of costs and benefits. International development banks also have a vital role to play in financing long-term projects.

Future research should focus on the energy needs of African cities as they face increasing urbanization, seeking sustainable solutions to improve living conditions. This study paves the way for further research on opportunities related to the energy transition in Sub-Saharan Africa.

6.2 Strategic Recommendations

6.2.1 Implications of Urban Growth

Rapid urbanization presents substantial economic opportunities for Sub-Saharan African countries. Expanding urban areas can become centres of economic activity, attracting investments and creating jobs. This transformation enhances infrastructure and services, both essential for sustainable development.

Despite its economic benefits, urban growth intensifies social challenges such as inequality and limited access to basic services. Many urban residents live in informal settlements with inadequate infrastructure, highlighting the urgent need for inclusive urban planning to bridge these disparities.

Rising energy consumption and CO_2 emissions are critical environmental challenges. Urban areas often face higher pollution and resource depletion levels. A shift toward renewable energy and sustainable urban practices is crucial to mitigate these impacts and promote a circular economy.

6.2.2 Economic Diversification and Energy Demand

Urban expansion is likely to foster new economic sectors, such as technology and services, which will drive increased energy consumption. As cities evolve into economic hubs, energy demand will rise due to both population growth and the diversification of energy-intensive economic activities.

6.2.3 Energy Infrastructure Challenges

Rapid urbanization may surpass the development of energy infrastructure, creating significant challenges in meeting rising energy demands. Urban areas, especially informal settlements, often lack reliable infrastructure, exacerbating inequalities in energy access, particularly for low-income populations.

6.2.4 Climate Resilience and Adaptation

As energy consumption increases, adopting climate-resilient strategies becomes critical to mitigating the adverse environmental effects of urbanization. Investments in sustainable urban planning and renewable energy technologies will help adapt to changing climate conditions while securing energy needs.

6.2.5 Social Equity Considerations

The rise in urban populations necessitates prioritizing social equity in energy access. Policymakers must adopt inclusive approaches to ensure marginalized communities benefit from urbanization and gain access to affordable, reliable energy, which is crucial for achieving broader sustainable development goals.

6.2.6 Policy Frameworks Recommendations for Sustainable Development

Policymakers must prioritize sustainable urban planning based on circular economy principles. This includes promoting green spaces, efficient public transportation, and sustainable waste management to reduce environmental impacts.

Meeting growing energy demands sustainably requires strong investment in renewable energy sources. This shift supports environmental goals, enhances energy security, and strengthens resilience to climate change.

Integrated policy frameworks are needed to align urban planning with energy policies. Collaboration among government agencies, private sector players, and civil society is essential for creating cohesive strategies to manage urban growth and its associated energy demands effectively.

Effective governance is essential for managing urban growth. Key actions include improving data collection on urbanization trends, engaging communities in decision-making, and ensuring policies are adaptable to changing demographics and environmental conditions.

References

Abassa, A. N. (2017). Urbanization and precariousness of electric energy in major cities of West Africa: The example of Niamey in Niger (Bibliographic analysis). *Master's Thesis, University Abdou Moumouni of Niamey*, p. 118.

African Development Bank (AfDB). (2021). Report on the energy situation in the Democratic Republic of Congo.

African Development Bank (AfDB). (2022). Dynamics of African urbanization: The economic influence of African cities. *Joint report by the African Development Bank, OECD Sahel and West*

Africa Club (SWAC/OECD), and United Nations Economic Commission for Africa (UNECA), p. 205

Agence Internationale de la Francophonie. (2001). City, energy, and environment. *In Proceedings of the conference held in Beirut, Lebanon*, 17–19, p. 232.

Anjali, S., et al. (2012). Access to electricity in sub-Saharan Africa: Experience feedback and innovative approach. p. 103.

Bartier, M., & Keller, A. (2016). Urban energy solutions in sub-saharan Africa: Toward sustainable development. *Environmental Science & Policy, 65*, 55–66. https://doi.org/10.1016/j.envsci.2016.07.004

Basu, S., Bale, S. E., Wehnert, T., & Topp, K. (2019). A complexity approach to defining urban energy systems. *Cities, 95*, 102358. https://doi.org/10.1016/j.cities.2019.102358

Blanco, H., et al. (2018). Trends in urban energy demand in Africa: A sustainable development perspective. *Renewable and Sustainable Energy Reviews, 92*, 1–10. https://doi.org/10.1016/j.rser.2018.04.007

Borchers, M., Euston-Brown, M., Bawakyillenuo, S., Ndibwami, A., & Batchelor, S. (2018). Sustainable energy transitions: Shifting from "business as usual" in sub-Saharan African urban areas. *International Journal of African Development, 5*(1), Article 4. Retrieved from https://scholarworks.wmich.edu/ijad/vol5/iss1/4.

Burhama, E. C. (2023). Challenges related to energy transition in sub-Saharan Africa: The case of the Democratic Republic of Congo. *Master's Thesis, University of Ottawa*, p. 68.

Bîrgovan, A. L., Lakatos, S. E., Szilagyi, A., Cioca, I. L., Pacurariu, L. R., Ciobanu, G., & Rada, C. E. (2022). How should we measure? Review of circular city indicators. Retrieved from https://ncbi.nlm.nih.gov.

Cajot, S., Peter, M., Bahu, J.-M., Guignet, F., Koch, A., & Maréchal, F. (2017). Obstacles in energy planning at the urban scale. *Sustainable Cities and Society, 30*, 223–236. https://doi.org/10.1016/j.scs.2017.02.003

Calitz, M. (2019). Residents' perspectives on Nelson Mandela Bay as a sustainable city. *Master's Thesis, Nelson Mandela University.*

Chen, Y., & Liu, Z. (2012). Energy transition and urban sustainability: A study of Cities in sub-saharan Africa. *Urban Studies, 49*(7), 1389–1403. https://doi.org/10.1177/0042098012442855

Cornillon, P. Y., et al. (2011). Urban energy demand and sustainability: A case study of sub-saharan Africa. *Energy Policy, 39*(6), 3478–3489. https://doi.org/10.1016/j.enpol.2011.03.010

Dandonougbon, I., & Hetchel, F. (2013). Spatial extension of Lomé and its impact on peri-urban agriculture. *Cultural Engineering, Biannual Scientific Review of IRES-RDEC, 002*, 115–136.

Duan, X., Li, X., Tan, W., & Xiao, R. (2022). Analysis of decoupling relationships between urbanization and carbon emissions in 33 African countries. Retrieved from https://ncbi.nlm.nih.gov.

Dziwonou, Y. (2001). Urbanization and urban planning challenges. Revue CAMES-Série B, Vol. 005. *Karthala, No., 1–2*, 281–291.

Fagbedji, K. G., Hetcheli, F., & Dandonougbon, I. (2017). Electrification of Lomé and its peripheries: Disparities and population adaptation. Ivorian Review of Geography of Savannahs, No. 2. *ISSN, 2521–2125*, 62–76.

Favennec, J. P., Adedjoumon, C., Duhamel, B., Giri, J., Gilles, H., & Tronche, A. (2009). Energy in Africa by 2050. Paris, p. 84.

Grimaldi, J. (2018). Models of growth and colonial implications in the renewable energy sector in Africa. *Claremont Colleges Library Undergraduate Research Award, 6*. Retrieved from https://scholarship.claremont.edu/cclura_2018/6/.

Grosset, F., & Van, P. N. (2016). Energy consumption and economic growth in sub-Saharan Africa. *Worlds in Development, 44*(4), 176.

Guerrero, L. A., Ger, M., & William, H. (2013). Solid waste management challenges for cities in developing countries. *Waste Management, 33*, 220–232.

Gwaleba, M. J. (2018). Urban growth in Tanzania: Exploring challenges, opportunities, and management. *International Journal of Social Science Studies, 6*(12), 47.

Halog, A., & Anieke, S. (2021). Circular economy and sustainability: A review of circular economy studies in developed countries and its potential adoption in developing countries. *Circular Economy and Sustainability, 1*, 209–230.

Harris, F. (1996). Intensification of agriculture in semi-arid areas: Lessons from the Kano Close-Settled Zone, Nigeria. Gatekeeper Series, 59. *International Institute for Environment and Development, London, UK*.

Hemidat, S., Achouri, O., El Fels, L., Elagroudy, S., & Hafidi, M. (2022). Solid waste management in the context of a circular economy in the MENA region. *Sustainability, 14*(1), 480. https://doi.org/10.3390/su14010480

ICLEI Africa & African Circular Economy Network (ACEN). (2023). Circular cities and the informal economy in Africa. *Renewable Matter.* Retrieved from https://renewablematter.eu.

International Energy Agency (IEA). (2014). Africa energy outlook: A focus on energy prospects in sub-Saharan Africa. *World Energy Outlook, Paris, November*.

Kibala, C. (2018). Energy demand in sub-saharan Africa's urban growth: Challenges and opportunities. *Energy for Sustainable Development, 42*, 45–56. https://doi.org/10.1016/j.esd.2017.12.003

Kondi-Akara, G., Hingray, B., Baptiste, F., & Arona, D. (2023). Recent trends in urban electricity consumption for cooling in West and Central African countries. *Energy, 276*, 127597. https://doi.org/10.1016/j.energy.2023.127597

Lamb, W. F., Wiedmann, T., Pongratz, J., Andrew, R., Crippa, M., Jos, G. J. O., Wiedenhofer, D., Mattioli, G., Khourdajie, A. A., House, J., Pachauri, S., Figueroa, M., Saheb, Y., Slade, R., Hubacek, K., Sun, L., Ribeiro, K. S., Khennas, S., de la Rue du Can, S., Chapungu, L., Steven, J. D., Bashmakov, I., Dai, H., Dhakal, S., Tan, X., Geng, Y., Gu, B., & Minx, J. (2021). A review of trends and drivers of greenhouse gas emissions by sector from 1990 to 2018. *Environmental Research Letters, 16*(7). 10.1088.

Magezi, T. J. (2015). Investigating the imbalance between rapid consumption culture and inadequate waste management infrastructure in sub-Saharan African cities. [PDF in English]

Mohammadi, N., & Taylor, J. E. (2016). Urban energy flows: Human mobility as a predictor of spatial changes. arXiv Preprint, arXiv:1609.01239.

Mutezo, G., & Mulopo, J. (2021). A review of Africa's transition from fossil fuels to renewable energy using circular economy principles. *Renewable and Sustainable Energy Reviews, 137*, 110609. https://doi.org/10.1016/j.rser.2020.110609

Nyassogbo, K. G. (2010). Constraints of urban sprawl or the absence of urban policy in Togolese cities: The example of Lomé. *Lomé Geography Review, 5*(1), 45–63.

Ogunmakinde, O. E., Egbelakin, T., & Sher, W. (2022). Contributions of the circular economy to the UN sustainable development goals through sustainable construction. Retrieved from

Oumar, H. N. (2018). Energy consumption and economic growth in Senegal: A causality and cointegration study. Doctoral Dissertation, University of Nîmes and University Cheikh Anta Diop (Dakar).

Owojori, O. M., & Okoro, C. (2022). The private sector role as a key supporting stakeholder towards circular economy in the built environment: A scientometric and content analysis. *Buildings, 12*(5), 695.

Pierre, X., et al. (2009). Urban energy systems and their role in sustainable urban development in Africa. *Energy Policy, 37*(4), 1401–1412. https://doi.org/10.1016/j.enpol.2008.11.004

Pourtier, R. (2001). *Black Africa* (pp. 151–188). Hachette Livre.

Rafitoson, K. (2017). The slow march towards energy transition in Madagascar: Status and prospects. *FES Madagascar*, p. 29.

Remøy, H., Wandl, A., Ceric, D., & van Timmeren, A. (2019). Facilitating circular economy in urban planning. *Urban Planning, 4*(3), 1–4. https://doi.org/10.17645/up.v4i3.2484.

Roques, P., & Roux, D. (2018). Energy consumption and practice theory: Pathways for energy transition. *Marketing Decisions, 90*, 35–54.

Saghir, J., & Santoro, J. (2018). Urbanization in sub-Saharan Africa: Meeting challenges by bridging stakeholders. Retrieved from https://csis.org.

Schröder, P., Lemille, A., & Desmond, P. (2020). Making the circular economy work for human development. *Environmental Research Letters, 15*(8), 083001.

Shanker, A., Clément, P., Tapin, D., & Buchsenschutz, M. (2012). Access to electricity in sub-Saharan Africa: Feedback and innovative approaches. French Development Agency, p. 103

Shepard, D. (1968). A two-dimensional interpolation function for irregularly spaced data. *In Proceedings of the 1968 ACM National Conference*, pp. 517–524. https://doi.org/10.1145/800186.810616.

Sotamenou, J., Ganry, F., Montange, D., Parrot, L., & Simon, S. (2010). Transfer stations for sustainable municipal solid waste management in Africa: Evidence from Cameroon. *Solid Waste Management & Environmental Remediation.* Nova Science Publishers, pp. 217–242.

Sreerupa, G., & K., & Andharia, J. (2023). Unpacking the circular economy: Highlighting contributions of informal economy workers in Delhi. *Social Science Research Network (SSRN).* https://doi.org/10.2139/ssrn.4622054

Todaro, M. P. (1997). Urbanization, unemployment, and migration in Africa: Theory and policy. Policy Research Division Working Paper No. 104, Population Council.

UN-Habitat. (2015). Sustainable urban development in Africa. *United Nations Human Settlements Programme* (UN-Habitat).

UN-Habitat. (2023). *Africa Urban Agenda Programme.* Retrieved from https://unhabitat.org.

United Nations Human Settlements Programme (UN-Habitat). (2020). *The State of African Cities 2020: The Geography of African Investment.* Retrieved from https://unhabitat.org.

United Nations Population Fund (UNFPA). (2007). State of the world population 2007: Unleashing the potential of urban growth. ISBN 978–0–89714–808–5.

Waite, M., Cohen, E., Torbey, H., Piccirilli, M., Tian, Y., & Modi, V. (2017). Global trends in urban electricity demands for cooling and heating. *Energy, 127*, 786–802.

Wang, Y., Cheng, K., & Wu, W. (2017). Atmospheric emissions of typical toxic heavy metals from open burning of municipal solid waste in China. *Atmospheric Environment, 152*, 6–15.

Winkler, H., Borchers, M., Hughes, A., Visagie, E., & Heinrich, G. (2006). Policies and scenarios for Cape Town's energy future: Options for sustainable city energy development. *Journal of Energy in Southern Africa, 17*(1), 28–41. https://doi.org/10.17159/2413-3051/2006/v17i1a3367.

World Bank. (2017). Doing Business: Equality of opportunity for all (4th edition). *World Bank Reports.* Retrieved from https://francais.ngbusiness.org/.

World Bank. (2022). Energy Report. *World Bank Publications.*

World Bank. (2021). *Africa's pulse, No. 24, October 2021: An analysis of issues shaping Africa's economic future.* The World Bank. https://doi.org/10.1596/978-1-4648-1805-9

World Bank. (2023a). Human Capital Indicator: An overview in the Democratic Republic of Congo. *World Bank Reports.*

World Bank. (2023b). Scaling up to phase down: Financing energy transition in the power sector. *World Bank Reports.*

Zoma, V., Kiemdé, A., & Sawadogo, Y. (2022). Key challenges of urban growth in Ouagadougou. Regard suds, 2(September), pp. 77–92. https://hal.archives-ouvertes.fr.

Energy Access Implications for Unserved Communities in Sub-Saharan Africa: Challenges and Opportunities

Nwabialu Biola Onyeanusi

Abstract Energy is the bedrock of development and a necessary derivative to improve the quality of life in unserved communities of sub-Saharan Africa. However, over 500 million people in rural areas of sub-Saharan Africa continue to lack access to energy. The extension of the National grid to sparsely populated rural communities incur high extension costs and is economically unviable. Renewable off-grid solutions such as solar mini-grids offer a faster and more feasible approach to energy access in rural communities. In addition to basic lighting, a range of social and economic benefits which include improved healthcare services, education, security, and economic prospects are expected to uplift existing limitations. Conceptually, the IEA's definition of energy access offers a holistic and fundamental insight into factors that influence effective energy access, entailing that a base connection is not sufficient to improve a household's quality of life because the dynamics of development and electrification in remote communities are complex and involve many underlying forces. Hence, a one-way linear assessment of impacts is insufficient as projections rarely match reality. Successful deployment of off-grid systems requires a conducive enabling environment, which includes supportive government policies, access to financing, and opportunities for further investment. This review paper revisits the current state of energy access in sub-Saharan Africa through the lens of a detailed outlook aimed at highlighting the factors that challenge socio-economic opportunities, thereby informing energy policies and investments in off-grid rural electrification. It provides an insight into the current energy situation of rural electrification across countries in SSA. The findings suggest that renewable off-grid technologies can be a transformative technology for rural electrification but must be accompanied by holistic development strategies to maximize its benefits for unserved communities.

N. B. Onyeanusi (✉)
Department of Global Environmental Studies, Sophia University, Tokyo, Japan
e-mail: nonyeanusi@gmail.com

M.-M. Narra and S. Narra (eds.), *African Green Transition Through Innovative Pathways*, World Sustainability Series, https://doi.org/10.1007/978-3-031-87043-9_9

1 Introduction

Energy is a natural, hidden, and invisible system within civilization. It is vital for economic growth and acceleration to the top of any country desiring progress. Clearly defined as the ability to do work; no work can be done without a form of energy. Access to energy is an essential stimulant for ensuring economic growth within a society. Civilization cannot exist as we know it without the consumption of vast quantities of energy, so energy is indeed a moral good which is fundamental to the creation and functioning of societies (Hefner III, 2009). While many countries have journeyed to attain a 100% electrification rate, countries in sub-Saharan Africa (SSA) have continued to lag with energy issues plaguing both urban and rural areas. Energy systems require infrastructures and are considerably capital-intensive through the generation, transmission and distribution stages. The high cost and technicalities involved in the extension of the main grid to dispersed rural communities remain a challenge, mainly due to their low demand profile, economic conditions, and proximity to the main grid (Babalola et al., 2022). Hence, these communities are forced to tend to their energy needs by utilizing alternate energy sources such as kerosene lamps and candles. Ironically, many urban areas despite connection to the main grid continually experience fluctuations and intermittent supply. Over the past decade, the Sustainable Development Goals (SDGs) launched by the United Nations have addressed development indicators inclusive of energy. Goal 7 aims to ensure access to affordable, reliable, sustainable and modern energy for all (UN, 2023). Yet, it is agreed that electricity consumption can bring about the needed economic development and is closely linked to other goals such as poverty elimination, education, health and well-being. Various studies corroborate these findings and emphasize that electricity consumption not only improves the quality of life but also propels the economy to greater heights (Agyei-Sakyi et al., 2021; Altinay & Karagol, 2005; Yoo & Kwak, 2010).

Energy access in rural communities of SSA have been perceived as a mere connection or lighting source to a household performing a bare minimum often referred to as basic lighting, however, this concept transcends these limitations. A more comprehensive definition of energy access from a household's perspective is stated by the IEA as "*a household having reliable and affordable access to both electricity and clean cooking facilities, which is enough to supply a basic bundle of energy services initially, and then an increasing level of electricity over time to reach the regional average*".

This definition sums up the key energy access indicators: *reliable and affordable access* ensures that electricity is available at a fair cost whenever needed without interruption, *supply* postulates that the electricity supplied must meet demand at the initial stage and *increasing level over time* this focuses on the energy consumption pattern of households which must grow in a sustainable way with the purchase of efficient appliances.

However, despite efforts to address energy access globally, a lot of collective work is required to boost energy access in SSA. Of the 48 SSA countries, only Mauritius

Table 1 Population without access by region

Regions	2010 (million people)	2021 (million people)	Population growth rate (%)
Central and Southern Asia	414	24	1.4
Eastern and Southeastern Asia	90	35	0.5
Northern Africa and Western Asia	37	30	1.5
Sub-Saharan Africa	566	567	2.5

Source Author's elaboration (IEA et al., 2023; Worldometer, 2024)

and Seychelles have achieved universal access since 2010 with Nigeria, Ethiopia and the Democratic Republic of Congo representing the largest access deficit worldwide (IEA et al., 2024).

This in fact could be attributed to a 2.5% population growth rate when compared to other regions, hence, efforts must match future population growth. Table 1 shows the progression of each region's population with access. Here, we see significant advancement in Asia, especially for the Central and Southern regions where over 90 percent of the unserved population gained access in 2021.

To rapidly achieve universal electrification by 2030, decentralized energy systems have become the preferred choice to accelerate energy access and contribute to the SDGs. These systems either as mini-grids or distributed individual energy services have enabled the development of innovative, community-driven energy solutions tailored to the specific needs and resources of communities (Kammen et al., 2015).

In consideration of environmental concerns such as pollution and resource depletion, decentralized renewable energy systems offer sustainable solutions to expand access while mitigating climate change (Nyarko et al., 2023). Particularly, solar-powered systems comprising of Solar Home Systems (SHS) and Solar Mini-grids (SMG) have been heralded as frontrunners among renewable energy sources in the context of rural areas. (Sarker et al., 2020; Wagner et al., 2021) describe it as the most viable option for rural areas due to its modularity and scalability.

2 Methodology

The central objective of this study is to examine energy access in sub-Saharan Africa (SSA) through factors that can undermine potential socio-economic development in off-grid communities. It reviews the existing literature related to off-grid energy systems applicable in the SSA region by synthesizing the current understanding of energy access with factors that challenge or play key roles in the attainment of sustainable development. The literature search targeted publications using keywords such as "energy access," "renewable energy," "electricity access, " "off-grid solutions, " "rural electrification," and "sustainable energy" that address the energy access landscape in SSA, the role of renewable energy sources, and the barriers and opportunities for improving energy access in the region. Collectively, academic peer-reviewed articles, reports from international organizations such as the World Bank and the International Energy Agency (IEA), policy documents, and grey literature with particular emphasis on rural areas where energy access is a critical issue have been utilized to capture various perspectives in the study. First, it discusses the current state of electrification in SSA highlighting the wide range of energy sources that shape the electricity mix as well as the percentage of the rural population with access to electricity. Then, it examines the key dimensions critical for understanding the complex nature of energy access and ensuring that energy solutions are designed and implemented to meet the diverse needs of communities in the region. Finally, the connection between energy access and sustainable development is established providing a comprehensive overview for evaluating and improving energy access in sub-Saharan Africa.

2.1 Current State of Electrification in SSA

The total electricity capacity installed in Africa reached 245GW in 2022, marking a 50% increase since 2015 equivalent to Germany's generation. This cumulative capacity comprises of fossil fuels, hydropower, and other renewables. Hydropower remains the most established of all renewable energy sources, accounting for 42GW of the total installed capacity (IHA, 2024). However, more than half of these hydro plants connected to the main grid have aged impairing the performance and reliability of the system. Infrastructural decay has limited the expansion of the grid to rural areas, with population growth further widening the gap between existing capacity and energy demand. To address this, attention has drifted to the use of solar-powered technologies to provide electricity to off-grid households. With an average global solar horizontal irradiance of over 2045 kWh/m^2/year, Africa accounts for 40% of the world's solar potential earmarked at 60,000,000 TWh/year making it the most sun-rich continent (Liu, 2015). Therefore, this presents a great opportunity to meet the energy needs of unserved communities while transitioning from non-renewable energy sources.

In SSA, addressing energy access is of utmost importance. Approximately 570 million individuals reside in areas without access to electricity, while an estimated 1 billion people lack the means to cook with clean energy sources. This problem exacerbates the low development rate prevalent in these areas and intensifies multidimensional poverty. Few communities have supported themselves in the past with off-grid fossil fuel-based technologies. However, with increasing fuel costs, sustenance has become impossible. For example, Nigeria removed its fuel subsidy in 2023, skyrocketing the price of a litre of fuel from N195 per litre ($0.12) to a current price of N855 per litre ($0.52), further increasing the number of people who can barely afford to electrify their homes, and plunging them deeper into energy poverty.

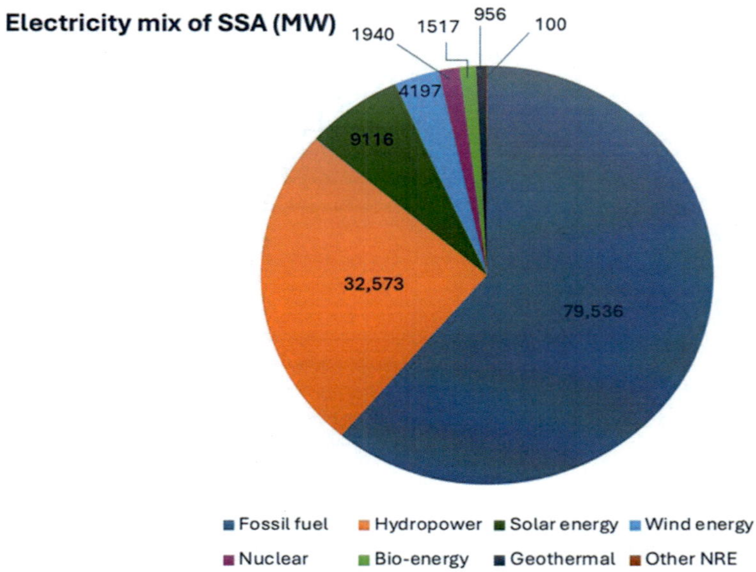

Fig. 1 Electricity mix of sub-Saharan Africa in MW. *Data source* (DTE, 2023)

Figure 1 shows a breakdown of the electricity mix across SSA, solar energy installations have continued to grow annually, signifying efforts to electrify 61% of unserved rural communities. Notably, the cost of solar PV systems has continued to decrease, falling by 90% as the cumulative global installed capacity increases, this can be attributed to the falling module prices and the installation system cost (Baurzhan & Jenkins, 2016).

The trend of other associated hardware components like batteries shows similar traits like the cost of lithium-ion batteries falling by 80% from $1000 per kWh to under $200 per kWh (Comello et al., 2018). Hence, it is expected that the total energy cost of solar PV will continue to decline given technological advancement, economics of scale, market maturity and the urgency to reduce emissions. Figure 2 shows the electrification rate of 48 countries within SSA where three categorizations of energy access achievers are identified. The first group are the high achievers with at least 80% of rural electrification rate made up of Mauritius, Seychelles, Cape Verde, Comoros, South Africa, and Eswatini, which have achieved great success towards electrification in the rural areas.

Next, the moderate achievers including Kenya, Ghana, Sao Tome and Principe have shown great tenacity to improve electrification rates to an average level of 50% to 70%. The remaining 39 countries fall within the range of underachievers with less than 50% rural electrification rate (World bank, 2023).

Specifically, countries with less than 10% rural access rate like South Sudan, Sierra Leone, Niger, Mozambique, Mauritania, Malawi, Equatorial Guinea, the Democratic Republic of Congo, Chad, the Central African Republic, Burundi, Burkina

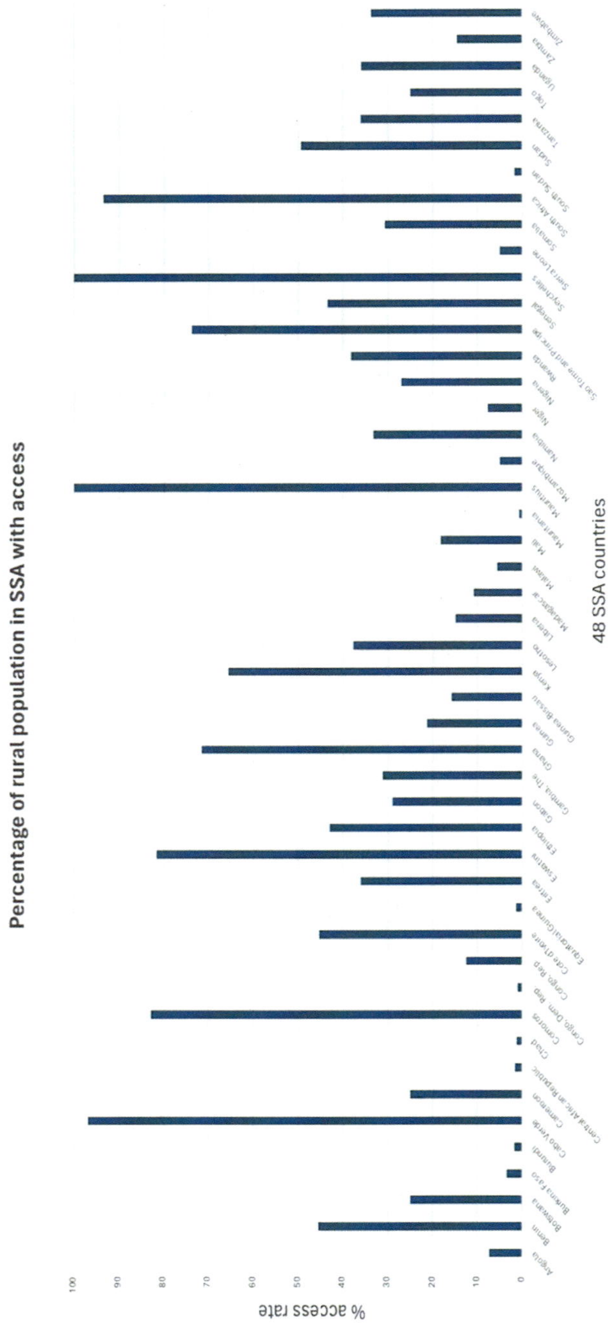

Fig. 2 Rural electrification rate in SSA. *Data source* (World bank, 2023)

Faso and Angola and a rural population of 79%, 56%, 83%, 61%, 82%, 26%, 53%, 76%, 56%, 85%, 67% and 31% respectively face significant challenges including limited infrastructure, financial constraints and political instability. Hence, sustainable off-grid systems, such as solar panels with battery storage have the potential to rapidly increase access to electricity in these areas not connected to the main grids. These systems can provide reliable and clean energy, improving lives in remote communities.

However, the challenge of electrification cannot be solely attributed to the lack of access in SSA. Another consideration is the disparity in the annual per capita consumption, which is evident not only in rural areas but also in urban areas. The annual per capita electricity consumption amongst SSA countries excluding South Africa remains low averaging about 180 kilowatt-hours (kWh) per person in comparison with other regions and considerably worse in rural communities with an annual per capita consumption of less than 100kWh (Falk et al., 2021; Groth, 2019; Tagliapietra, 2018). This could hinder the potential economic and social development necessary to alleviate poverty in the region.

2.2 *Energy Access as a Development Strategy*

Typically, basic lighting has emerged as the most recognized benefit of energy access in sub-Saharan Africa. Numerous studies have highlighted its significance as a crucial indicator when assessing socio-economic impacts (Aklin et al., 2017; Groth, 2019; Wagner et al., 2021). The presence of reliable lighting not only enhances daily living conditions but also plays a vital role in improving educational opportunities, boosting local economies, and fostering community development. In this context, the availability of energy for lighting serves as a foundational element that supports broader advancements in quality of life across the region. In alignment with the strategic priorities set forth by the International Energy Agency (IEA), this paper undertakes a comprehensive examination of energy access. This concept extends far beyond the provision of basic lighting; it encompasses the full spectrum of benefits that electricity can offer. However, before these transformative benefits can be fully harnessed, it is essential to gain a deep understanding of the various factors that critically influence the standardization of methods for achieving energy access. In this regard, this paper delineates the key components essential for effective energy access which are Reliability, Accessibility, Affordability, and Usability (collectively referred to as RAAU) as shown in Fig. 3, specifically in the context of rural communities. These components not only serve as the foundational pillars that determine how effectively energy can be integrated into everyday life ultimately fostering resilience but also unlocking myriad opportunities for economic growth and vitality within these communities.

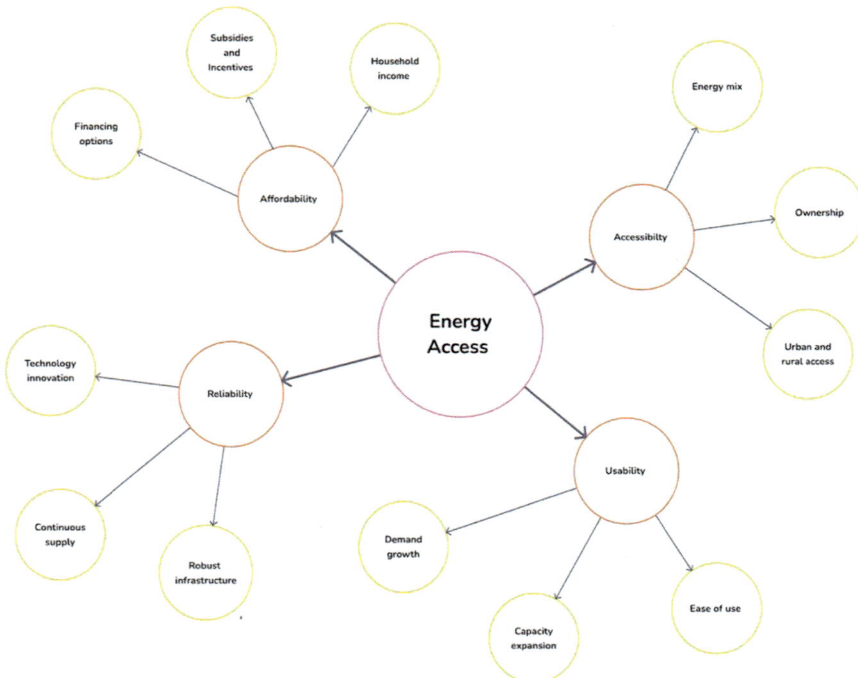

Fig. 3 Key components of effective energy access

2.2.1 Accessibility

The consideration of accessibility is pivotal in conversations about energy access because it describes the extent to which individuals, communities, and industries can obtain and utilize energy for a wide range of needs and activities. Rural electrification can be achieved using decentralized energy systems to curb the inadequacies of grid extension, of which, three categories are prevalent in rural SSA. The Home Pico systems are compact photovoltaic (PV) panels, generating just a few watts of power within a 1–10 W capacity range, supporting essential home services. This small yet powerful device can illuminate spaces with dependable lighting, charge mobile phones to keep you connected, and power a radio, allowing you to stay informed and entertained. They can be directly purchased at the marketplaces or through credit financing with no distribution control and are best suited as short-term backups for emergency and unprecedented situations. The second category is the stand-alone systems which are wired directly in individual homes. Originally, these systems have operated as generator sets using fuel with varied capacities depending on needs and income. However, air pollution accidents, noise pollution, and fluctuating fuel prices across regions have made them unsuitable in this current energy transition. Alternatively, Solar Home Systems (SHS) and other renewable energy options offer a clean, readily available energy source at the base of use and are tailored to match the household's existing demand. Evidence from countries like Kenya shows that installers additionally combine these systems with various appliances to improve livelihoods and boost demand (Wagner et al., 2021). In sub-Saharan Africa, SHS are generally purchased through a payment plan, allowing consumers to pay in manageable instalments over an extended period. Alternatively, customers have the option to pay the total amount upfront or rent the system from a third-party operator. The last category which is the mini grid offer a more advanced capacity between 0.2 kW and 2 MW to serve entire communities. This system size acts as an alternating current (AC) system which imitates the centralized grid system and is installed at a discrete location within the community, eluding the high cost of grid extension and transmission losses to unserved communities (Pedersen, 2016). The ownership model for the mini-grid varies from place to place and depends on the community maturity level though the utility-based model offers a lower tariff due to government-induced subsidies (Nyarko et al., 2023), see Table 2.

The private developer model takes centre stage in SSA since the developers can independently scout for viable sites and manage the setup from design to operation after obtaining the required permits. This allows for faster implementation of projects, which can be a significant advantage. However, one of the key challenges faced in this model is establishing a tariff that aligns closely with the existing rates of the main electricity grid. This alignment is crucial to ensure competitiveness and fairness, as it affects both consumer acceptance and the overall feasibility of the project. (Palit & Bandyopadhyay, 2016) highlight the community-based model in which local representatives play a crucial role in managing the mini-grid. They are responsible for overseeing both the installation and ongoing maintenance processes, ensuring that the system meets the needs of the community effectively. This model has been effectively implemented in SSA countries like Cape Verde and Liberia but the technical know-how of the representatives and the financial implications could challenge the system's sustainability (Nyarko et al., 2023).

Table 2 Accessibility options for rural electrification

	Mini grid systems	Isolated/home systems	Portable/pico systems
Type	Renewable energy dependent grid e.g. solar mini-grids, Hybrid energy systems[1]	Renewable energy sources like SHS, wind, bioenergy	LED lighting, Mobile chargers, Radio
Ownership	Utility-based, Private developer, Community-based, Hybrid	Direct purchase model, Third-party ownership, Lease -to-own model	Direct purchase model, Dealer credit model
System size	0.2 KW–2 MW	10–100 W	1–10 W

[1]Other systems reliant on conventional energy sources exist, but in consideration of a clean energy transition, this review paper has focused on renewable energy sources for rural electrification. Source: Adapted from (Nyarko et al., 2023; Palit & Bandyopadhyay, 2016; Pedersen, 2016; Ugwoke et al., 2020)

In practical applications, these models can be effectively combined to facilitate the electrification of communities, all while adhering to specific objectives and timelines. This hybrid model allows for a tailored strategy that meets the unique energy needs of different communities, ensuring a sustainable and efficient transition to reliable electricity access.

2.2.2 Affordability

The affordability indicator is determined by the household's income, financing options, subsidies, and incentives. In rural areas of Sub-Saharan Africa (SSA), households are categorized as low-income earners, mainly relying on agriculture, particularly subsistence farming (Bensch et al., 2011; Peters & Sievert, 2016; Wamukonya & Davis, 2001). Consequently, the cost of accessing electricity is a significant concern for these households, as they are more likely to prioritize basic needs such as food, shelter, and clothing due to their limited income. The availability of several financing options can augment the strain electrification could cause on a household's income and mobilize options for vulnerable households. (Tinta et al., 2024) shows in his study, a connection between income and solar PV adoption in rural areas further highlighting that access to financial services impacts the adoption of off-grid systems. These financial services in the form of rebates, loans, grants, green bonds, power purchasing agreements (PPA) and community-based financing are among the various options available to energy developers and investors (Khellaf, 2018).

Also, policy measures that provide funding and subsidies strongly promote the large-scale adoption and utilization of solar PV systems considering the high cost of investment (Wassie & Adaramola, 2021). In (Mondal & Klein, 2011) study of rural Bangladesh, users who purchased SHS experienced difficulties in paying both the up-front costs and the monthly installation fee.

The financial burden of investing in a SHS poses a significant challenge for households. Unlike their previous sporadic energy procurement practices, they are faced with the necessity of adhering to a specific payment schedule for these systems. This shift in financial responsibility requires careful planning and budgeting for many households.

Alternatively, the usage-based payment model offered by mini-grids allows consumers to charge their prepaid meters and access the consumption capacity equivalent to the amount deposited. One drawback of this system is that customers cannot control or own it, and they are charged a fixed rate for each kWh used. This stands in contrast to the SHS, which offers more control to the customers. In a Tanzanian study, (Falk et al., 2021) finds that households were willing to pay higher than their average income, a maximum amount of EUR 0.40, three times less than the fixed local utility rate of EUR 1.39. Considering the energy cost is crucial for ensuring that energy access is effective, particularly because there is a significant population in these areas living below the poverty line of $2.15. This underscores the need for affordable energy solutions to cater to the specific economic conditions of the community. Again, the economics of scale can be applied to overturn the high cost of electricity ensuring that as more connections are made within a locality, the rate per kWh decreases, unburdening connected users.

2.2.3 Reliability

The primary objective of ensuring access to electricity is to guarantee a continuous and dependable supply that can be accessed at any time of the day as per requirement. Historically, reliability of supply has remained a barrier to electricity access in SSA. Despite connection to the main grid in urban areas, the actual duration of supply remains limited. Investors and developers of SHS and mini-grid systems should guarantee their reliability to circumvent these challenges. In rural areas, an inconsistent and unreliable supply of electricity necessitates that households rely on alternative sources such as candles, kerosene, and other traditional energy sources for their lighting and energy needs (Wassie & Adaramola, 2021).

This constant dependence on these conventional energy sources could affect the daily lives and well-being of the residents. Several factors have been identified to impair the reliability of energy systems. For renewable energy sources like solar which are intermittent in nature, they are prone to fluctuations caused by the day-night cycle, weather conditions, and seasonal variations within a locality.

To address this issue, integrating a backup energy source to form a hybrid system ensures continuity of supply. These backup sources depending on the area could be an alternate renewable energy source or a conventional generator ensuring a reliable and steady supply of electricity. Also, energy storage at the solar peak hours using batteries can support the reliability of off-grid systems, especially for domestic use in the evenings.

Batteries typically have a shorter lifespan than solar panels, and most off-grid systems in SSA have been observed to rely on lead-acid batteries instead of more efficient lithium-ion batteries (Antonanzas-Torres et al., 2021). Moreover, an increase in temperature can exacerbate this issue, leading to a shortened operational life for the batteries (Adu-Poku et al., 2023). Technological innovation could play a role in enhancing solar panel conversion efficiency enabling more efficient conversions from irradiation capturing and advanced cost-effective battery storage devices. (Numminen & Lund, 2019) identified the instability and the low quality of the system component unit as the central point that diminishes the reliability of supply.

Various indicators like the Loss of Load Probability (LOLP) and Mean Down Time (MDT) can be used to measure the system's reliability. For example, (Yibeltal T Wassie & Ahlgren, 2023) has applied the System Average Interruption Duration Index (SAIDI) in his study of southern Ethiopia. It was observed that the interruptions hour (SAIDI) increased from 15.5 h to 402.38 h within 8 months, attributing the interruption to a rise in the number of new customer connections despite a static supply. This asserts the position that many mini-grid developers are unable to expand on their initial grid capacities, hence further addition of customers impends the system's reliability. Overall, the system components' quality and sizing, the energy source type, high temperatures experienced in the region, and the growth in electricity demand over time can affect the reliability of off-grid systems.

2.2.4 Usability

Usability as a component of energy access is used to determine how easily people adopt the systems, enhance their lives through its operation, with an added opportunity to scale up the system to meet the evolving demand. In communities with limited technical knowledge and infrastructure, energy solutions are likely to be successfully accepted and sustained if they are user-friendly. One of such examples is the prepaid metering system commonly referred to as the Pay-as-you-go (PAYGO) meters in SSA. These meters are recharged by either purchasing credits through a local agent or a mobile app banking platform in advance thereby simplifying the rather complex process for residents.

Additionally, it gives each household the independence to manage their energy consumption as necessary within monetary constraints. For SHS, households must understand the basic technicalities involved in operating the system since they are managed individually per household, hence a simplified interface control could aid maintenance and durability. Usability mirrors the potential socio-economic benefits within a household through different indicators like education and productivity. For example, lighting could offer students the opportunity to extend study time after dusk while utilization of machinery enhances productivity which could generate alternate household income. The Multi-Tier Framework (MTF) categorizes households in rural Sub-Saharan Africa (SSA) as Tier 1, indicating that their access to electricity and modern appliances is extremely limited. In this tier, the primary devices found are basic lighting sources and radios, reflecting a significant lack of more advanced electrical technologies and amenities. Other additional appliances that can be found in rural households are televisions for entertainment and fans for cooling, placing a limited number of households within the Tier 2 category (Bertheau, 2020).

These classifications hinder an improvement in their quality of life as the benefits of utilization are restricted. Ideally, the scope of energy access entails a growth in demand over time for social and productive uses such as water pumping for irrigation and agricultural processing machinery, these projected increase in demand will eventually lead to a capacity expansion phase (Bannor et al., 2023; Bhandari et al., 2017).

On the technology side, renewable energy systems like solar systems are flexible and can be scaled to meet the existing demand. However, the cost of expanding an existing system is a delimiting factor for isolated systems and mini grids. The implementation of these system upgrades is significantly influenced by the income levels of rural households, which often renders them financially unfeasible. Many families in these areas may struggle with limited resources, making it challenging to afford the necessary improvements (Mazzone, 2019).

2.3 *Socio-Economic Benefits of Effective Energy Access*

Effective energy access in rural sub-Saharan Africa improves an individual's quality of life by unlocking a wide range of socio-economic benefits leading to further community development. These socio-economic benefits impact various sectors of the economy including better lighting, education, health services, security, productivity, and businesses through job creation and income-generating activities. Academic work on the socio-economic impact of off-grid rural electrification in developing countries has continued to grow in recent years, although indicators have varied from significant to insignificant (Bhandari et al., 2017). Energy access improves healthcare services in communities and enables clinics to operate efficiently by offering refrigeration of medicines and vaccines, lighting for nighttime emergencies, and the operation of medical equipment. Falk in his study of rural Tanzania finds respondents acknowledged improved health care at the local hospital after the installation of the mini grid (Falk et al., 2021). This also aligns with (Carabajal et al., 2024) cohort study of Kenya and Nigeria. Furthermore, the health status of household members improves over time after establishing access, this can be attributed to a reduction in hazardous energy use (Wamukonya & Davis, 2001).

Another strong indicator of the socio-economic benefits of energy access is its impact on education. Several studies have shown linkages between energy access and educational achievement (Aguirre, 2017). In Sierra Leone, students are able to study and finish their homework with better lighting at nighttime increasing study hours per day, thereby improving their academic performance (Sadik-Zada et al., 2023). The estimated nighttime study hours could reach 0.94 h/day (Beyene et al., 2024). Other benefits of energy access in education include access to online educational resources and technology-assisted learning. (Sadik-Zada et al., 2023) suggests this interconnection could potentially drive agricultural modernization of rural Sierra Leone in the future. Furthermore, effective energy access can stimulate business development and economic growth for the region. This could be achieved through job creation, productivity, or income generation. For instance, it can increase agricultural productivity by using energy for irrigation and food processing, which enhances food security. However, a great disparity exists in the assessment of economic impacts. While (Azimoh et al., 2017; Carabajal et al., 2024; Falk et al., 2021; Tinta et al., 2024) indicates positive impacts can be achieved by energy access, (Baurzhan & Jenkins, 2016; Bensch et al., 2011; Beyene et al., 2024) demonstrates no significant effect is observed in one or more sub-indicators from their assessment.

This could be attributed to the capacity range of available energy sources sufficient to bring about economic opportunities for households and communities. Also, enabling policies can further enhance the prospect of off-grid systems in developing countries as seen in Fig. 4. For example, the affordability of service can be improved with policies that support subsidies and financial loans to residents. Overall, these three indicators highlighted above reflect the most studied for socio-economic impacts, other emerging empirical evidence studies of socio-economic

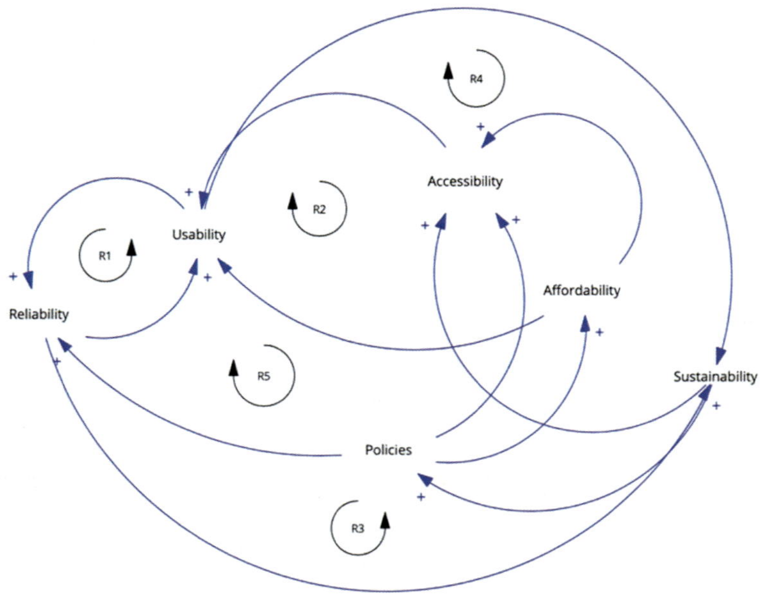

Fig. 4 Interconnections of policy and sustainability on effective energy access

impacts would include impacts on migration, security, and the reduction in battery waste and alternate energy sourcing of rural households.

3 Conclusion

This review study aims to contribute to the literature on energy access for rural communities in sub-Saharan Africa (SSA). available capacity. The current evaluation of electrification indicates that more than half of SSA countries' energy is derived from conventional energy sources, although scalable renewable energy sources like solar are utilized for rural electrification, hence a growing trend. Rural communities in Sub-Saharan Africa have been categorized based on their level of access to electricity, highlighting the disparities in rural electrification efforts across the region. This classification serves to identify areas that are well-served by energy infrastructure and those that remain under-electrified, which is crucial for implementing targeted policies and interventions aimed at improving energy access and supporting sustainable development in the region. However, the use of renewable energy sources such as solar can provide a sustainable energy system that supports essential services and development in rural areas. To achieve this, energy access must transcend from a minimal lighting source to an all-inclusive equipment utilization for sustainable development influenced by the reliability, accessibility, affordability, and usability of the energy system.

Integrating these factors with enabling policies as shown in Fig. 4 indicates the positive reinforcement of one variable on the other. As the factors influence the sustainability of the community, sustainable development enhances the expansion of supportive policies through regulatory frameworks and subsidies thereby attracting both domestic and international investors. Therefore, effective energy access is a key driver of social and economic development, empowering rural communities to improve their living conditions and unlocking new opportunities.

The task of attaining universal energy access in Sub-Saharan Africa presents a considerable challenge, ranging from inadequate infrastructure to funding shortages and regulatory complexities. However, the vast socio-economic advantages that come with overcoming these hurdles are enormous. Addressing energy access holistically in SSA is imperative for policymakers, development organizations, and energy professionals to set the foundation for a more stable, prosperous, and equitable society.

References

Adu-Poku, A., Koku Aidam, G. s. S., Jackson, G. A., N'Tsoukpoe, K. E., Kponyo, J. J., Messan, A., Kemausuor, F. (2023). Performance assessment and resilience of solar mini-grids for sustainable energy access in Ghana. *Energy, 285*, 129431. https://doi.org/10.1016/j.energy.2023.129431.

Aguirre, J. (2017). The impact of rural electrification on education: A case study from Peru. *The Lahore Journal of Economics, 22*(1), 91.

Agyei-Sakyi, M., Shao, Y., Amos, O., & Marymargaret, A. (2021). Determinants of electricity consumption and volatility-driven innovative roadmaps to one hundred percent renewables for top consuming nations in Africa. *Sustainability, 13*(11), 6239.

Aklin, M., Bayer, P., Harish, S., & Urpelainen, J. (2017). Does basic energy access generate socioeconomic benefits? A field experiment with off-grid solar power in India. *Science Advances, 3*(5), e1602153.

Altinay, G., & Karagol, E. (2005). Electricity consumption and economic growth: Evidence from Turkey. *Energy Economics, 27*(6), 849–856.

Antonanzas-Torres, F., Antonanzas, J., & Blanco-Fernandez, J. (2021). State-of-the-art of mini grids for rural electrification in West Africa. *Energies, 14*(4), 990.

Azimoh, C. L., Klintenberg, P., Mbohwa, C., & Wallin, F. (2017). Replicability and scalability of mini-grid solution to rural electrification programs in sub-Saharan Africa. *Renewable Energy, 106*, 222–231.

Babalola, S. O., Daramola, M. O., & Iwarere, S. A. (2022). Socio-economic impacts of energy access through off-grid systems in rural communities: A case study of southwest Nigeria. *Philosophical Transactions of the Royal Society A, 380*(2221), 20210140.

Bannor, R. K., Amfo, B., Sarquah, K., Oppong-Kyeremeh, H., & Kyire, S. K. C. (2023). Unwrapping the nexus between the off-grid system and its impact on the islands and remote villages of Ghana. *International Journal of Energy Sector Management, 17*(1), 82–105.

Baurzhan, S., & Jenkins, G. P. (2016). Off-grid solar PV: Is it an affordable or appropriate solution for rural electrification in Sub-Saharan African countries? *Renewable and Sustainable Energy Reviews, 60*, 1405–1418.

Bensch, G., Kluve, J., & Peters, J. (2011). Impacts of rural electrification in Rwanda. *Journal of Development Effectiveness, 3*(4), 567–588.

Bertheau, P. (2020). Assessing the impact of renewable energy on local development and the sustainable development goals: Insights from a small Philippine island. *Technological Forecasting and Social Change, 153*, 119919.
Beyene, A. D., Mekonnen, A., Jeuland, M., & Czakon, S. (2024). Socioeconomic impacts of solar home systems in rural Ethiopia. *Renewable and Sustainable Energy Reviews, 192*, 114197.
Bhandari, B., Lee, K.-T., Chu, W.-S., Lee, C. S., Song, C.-K., Bhandari, P., & Ahn, S.-H. (2017). Socio-economic impact of renewable energy-based power system in mountainous villages of Nepal. *International Journal of Precision Engineering and Manufacturing-Green Technology, 4*, 37–44.
Carabajal, A., Orsot, A., Elimbi Moudio, M. P., Haggai, T., Okonkwo, C. J., Jarrard, G. T., & Selby, N. S. (2024). Social and economic impact analysis of solar mini-grids in rural Africa: a cohort study from Kenya and Nigeria. Environmental Research: Infrastructure and Sustainability.
Comello, S., Reichelstein, S., & Sahoo, A. (2018). The road ahead for solar PV power. *Renewable and Sustainable Energy Reviews, 92*, 744–756.
DTE. (2023). 30 sub-Saharan Africa countries among 38 that fail to provide basic electricity access. Retrieved January 09, 2025, from https://www.downtoearth.org.in/africa/30-sub-saharan-africa-countries-among-38-that-fail-to-provide-basic-electricity-access-91045.
Falk, J., Angelmahr, M., Schade, W., & Schenk-Mathes, H. (2021). Socio-economic impacts and challenges associated with the electrification of a remote area in rural Tanzania through a mini-grid system. *Energy, Ecology and Environment, 6*(6), 513–530.
Groth, A. (2019). Socio-economic impacts of rural electrification in Tanzania. *International Journal of Sustainable Energy Planning and Management, 21.*
Hefner III, R. A. (2009). *The grand energy transition: the rise of energy gases, sustainable life and growth, and the next great economic expansion.* John Wiley & Sons.
IEA, Irena, UNSD, World Bank, & WHO. (2023). *Tracking SDG 7: The energy progress report 2023.* World Bank.
IEA, IRENA, UNSD, World Bank, & WHO. (2024). Tracking SDG 7: The energy progress report, 2024. World Bank, Washington DC. Retrieved from https://www.iea.org/reports/tracking-sdg7-the-energy-progress-report-2024, Licence: CC BY NC 4.0.
IHA. (2024). *World hydropower outlook: Opportunities to advance net zer*o. Retrieved from
Kammen, D. M., Alstone, P., & Gershenson, D. (2015). Energy access and sustainable development. *Physics of sustainable energy III (PSE III): Using energy efficiently and producing it renewably, 1652*(1), 14–32.
Khellaf, A. (2018). *Overview of economic viability and social impact of renewable energy deployment in Africa*, pp 59–70
Liu, Z. (2015). *Global energy interconnection.* Academic Press.
Mazzone, A. (2019). Decentralised energy systems and sustainable livelihoods, what are the links? Evidence from two isolated villages of the Brazilian Amazon. *Energy and Buildings, 186*, 138–146. https://doi.org/10.1016/j.enbuild.2019.01.027.
Mondal, A. H., & Klein, D. (2011). Impacts of solar home systems on social development in rural Bangladesh. *Energy for Sustainable Development, 15*(1), 17–20.
Numminen, S., & Lund, P. D. (2019). Evaluation of the reliability of solar micro-grids in emerging markets–Issues and solutions. *Energy for Sustainable Development, 48*, 34–42.
Nyarko, K., Whale, J., & Urmee, T. (2023). Drivers and challenges of off-grid renewable energy-based projects in West Africa: A review. *Heliyon, 9*(6).
Palit, D., & Bandyopadhyay, K. R. (2016). Rural electricity access in South Asia: Is grid extension the remedy? A critical review. *Renewable and Sustainable Energy Reviews, 60*, 1505–1515.
Pedersen, M. B. (2016). Deconstructing the concept of renewable energy-based mini-grids for rural electrification in East Africa. *Wiley Interdisciplinary Reviews: Energy and Environment, 5*(5), 570–587.
Peters, J., & Sievert, M. (2016). Impacts of rural electrification revisited–the African context. *Journal of Development Effectiveness, 8*(3), 327–345.

Sadik-Zada, E. R., Loewenstein, W., & Dumbuya, F. S. (2023). Modernization through solar off-grid electrification? A mixed picture for rural Sierra Leone. *The Electricity Journal, 36*(7), 107316.
Sarker, S. A., Wang, S., Adnan, K. M., Anser, M. K., Ayoub, Z., Ho, T. H., & Hoque, M. M. (2020). Economic viability and socio-environmental impacts of solar home systems for off-grid rural electrification in Bangladesh. *Energies, 13*(3), 679.
Tagliapietra, S. (2018). Electrification in sub-Saharan Africa: the role of international institutions. The oxford institute for Energy studies (115).
Tinta, A. A., Okoko, A., & Lankouande, E. (2024). From light to income: Solar PV technology. *Journal of the Knowledge Economy*, 1–23.
Ugwoke, B., Adeleke, A., Corgnati, S. P., Pearce, J. M., & Leone, P. (2020). Decentralized renewable hybrid mini-grids for rural communities: Culmination of the IREP framework and scale up to urban communities. *Sustainability, 12*(18). https://doi.org/10.3390/su12187411.
UN. (2023). *Progress towards the sustainable development goals: towards a rescue plan for people and planet*. Retrieved from https://unstats.un.org/sdgs/files/report/2023/SDG_Progress_Report_Special_Edition.pdf.
Wagner, N., Rieger, M., Bedi, A. S., Vermeulen, J., & Demena, B. A. (2021). The impact of off-grid solar home systems in Kenya on energy consumption and expenditures. *Energy Economics, 99*, 105314.
Wamukonya, N., & Davis, M. (2001). Socio-economic impacts of rural electrification in Namibia: Comparisons between grid, solar and unelectrified households. *Energy for Sustainable Development, 5*(3), 5–13. https://doi.org/10.1016/S0973-0826(08)60272-0.
Wassie, Y. T., & Adaramola, M. S. (2021). Socio-economic and environmental impacts of rural electrification with solar photovoltaic systems: Evidence from southern Ethiopia. *Energy for Sustainable Development, 60*, 52–66. https://doi.org/10.1016/j.esd.2020.12.002
Wassie, Y. T., & Ahlgren, E. O. (2023). Performance and reliability analysis of an off-grid PV mini-grid system in rural tropical Africa: A case study in southern Ethiopia. *Development Engineering, 8*, 100106.
World bank. (2023). *Rural population (percentage of total population)-sub-Saharan Africa access to electricity, rural (% of rural population)*. Retrieved from. https://data.worldbank.org/indicator/EG.ELC.ACCS.RU.ZS?locations=ZG. https://data.worldbank.org/indicator/EG.ELC.ACCS.RU.ZS?locations=ZG.
Worldometer. (2024). Population growth rate. Retrieved from https://www.worldometers.info/.
Yoo, S.-H., & Kwak, S.-Y. (2010). Electricity consumption and economic growth in seven South American countries. *Energy Policy, 38*(1), 181–188.

Comparative Study of Conventional and Biofuels for Sustainable Mobility: A Case Study of Ghana

Janet Appiah Osei, Satyanarayana Narra, Rabani Adamou, and Amos T. Kabo-Bah

Abstract As means to promote energy security and mitigate climate change, utilization of biofuels has become attractive in recent years. Their local production from biomass resources etc. isolate them from political disruptions in the international market and boost the national security by lowering a country's reliance on petroleum imports. Biofuel utilization in Ghana is still in the nascent stage, though policies like the Renewable Energy Act encourages its application in the energy sector. Hitherto, the transport sector is void of biofuels and remain entirely dependent on conventional fuels. To elucidate the benign effects of biofuels over conventional fuels under the ambient conditions in Ghana, empirical analysis was conducted using a physics-based vehicle model. The model incorporated both the clean fuels and the fossil fuels to assess their respective impacts. The model output highlighted the environmental benefits of using biofuels as increasing bioethanol blend from 10 to 85% and 100% reduced fossil-based CO_2 emissions by 8% and 87% respectively with E100 having no fossil emissions when gasoline was used as the base fuel. Similarly, use of B10 and B50 biodiesel blends instead of pure diesel fuel downsized fossil CO_2 emissions by 6% and 48% respectively. Biofuels can therefore ameliorate the transport sector by downsizing the total emissions associated with this sector.

Keywords Analytical model · Bioethanol · Biodiesel · Decarbonization

J. A. Osei (✉) · R. Adamou
WASCAL Climate Change and Energy, Abdou Moumouni University, Niamey, Niger
e-mail: appiah.j@edu.wascal.org

S. Narra
Material and Energy Valorisation, University of Rostock, Rostock, Germany

A. T. Kabo-Bah
University of Energy and Natural Resources, Sunyani, Ghana

M.-M. Narra and S. Narra (eds.), *African Green Transition Through Innovative Pathways*, World Sustainability Series, https://doi.org/10.1007/978-3-031-87043-9_10

1 Introduction

Globally, transportation relies heavily on combustion engines, with over 90% of vehicles using petroleum fuels. A 2015 report from the European Union indicated that 56% of the vehicle fleet in Europe operated on gasoline, 41% on diesel and only 3% on alternative fuels (Kalghatgi et al., 2018). In Ghana, the transport sector is entirely dependent on petroleum fuels, experiencing an annual growth rate of 5.5% (Ghana Energy Commission, 2022). This makes it the largest emission sector within the energy sector, accounting for approximately 43% of the total emissions (Kemausuor et al., 2015). Moreover, the absolute reliance on imported petroleum fuels threatens energy security, depreciates the Ghanaian cedis and deteriorates the balance of payment and the country's foreign currency reserves (Pelizan, 2019).

In alignment with the Paris Agreement goals, many developed countries have committed to over 50% GHG emission reduction by 2030. Ghana, through its Nationally Determined Contribution (NDC) announced in 2015, aimed to "*lower GHG emissions by 15% (11.1 $MtCO_{2e}$) relative to a Business- as- usual scenario emission of 73.95 $MtCO_{2e}$ by 2030 and to further minimize emissions by 30% in case of available external support*" (MESTI, 2021).

To drive the decarbonization process, the Renewable Energy Act was enacted in 2011 under the auspices of the government to promote the use of renewable energy. The Act called for the extensive use of bioenergy technologies and the utilization of biomass resources, setting a target of 10% biofuel in the transportation fuel mix by 2020 and 20% by 2030.

Nevertheless, the 10% blend has not been actualized whilst the 20% remain uncertain (Kemausuor et al., 2015). In 2018, Ghana Standard Authority included biofuel standards in its catalogue, but these standards had no practical impact or application, as no biofuel mandate was ever implemented and additionally, biofuels were never classified as commercial products (Pelizan, 2019).

According to the research by Kemausuor et al. (2014), approximately 9.7×10^{16} J of bioenergy can be generated annually from organic materials. Additionally, (Duku et al., 2011; Mohammed et al., 2013) discovered bioenergy potential of 7.5×10^{16} J and 1×10^{17} J per year respectively. This indicates a significant availability of substrates for biofuel production in Ghana. Conversely, existing energy policies have failed to capture the degree to which biofuels could complement the transport fuels in Ghana. A study by Saisirirat et al. (2022), concluded that integrating biofuels into the transport sector presents a more viable decarbonization strategy compared to electricity. This conclusion was drawn from the challenges posed by intermittent power supply and the dependence on non-renewable energy sources for electricity generation, which make electric vehicles an unsustainable option in Ghana at this time, unless significant reforms are implemented in the power sector.

In Ghana, the potential of biofuels as decarbonizing fuel in the transport sector has not been adequately explored in literature. It is therefore imperative to increase research in this area to facilitate the implementation of sustainable biofuel policies. To address this gap, the study conducts a comparative analysis of the fuel consumption and emission rates of biofuels versus conventional fuels, taking into account the local environmental conditions in Ghana. An empirical analysis was performed using an analytical model that applies the physical principles of a moving vehicle.

2 Materials and Methods

2.1 *Study Data*

Data on the prevailing temperature and relative humidity in Accra were provided by the Ghana Meteorological Agency whilst vehicle selection was based on the prevalent taxis and minivans (*trotros*) available in the central business districts of Accra.

2.2 *Methodology*

The study employed an analytical model based on physical principles to ascertain the varying rate of fuel consumption between conventional and biofuels. To attain the aim of the study, three scenarios were employed—Business as Usual (BAU), Alternative and Extreme scenarios. BAU defined the current situation of gasoline and diesel fuel usage whilst Alternative and Extreme are the mitigation scenarios involving biofuel use. The alternative mitigation measure considered E10 (10% bioethanol 90% gasoline) and B10 (10% biodiesel 90% diesel), the extreme measure focused on E85 (85% bioethanol 15% gasoline), E100 (100% bioethanol), B50 (50% biodiesel 50% gasoline) and B100 (100% biodiesel).

The design of the model involved five key stages: (1) Identification of the physical system (2) Modelling by creating mathematical formulas (3) Simulation of mathematical equations (4) Interpretation of results (5) Evaluation (Redish & Smith, 2008). These five steps are summarized in Fig. 1.

2.3 *Physical System Identification*

Three different taxis and minibuses (*trotros*)—Daewoo Matiz, Toyota Matrix, Toyota Yaris, Hyundai Grace, Toyota HiAce and Mercedes Sprinter were selected among the prevalent commercial vehicles in Accra. The criteria for selection were based on their distinctive engine capacities to enable the determination of fuel consumption and emissions based on different vehicle sizes. This will increase the scope and applicability of the study, with these vehicles serving as proxies. Their specifications of interests needed for efficient development of the model were identified and enlisted in Table 1.

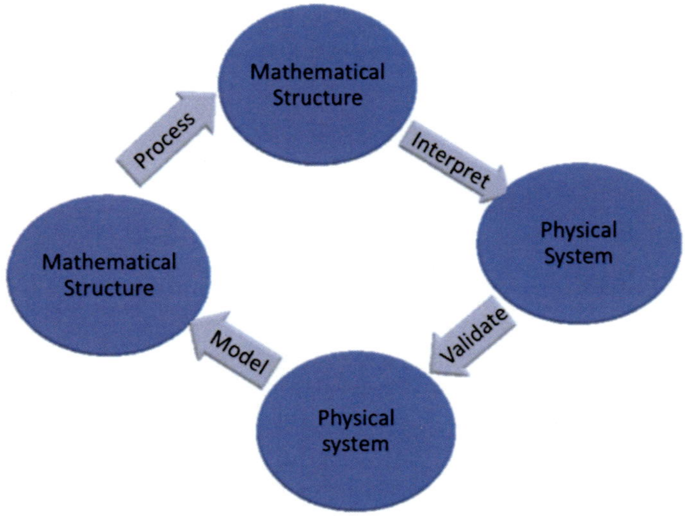

Fig. 1 Structure of a physical model

Table 1 Vehicle model and specifications

Vehicle make	Model	Engine capacity (l)	Mass (kg)	Frontal area (m^2)	Drag (Kw)	Maximum power (Kw)	Acceleration (m/s^2)
Chevrolet	Matiz	0.8	775	1.91	0.34	38	1.63
Toyota	Yaris	1.4	1000	2.09	0.29	64	2.12
Toyota	Matrix	1.8	1300	2.24	0.33	99	2.79
Hyundai	Grace	2.5	1604	2.82	0.35	59	2.36
Toyota	HiAce	3.0	1780	2.84	0.36	96.5	2.53
Mercedes	Sprinter	3.5	2546	4.46	0.43	115	2.31

2.4 Modelling and Simulation Approach

The specifications of the selected vehicles alongside the fuel properties in Table 2 were used to construct the physics-based model involving three critical forces that acts on a vehicle during motion- force of aerodynamic drag, force of inertia acceleration and force of rolling resistance. These forces interact to influence the fuel consumption of a vehicle in motion (Ben-Chaim et al., 2013).

Alongside the vehicle components, the model was built using the environmental conditions specifically temperature and relative humidity. The model used a bottom-up approach as illustrated in Eqs. 1 to 5.

Table 2 Conventional and Biofuel Properties

Fuel type	Brake specific fuel consumption (BSFC) (g/kWh)	Density (kg/l)
Gasoline	250	0.748
E10	259	0.752
E85	327	0.784
E100	340	0.790
Diesel	200	0.845
B10	203	0.849
B50	215	0.863
B100	230	0.880

Source (Mehra et al., 2020; Patel et al., 2019 and Usman et al., 2023)

$$Q = bsfc * 10^{-2}\left[\frac{F_{roll} * F_{inertia} * F_{drag} * (v)^3}{\eta \mathrm{T} * \rho \mathrm{f} * \mathrm{vh}}\right] \quad (1)$$

$$F_{roll} = f\langle 1 + \frac{v}{147}\rangle mg \quad (2)$$

$$F_{inertia} = ma \quad (3)$$

$$F_{drag} = 0.5 * d * A * \rho \mathrm{a} * \mathrm{v}^2 \quad (4)$$

$$\eta \mathrm{T} = \frac{F_{inertia} v}{h} \quad (5)$$

where; Q = fuel consumption (l/100 km), bsfc = brake specific fuel consumption (gkW/h), F_{roll} = rolling resistance force, $F_{inertia}$ = force of inertia acceleration, F_{drag} = force to overcome aerodynamic drag, ηT = transmission efficiency, ρf = fuel density (kg/l), v = velocity in (m/s), vh = velocity in (km/h), m = mass of vehicle (kg), a = vehicle acceleration, g = acceleration due to gravity, d = vehicle drag, A = frontal area, ρa = density of air h = maximum power (kW).

Simulations were run in python under SimPy environment. The different fuel types were subjected to variations whilst the weather factors and the vehicle specifications were kept constant depending on the vehicle type. Additionally, the driving speed and road surface were kept constant, with an asphaltic road and an average speed of 44.2 km/h considered in the model simulation. Figure 2 depicts the framework of the simulation process which describes variations in the model inputs using the proposed physical model to attain the target variable—fuel consumption of the vehicles.

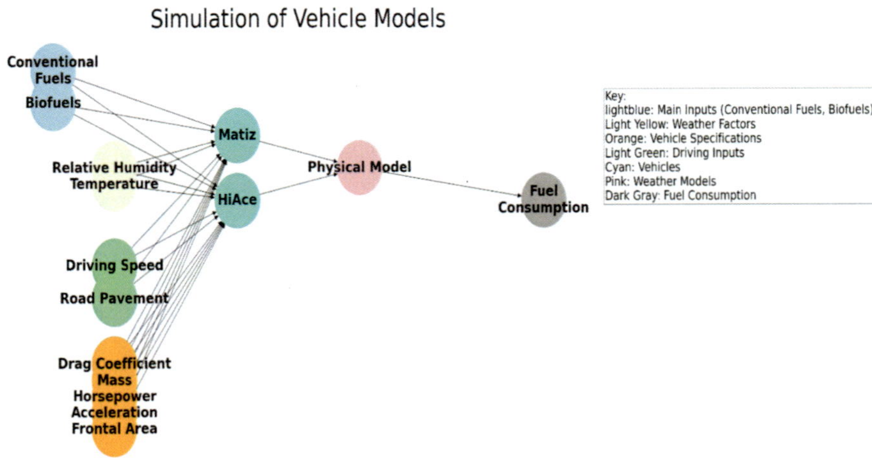

Fig. 2 Vehicle Simulation process

2.4.1 Model Evaluation

Uncertainty analysis was conducted by the use of sensitivity analysis in order to determine the logical behaviour and robustness of the analytical model. Sensitivity analysis was performed on the key input factors used in the model construction. One-factor-at-a time variation approach which requires varying one input at a time whilst keeping the others constant was used to perform the sensitivity analysis. The analysis was interpreted using the uncertainty metrics—margin of error (MOE), confidence interval (CI) and Standardized regression coefficient (SRC) to aid in describing the robustness of the model.

2.5 *CO_2 Emission Determination*

Using the model output and the IPCC default CO_2 emission factors, the total CO_2 emissions were ascertained. Default factors were applied due to the lack of country-specific emission factors in Ghana per authors' knowledge. Emissions were determined in two parts (a) fossil CO_2 emissions which focused mainly on the fossil fractions of the biofuel blends and the conventional fuels (b) biogenic CO_2 emissions that emphasized solely on the biological fraction of the biofuel blends.

The biogenic and fossil portions in the blends were computed by multiplying their respective fractions by the emission factors provided by the IPCC (Narayanan, 2006).

Equations 6 and 7 depict the IPCC approach used to determine the tailpipe emissions in this study.

$$\text{Fossil } CO_2 = F_f \times E_{CO2,f} \times GWP_{CO2} \tag{6}$$

$$\text{BioCO}_2 = F_b \times E_{CO2,b} \times GWP_{CO2} \tag{7}$$

where Fossil CO_2 and Bio-CO_2 are fossil and biogenic carbon dioxide emissions (kg) respectively; F_f = fuel consumption (fossil) (TJ); F_b = fuel consumption (biogenic) (TJ); $E_{CO2,f}$ = carbon dioxide (fossil) emission factor (kg/TJ); $E_{CO2,b}$ = carbon dioxide (biogenic) emission factor (kg/TJ), GWP is global warming potential given as 1.

3 Results and Discussion

Figures 3 and 4 and Tables 3 and 4 display the rate of fuel consumption and CO_2 emissions between the conventional fuels and the different biofuel fractions serving as mitigation measures. The biogenic portion of biofuels played significant role in the amount of fuel consumed and CO_2 emitted. The greater the bio-content the higher the fuel consumption and vice versa. With the fuel increment notwithstanding, fossil CO_2 emissions decreased substantially as presented in the tables and figures when gasoline and diesel were used as base fuels respectively. Bio-CO_2 emissions on the other hand increased significantly as the fossil emissions reduced. Downsizing the bio-content from E100 to E85, E10 and gasoline resulted in 13%, 88% and 100% decrease in Bio-CO_2 while B100 to B50 and BI0 blends as well as pure diesel led to 90%, 98% and 100% reduction in Bio-CO_2 emissions. Biogenic emissions are considered less harmful because biofuels are carbon–neutral—the emitted CO_2 is reabsorbed by plants during photosynthesis (Osei et al., 2020). Similarly, the study by Takagi et al. (2018) observed no fossil-CO_2 emission in 100% biofuels. They posited further that the environmental impacts of biofuels are due entirely to the conventional fuel share in the blend.

Calorific value is a primary determinant of fuel consumption. It describes the energy content of the fuel and represent the total heat released when a unit amount of fuel combusts in the presence of oxygen. Therefore, the higher the calorific value, the higher the fuel efficiency due to the release of greater heat that enhance engine performance during combustion (Wan Ghazali et al., 2015). Biofuels generally have lower calorific values, a primary factor leading to their higher fuel consumption, as observed in this study. This limitation in energy content presents a notable challenge for biofuel adoption, as projections by (Kalghatgi et al., 2018) indicated a substantial use of petroleum products by 2050, despite efforts to shift towards alternative or cleaner fuels. Nevertheless, many European biofuel-specific vehicles are currently designed with improved fuel economy compared to standard conventional cars. This enhanced fuel efficiency is not typically achievable in retrofitted gasoline and diesel vehicles, which were assumed in this study. Rimkus et al. (2024) discovered reduction in efficiency of bioethanol during combustion as the gasoline-

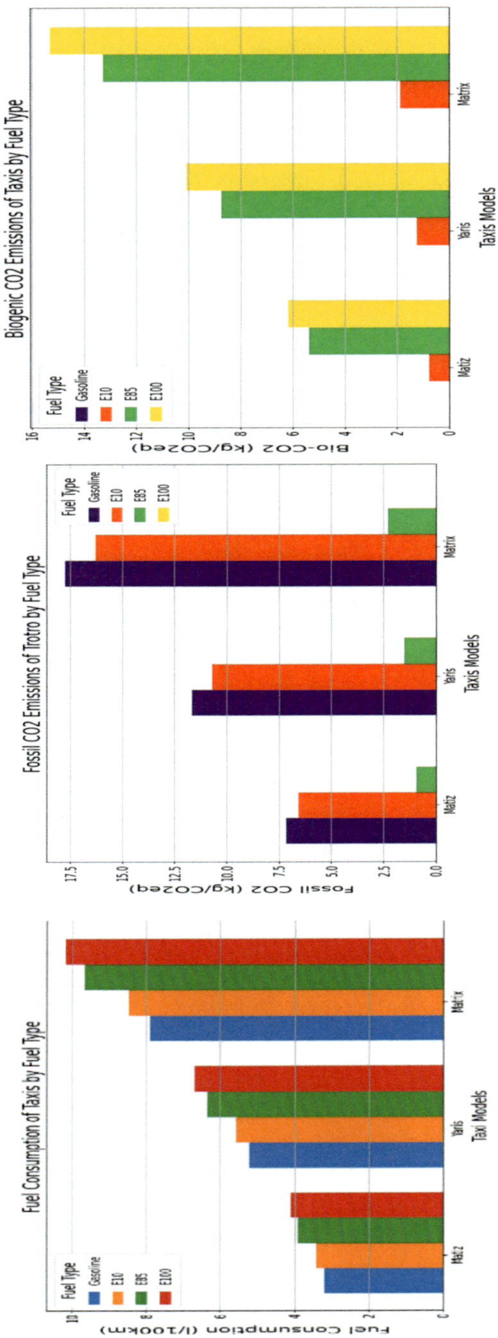

Fig. 3 Bioethanol and gasoline consumption and CO_2 emissions of taxis

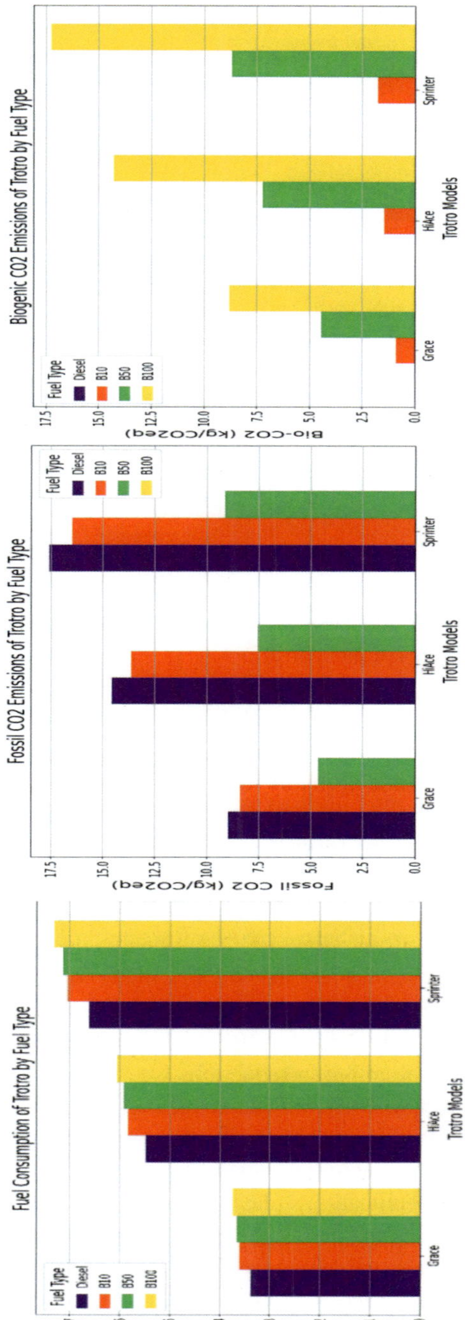

Fig. 4 Biodiesel and diesel consumption and CO_2 emissions of minivans

Table 3 Percentage Bioethanol Increase and FossilCO_2 emission savings

Bioethanol	Fuel increment (%)	Fossil CO_2 savings (%)
E10	7.14	8.22
E85	22.12	87.09
B100	28.78	100

Table 4 Percentage biodiesel increment and Fossil CO_2 emission savings consumption and

Biodiesel	Fuel increment (%)	Fossil CO_2 savings (%)
B10	6.5	6.35
B50	7.82	48.23
B100	10.43	100

retrofitted engine was not optimized to run on fuels with higher octane number—a beneficial characteristic of biofuels. This mismatch ultimately resulted in increased fuel consumption as bioethanol concentrations rose, compared to pure gasoline. Wei et al. (2024) conducted tests on a diesel engine retrofitted for biodiesel use and noticed a substantial increase in fuel consumption as biodiesel concentration increased. This was attributed to the decline in the engine's thermal efficiency resulting from the lower calorific values associated with increasing the percentage of biodiesel blends.

Besides the energy content, the fuel properties as denoted in Table 1 also significantly influence the differences in fuel consumption rates between conventional fuels and biofuels. As shown in Table 1, BSFC is higher in biofuels relative to the fossil-based fuels due to the inverse relationship with calorific value—higher heating value reduces BSFC and vice versa (Thangavelu et al., 2016). BSFC is defined as the ratio of fuel consumed per unit power, and normally used to assess engine efficiency in terms of fuel consumption. According to the study by (Usman et al., 2023) on the effects of bioethanol blends on engine performance, a notable increase in fuel consumption was observed at higher BSFC values, associated with higher blending ratios. Similarly, Guo et al. (2016) accrued the elevated BSFC associated with biodiesel as the key factor in its fuel consumption rate. Additionally, density impacts engine performance characteristics.

As depicted in Table 1, the gradual increase in density with rising biogenic content contributed to the elevated fuel consumption which was also observed by (Guo et al., 2016). In terms of fossil emission reductions, studies like (Guo et al., 2016; Rimkus et al., 2024; Solaymani, 2023; Usman et al., 2023) all reported lower fossil CO_2 emissions with increasing biofuel blending ratios irrespective of the higher fuel consumption which is consistent with findings in this study.

Moreover, the vehicle specifications played crucial roles in the rate of fuel consumption and emissions. From the table, vehicles with smaller engine capacities and mass exhibited better fuel economy. Vehicle mass affects two important forces in physical models: inertia acceleration and rolling resistance forces. Increase in these two forces ultimately augment fuel consumption and its concomitant CO_2 emissions. Using Matiz and Yaris as taxis resulted in 60% and 34% fuel savings when compared to Matrix while Hyundai Grace and HiAce saved 49% and 38% fuel respectively using Sprinter as base vehicle.

3.1 Model Validation

Table and Fig. 5 depict the results from the error analysis performed on the proposed model. The SRC values indicate a strong correlation between the input factors and fuel consumption especially the rolling resistance coefficient which had a perfect relationship with the output. With the exception of the rolling coefficient which was linearly related with the model output (fuel consumption), density and driving speed had inverse relation with the output—increase in these two factors decrease fuel consumption as explicitly illustrated by Fig. 5. Comparably, variability and uncertainty in fuel consumption were higher in driving speed followed accordingly by rolling coefficient with density showing higher level of precision as indicated by the MOE and CI values.

Overall, based on the higher correlation and relatively lower variability and uncertainty, the proposed model is described as robust and effective in attaining the aim of the study.

4 Conclusions and Policy Recommendations

This study aimed to elucidate the impacts of biofuel in the transport sector of Ghana. A physics-based model, incorporating specific vehicle and fuel characteristics were employed to identify the distinctive difference between conventional and biofuels in terms of fuel consumption and CO_2 emissions under the environmental settings of Ghana. The findings depicted the ecological benefits of biofuels relative to standard fuels through the reduction of fossil CO_2 emissions by 8% and 87% in E10 and E85 bioethanol blends as well as 6% and 48% in B10 and B50 biodiesel blends respectively. For 100% bioethanol and biodiesel, fossil CO_2 emissions were null. Albeit, biofuels recorded higher fuel consumption due to their lower energy content compared to conventional fuels, they are still considered as viable fuel alternatives due to their decarbonizing nature. Recent advancement in vehicle technology has improved the fuel efficiency of biofuel-specific vehicles, making them more economical in terms of fuel saving and thus increasing their advantages over standard fuels.

Results also depicted minimal biogenic CO_2 emissions but relatively higher fossil-based emissions in the alternative scenario blends (E10 and B10), limiting their impact on climate change mitigation. Consequently, policies should promote high biofuel blends or pure biofuels (B100 and E100) to maximize reductions in fossil CO_2 emissions. Given that, current technology presents challenges for high-blend biofuels (extreme scenarios), as noted by (Saisirirat et al., 2022), international partnerships

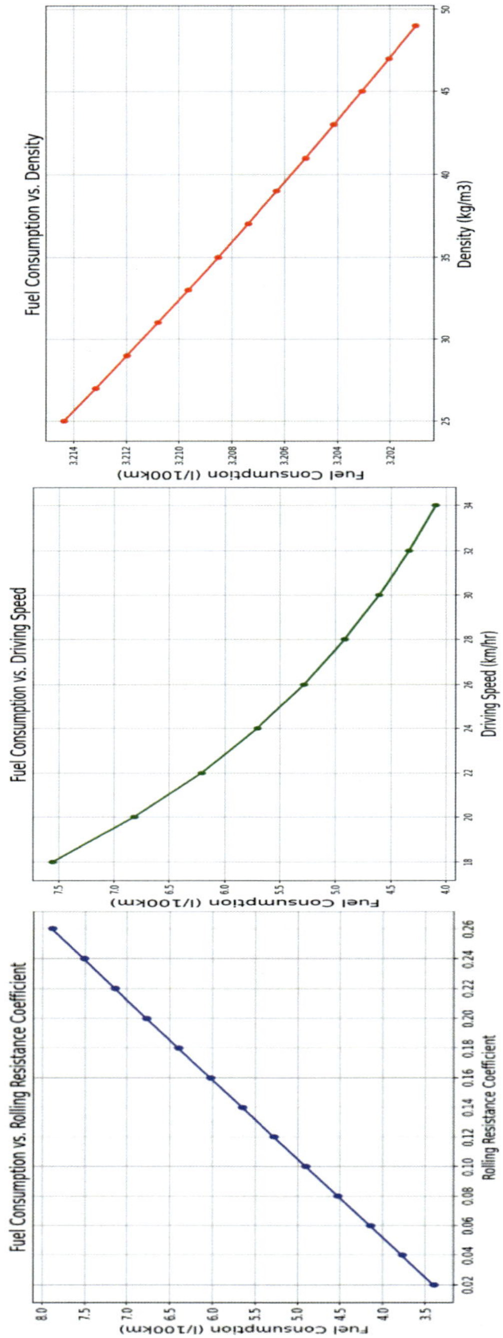

Fig. 5 Uncertainty results of model input factors

Table 5 Error analysis results

Key factors	SRC	MOE	CI
Density	−0.9994	0.0021	3.1810 to 3.1851
Driving speed	−0.906	1.2580	4.4117 to 6.9278
Rolling resistance coefficient	1.00	0.9925	6.1465 to 8.1316

with biofuel-leading countries like Brazil and the USA should be encouraged to support effective biofuel implementation in the transport sector.

Government policies in the form of incentives, such as tax credits for biofuel vehicles, carbon taxation and meager import duties should be promoted to augment the purchase of biofuel-specific vehicles while supporting massive price reductions for biofuels to substantiate the attractiveness of biofuels to consumers. Moreover, lowering taxes for commercial drivers using biofuel-retrofitted vehicles will boost biofuel use in public transport. By incorporating these policies, biofuels can sustainably be implemented within the transport sector to facilitate the attainment of the net-zero emission goals outlined in the Ghana Transition Plan for 2060.

Acknowledgement The authors appreciate the German Federal Ministry of Education and Research (BMBF) and the West African Science Centre on Climate Change and Adapted Land Use (WASCAL) for financially supporting this project.

References

Ben-Chaim, M., Shmerling, E., & Kuperman, A. (2013). Analytic modeling of vehicle fuel consumption. *Energies, 6*(1), 117–127. https://doi.org/10.3390/en6010117.

Duku, M. H., Gu, S., & Hagan, E. B. (2011). A comprehensive review of biomass resources biofuels potential in Ghana. *Renewable and Sustainable Energy Reviews, 15*(1), 404–415.

Ghana Energy Commission. (2022). *National Energy Statistics 2000–2022 edition* (Issue April). http://www.energycom.gov.gh/old/downloads/Energy_Statistics_2011.pdf.

Guo, S., Yang, Z., & Gao, Y. (2016). Effect of adding biodiesel to diesel on the physical and chemical properties and engine performance of fuel blends. *Journal of Biobased Materials and Bioenergy, 10*(1), 34–43. https://doi.org/10.1166/jbmb.2016.1566.

Kalghatgi, G., Levinsky, H., & Colket, M. (2018). Future transportation fuels. *Progress in Energy and Combustion Science, 69*, 103–105. https://doi.org/10.1016/j.pecs.2018.06.003.

Kemausuor, F., Kamp, A., Thomsen, S. T., Bensah, E. C., & Østergård, H. (2014). Assessment of biomass residue availability and bioenergy yields in Ghana. *Resources. Con Serv Recycl, 86*, 28–37.

Kemausuor, F., Nygaard, I., & Mackenzie, G. (2015). Prospects for bioenergy use in Ghana using Long-range Energy Alternatives Planning model. *Energy, 93*, 672–682.

Mehra, K. S., Singh, S., Singh, A. K., Kharkwal, H., & Avikal, S. (2020). Performance, energy, emission and cost analysis of Jatropha (Jatropha Curcas) oil as a biofuel for compression ignition engine. *Materials Today: Proceedings, 43*, 348–354. https://doi.org/10.1016/j.matpr.2020.11.675.

MESTI. (2021). Ghana: Updated Nationally Determined Contribution under the Paris Agreement (2020–2030). *Environmental Protection Agency (EPA) and the Ministry of Environment, Science,*

Technology and Innovation (MESTI), September, 10. https://www4.unfccc.int/sites/ndcstaging/PublishedDocuments/GhanaFirst/Ghana%27s Updated Nationally Determined Contribution to the UNFCCC_2021.pdf

Mohammed, Y. S., Mokhtar, A. S., Bashir, N., & Saidur, R. (2013). An overview of agricultural biomass for decentralized rural energy in Ghana. *Renewable and Sustainable Energy Reviews, 20*, 15–22.

Narayanan, H. (2006). Chaper 1 Introduction. *Annals of Discrete Mathematics, 54*(C), 1–13. https://doi.org/10.1016/S0167-5060(08)70670-8.

Osei, J. A., Manohar, S., & Kitur, E. (2020). Effects of different incubation methods on Ethanol production from selected food wastes products. *Indonesian Journal of Environmental Management and Sustainability, 4*(3), 64–69.

Patel, S., Azad, A. K., & Khan, M. (2019). Numerical investigation for predict diesel engine performance and emission using different fuels. *Energy Procedia, 160,* 834–841. https://doi.org/10.1016/j.egypro.2019.02.150.

Pelizan, L. (2019). *Biofuel Production in Ghana: Exploring the Opportunity. November*. https://doi.org/10.15581/018.OP-320.

Redish, E. F., & Smith, K. A. (2008). Looking beyond content: *Journal of Engineering Education*, 295–307.

Rimkus, A., Pukalskas, S., Mejeras, G., & Nagurnas, S. (2024). Impact of bioethanol concentration in gasoline on SI engine sustainability. *Sustainability, 16*(6). https://doi.org/10.3390/su16062397

Saisirirat, P., Rushman, J. F., & Silva, Kampanart Nuwong, C. (2022). Contribution of road transport to the attainment of Ghana's nationally determined contribution (NDC) through biofuel integration. *Energies, 15,* 880. https://doi.org/10.3390/en15030880

Solaymani, S. (2023). Biodiesel and its potential to mitigate transport-related CO2 emissions. *Carbon Research, 2*(1). https://doi.org/10.1007/s44246-023-00067-z.

Takagi, H., Takahashi, T., & Suzuki, N. (2018). *Welfare Effects of the US Corn-Bioethanol Policy.* https://doi.org/10.1007/978-4-431-54895-9_5.

Thangavelu, S. K., Ahmed, A. S., & Ani, F. N. (2016). Review on bioethanol as alternative fuel for spark ignition engines. *Renewable and Sustainable Energy Reviews, 56*, 820–835. https://doi.org/10.1016/j.rser.2015.11.089.

Usman, M., Ijaz Malik, M. A., Chaudhary, T. N., Riaz, F., Raza, S., Abubakar, M., Ahmad Malik, F., Muhammad Ahmad, H., Fouad, Y., Abbas, M. M., & Kalam, M. A. (2023). Comparative assessment of ethanol and methanol–ethanol blends with gasoline in SI engine for sustainable development. *Sustainability, 15*(9). https://doi.org/10.3390/su15097601.

Wan Ghazali, W. N. M., Mamat, R., Masjuki, H. H., & Najafi, G. (2015). Effects of biodiesel from different feedstocks on engine performance and emissions: A review. *Renewable and Sustainable Energy Reviews, 51*, 585–602. https://doi.org/10.1016/j.rser.2015.06.031.

Wei, C., Jiang, G., Wu, G., Zhou, Y., & Liu, Y. (2024). Effects on of Blended Biodiesel and Heavy Oil on Engine Combustion and Black Carbon Emissions of a Low-Speed Two-Stroke Engine. *Polish Maritime Research, 31*(1), 94–101. https://doi.org/10.2478/pomr-2024-0010.

Renewable Energy and Hydrogen Pathways

Green Hydrogen Economy: Paving the Way for a Just and Inclusive Energy Transition

Amy Kebe

Abstract The shift towards a green hydrogen economy, which involves producing hydrogen using renewable energy sources, is more than just a response to climate change. It presents a significant and timely opportunity to promote sustainable development and create widespread economic prospects, paving the way for a sustainable future for Africa. By 2030, up to 139 million new jobs will be created in the global energy sector, with 38.2 million in the renewable energy sector alone (IRENA & ILO in Renewable energy and jobs: Annual review 2022, International renewable energy agency. Abu Dhabi and International Labour Organization, Geneva, 2022). The transition to renewable energy and the circular economy could generate over 100 million jobs by 2050 (OECD in Green economy and energy transition in emerging markets, 2021). This growth is not limited to the sector but also has the potential to promote gender equality significantly. In 2019, the renewable energy sector employed 11.5 million people (Otanicar in A review of solar hybrid photovoltaic-thermal (PV-T) collectors and systems, 2023), with women making up 32% of the workforce, compared to 22% in the overall energy labour force (IRENA in Renewable capacity statistics 2020 international renewable energy agency, 2020). This data underscores the potential for a just energy transition to bridge gendered gaps that have become systemized (UN in The sustainable development goals report, 2021). However, without immediate and inclusive policy interventions, these opportunities may perpetuate gender disparities in the energy sector, and it is essential to address the gender gap in the energy sector, especially as many new jobs in the green economy will be in traditionally male-dominated sectors and occupations (ILO in World employment and social outlook. Trends, 2023). This paper delves into the intersections between the burgeoning green hydrogen sector and gender equality, proposing actionable strategies to ensure women play an integral role in this transformative industry. It not only scrutinizes current gender dynamics within the renewable energy sector and pinpoints barriers that impede women's participation and advancement but also highlights the significant role women can play in the green hydrogen sector. Through an in-depth literature review and case studies from various countries, the paper underscores best practices and successful models of gender integration in

A. Kebe (✉)
Faculty of the Humanities, English Department, Cheikh Anta Diop University, Dakar, Senegal
e-mail: amy.kebe@ucad.edu.sn

M.-M. Narra and S. Narra (eds.), *African Green Transition Through Innovative Pathways*, World Sustainability Series, https://doi.org/10.1007/978-3-031-87043-9_11

energy projects. A noteworthy example is Solar Sister in Kenya, a social enterprise that has empowered women through clean energy entrepreneurship in East Africa. Solar Sister has trained over 5,000 women entrepreneurs, equipping them with the skills and resources to market solar-powered products and clean cookstoves. This success story not only aids in alleviating energy poverty but also fosters economic empowerment among women, instilling confidence in the potential of gender integration in the green hydrogen sector. Furthermore, the paper presents a comprehensive framework for policymakers and industry stakeholders to embed gender equality principles in developing the green hydrogen economy. This analytical framework ensures a comprehensive approach to gender integration based on a theory of change and intersectionality. Instead of portraying women solely as victims, the approach emphasizes women's pivotal role as protectors of ecosystems and climate initiatives. This line of discussion explores the potential for a just energy transition and highlights ongoing innovative initiatives. Furthermore, it provides insights into alternative pathways and solutions for policymaking.

1 Introduction

1.1 How Are We Doing in Terms of Gender Equality?

1.1.1 Gender Equality and the Energy Transition

The 2023 UN Women and UN DESA report on the progress of gender equality across the 17 Sustainable Development Goals (SDGs) presents a sobering projection. In a worst-case climate scenario, up to 158 million women and girls could be pushed into poverty by 2050 due to climate change (DESA, 2023). Food insecurity, driven by these environmental shifts, may affect 236 million more women and girls. In sub-Saharan Africa, over 340 million women and girls are expected to remain in extreme poverty by 2030 unless significant progress is made (DESA, 2023). These figures reflect an apparent setback in achieving the 2030 Agenda for Sustainable Development and its "Leave No One Behind" (LNOB) commitment.

Despite numerous policies and frameworks aimed at promoting gender equality, systematic inequalities persist. The transition towards green and renewable energy offers a critical opportunity to address these disparities. With 139 million new jobs projected by 2030, including 38.2 million in renewable energy (IRENA & ILO, 2022), the green economy holds the potential to reshape labour markets. Furthermore, the circular economy could create over 100 million jobs by 2050 (OECD, 2021). However, 80% of these jobs will emerge in male-dominated sectors like engineering, manufacturing, and construction (ILO, 2023). This underscores the importance of integrating gender equality into designing and implementing new energy policies to prevent existing inequalities from being perpetuated.

Currently, 32% of the workforce in renewable energy is composed of women, compared to 22% in the broader energy sector (IRENA, 2020). While this indicates progress, barriers to leadership roles and technical jobs remain significant. If a gender-responsive approach is not adopted, the transition to renewable energy risks deepening gender-based disparities, especially for marginalized groups such as youth, displaced persons, and individuals with disabilities.

1.1.2 The Differentiated Impact of the Green Hydrogen Economy

The gendered dimensions of climate change underscore the differentiated needs and roles of men and women in climate protection efforts. Climate change impacts are not neutral—they disproportionately affect regions, age groups, and sexes due to social roles and unequal access to resources. Women's responsibilities within households and communities often increase their vulnerability. For example, droughts and floods place additional burdens on women and girls, often forcing them to travel long distances to fetch water and exposing them to harassment and violence (Afridi et al., 2022).

Natural disasters also reveal gendered vulnerabilities. In Bangladesh, women are more likely to die in floods because social norms discourage them from seeking refuge in emergency shelters where men are present (Ayeb-Karlsson, 2020). Similar patterns exist across Africa, where gender disparities impact women's ability to adapt and build resilience in the face of climate change. Integrating gender-responsive policies becomes crucial as the world shifts towards renewable energy sources such as green hydrogen. Green hydrogen is increasingly seen as a critical pathway to reduce greenhouse gas emissions and achieve sustainable development goals. However, energy transitions alone cannot guarantee gender and social equity unless they address structural inequalities in energy access, employment, and decision-making.

A just energy transition requires more than creating new jobs—it demands that women and marginalized groups have fair access to these opportunities. Without deliberate interventions, the transition risks reproducing the same power dynamics that have excluded women from economic growth in the past. Energy access and usage must reflect the diverse needs of all individuals, ensuring an equitable transition for both women and men.

1.1.3 Rationale for a Gender-Just and Equitable Energy Transition

Achieving a gender-just energy transition is essential for sustainable development. Clean and renewable energy, combined with gender equality, forms the foundation for achieving the 2030 Agenda for Sustainable Development, particularly SDG 5 (Gender Equality) and SDG 7 (Affordable and Clean Energy). Without access to sustainable energy, it will be impossible to empower all women and girls and promote economic and social equity (UNDP, 2016; UN Women, 2017).

Gender equity must be embedded within energy policies to mitigate women's vulnerabilities and leverage their capacity for positive change. Studies show that energy poverty disproportionately affects women and girls, limiting their health, education, and economic opportunities. For instance, 759 million people still live without access to electricity, and 2.6 billion lack clean cooking solutions. These gaps not only impact public health—with 3.2 million deaths annually from household air pollution—but also restrict women's participation in economic activities. Expanding clean energy technologies can significantly reduce these burdens and promote gender equity. Initiatives like Solar Sister in East Africa demonstrate the transformative potential of women's leadership in clean energy entrepreneurship. Such projects highlight the importance of gender inclusion in the green economy, showing that women can drive sustainable solutions and create economic opportunities for themselves and their communities.

1.1.4 Towards an Inclusive Energy Transition

A gender-just energy transition requires a comprehensive framework that addresses the barriers women face in accessing and benefiting from new energy systems. This includes targeted recruitment, mentorship programs, and leadership development initiatives to support women's participation in technical fields and decision-making processes. By adopting intersectional approaches—considering poverty, race, disability, and socioeconomic status—energy policies can ensure that marginalized groups are not left behind. Embedding gender-responsive policies in developing the green hydrogen economy will foster resilient, inclusive societies. The transition to green energy can catalyse sustainable development by prioritizing social equity and economic empowerment. Policymakers, industry stakeholders, and community leaders must work together to ensure that the shift to renewable energy reflects the needs and aspirations of all individuals, particularly women and other vulnerable groups.

2 Methodology

2.1 Gender-Just Energy Transition: Definitions

A gender-just energy transition embeds gender-responsive approaches into energy policies and practices. According to UN Women and UN Environment, such policies:

Identify and address gender gaps in the energy sector.

Encourage women's participation in decision-making processes.

Implement targeted actions to tackle gender-related challenges.

The ILO adds that ensuring a just transition—particularly for individuals facing intersecting discrimination—has the potential to transform gender roles (ILO,

2022e). In the shift to low-carbon economies, gender equality plays a critical role, requiring:

Fair access to energy and decision-making platforms.

Gender-specific policies to ensure equitable outcomes.

Addressing intergenerational justice concerns and considering how policy decisions impact future generations (IPCC, 2014).

The IPCC's sixth assessment report outlines essential components for a just transition:

Investments in low-emission, labour-intensive sectors.

Early assessment of the social and employment impacts of climate policies.

Social dialogue and consultation with stakeholders.

Creation of decent jobs with labour policies and workers' rights.

Fairness in energy access and use.

Training and retraining programs for workforce development.

Promotion of gender-specific policies to foster equitable outcomes.

Encouragement of international cooperation and multilateral actions.

2.2 *Research Focus: Gender Equality and the Green Hydrogen Sector*

This paper draws from over 12 years of experience with World Bank and UNEP projects in the Senegalese River Basin. It explores how gender equality and the green hydrogen economy intersect, focusing on ensuring women's active participation in this transformative industry. The discussion highlights the gendered impacts of climate change and the barriers that impede women's participation in the energy transition. Rather than presenting women as passive victims, the paper emphasizes their role as ecosystem protectors and critical contributors to climate initiatives.

2.2.1 Critical Areas of Analysis and Integration

The paper investigates the differentiated impacts of the energy transition on men and women and examines the barriers women face in participating in the green hydrogen sector. The research identifies:

- Practical needs and strategic interests of both genders.
- A framework for gender integration encompassing:
- Education and workforce development
- Policy and regulatory measures
- Support for women-led enterprises
- Inclusive work environments
- Monitoring and evaluation systems.

2.2.2 Case Studies and Best Practices

The research draws on successful models of gender integration, such as the Solar Sister initiative in Kenya, which empowers women through clean energy entrepreneurship. These case studies highlight:

The essential role of women in the renewable energy sector.

The economic and social benefits of gender-inclusive approaches offer replicable models for the green hydrogen economy.

2.2.3 Intersectionality and Structural Barriers

The study emphasizes the importance of recognizing intersecting factors that impact women, such as poverty, race, indigeneity, disability, and socioeconomic status.

Women in poverty face challenges joining decision-making bodies like energy-related committees.

Limited time and labour resources lead to poorer health outcomes for women and their children.

Aligning with a human rights-based approach and SDG 7 (universal energy access), the paper advocates for equal access to resources and opportunities for all individuals.

2.2.4 Alternative Pathways and Policy Recommendations

This section offers policymakers and industry stakeholders solutions to integrate gender equality principles into the green hydrogen economy. This approach ensures that the transition focuses on sustainability, social equity, and economic empowerment, particularly for women and marginalized groups.

2.3 Analytical Framework

2.3.1 A Bottom-Up Approach Mixed with Intersectionality and a Theory of Change

Addressing gender and social inequalities in African development policies, particularly in energy transition, is essential to achieving an inclusive renewable energy sector. A gender-responsive approach involves closing gender gaps, promoting women's empowerment, and ensuring inclusive decision-making processes. The Gender Expert Group highlights that a transformative approach goes beyond providing access to resources, challenging sociocultural norms, and discriminatory legal systems that perpetuate inequality (GGKP, 2022). This methodology fosters shared power, equitable resource control, and meaningful participation.

A grassroots, intersectional approach is fundamental for ensuring a fair energy transition, recognizing discrimination not only based on gender but also on age, ethnicity, and socioeconomic status. This strategy reshapes gender roles and norms while addressing power imbalances, contributing to a diverse and inclusive workplace. According to the Green Growth Knowledge Partnership (GGKP, 2023), such an intersectional approach draws attention to overlapping inequalities when implementing low-carbon energy systems, ensuring that marginalized groups are not left behind.

2.3.2 Energy Justice and Accessibility

Energy justice promotes equal access to energy while creating new economic opportunities in the sustainable energy sector. This framework emphasizes addressing social concerns such as energy poverty, affordability, distribution, and the needs of vulnerable populations (Tucho, 2019). It aims to identify where injustice is prevalent, understand which segments of society are excluded, and explore ways to correct these disparities (Müller & Neuman, 2021; Jenkins et al., 2016).

A key aspect of energy justice is linking income-generating activities to poverty reduction and ensuring women's participation in energy initiatives. Comprehending local circumstances and prioritizing energy accessibility can improve livelihoods and empower marginalized groups.

2.3.3 Human Rights-Based Approach to Energy Access

Energy access should not be seen as a matter of charity but as a fundamental human right. This research draws on international and regional human rights frameworks, focusing on the UN's Sustainable Energy for All Initiative (launched in 2011). The initiative advocates for universal access to modern energy services, improved energy efficiency, and an increased share of renewable energy by 2030. The rights-based approach ensures that energy policies are designed to meet the diverse needs of marginalized groups, including women, youth, and displaced persons. Governments and international organizations must uphold the rights of individuals by ensuring access to essential energy services. This approach not only promotes gender equity but also addresses the marginalized status of women during the energy transition.

2.3.4 Theory of Change and Intersectionality

This research rejects portraying women as passive climate change victims instead of recognizing them as resilient change agents. Women play critical roles as entrepreneurs, innovators, and decision-makers in the energy sector (Huyer, 2016). Reports from USAID and IUCN highlight how acknowledging women's contributions—rather than viewing them solely as consumers—can unlock new economic opportunities for both men and women in the energy industry. While considering the different needs and circumstances of men and women is essential, the language of victimhood must be avoided. Women should be recognized as vulnerable individuals and active participants in transitioning to sustainable energy.

2.3.5 Intersectionality and Specific Barriers for Marginalized Groups

Gender encompasses a spectrum of identities beyond traditional male–female dynamics. It is essential to account for intersectional factors that compound inequality, particularly for older women, Indigenous communities, and women with disabilities. Research indicates that women in poverty face barriers in joining decision-making bodies, such as village development committees, and experience poorer health outcomes due to time and labour constraints. The UN Women and UNDP 2023 Progress Report (DESA, 2023) shows that older women are more likely to experience poverty and violence than older men. Discrimination based on gender, race, disability, and indigeneity further deepens their marginalization. These intersecting factors must be addressed through targeted project investment and policy interventions to ensure all individuals benefit equally from the energy transition.

2.3.6 Gender-Just Energy Transition Policy and Legal Framework

This section surveys the international and regional human rights frameworks relevant to achieving a gender-just energy transition. As outlined in the background, the unequal impact of energy production, distribution, and consumption across genders is widely acknowledged. Women face challenges accessing and managing energy resources, contributing to energy poverty and more profound socioeconomic inequalities. For instance, households headed by women are disproportionately affected by rising energy costs, which further increases poverty levels. Additionally, the reliance on biomass fuel due to unaffordable energy sources heightens the unpaid care burden on women and girls, negatively impacting their health and livelihoods. A fair transition framework, aligned with global labour standards, can reduce these disparities by ensuring that marginalized groups benefit equitably from the shift to renewable energy. The GGKP Gender Report emphasizes the importance of inclusive frameworks that prevent exclusion during the transition away from fossil fuels and ensure that vulnerable communities can access the benefits of clean energy.

2.4 Key Legal and Policy Frameworks Supporting Gender Equality in the Energy Transition

2.4.1 Convention on the Elimination of All Forms of Discrimination Against Women (CEDAW)

The CEDAW convention is one of the most ratified international treaties, with provisions explicitly addressing equal access to modern energy services. Article 14 of CEDAW focuses on rural women, acknowledging the distinct challenges they face in energy access (UNWomen, 1979). Integrating energy access into gender equality frameworks ensures that rural women can overcome these barriers and participate fully in the energy transition.

2.4.2 Paris Agreement and Gender Action Plan

The 2015 Paris Agreement emphasizes the need for a just transition that creates high-quality employment opportunities while addressing climate change impacts. Under this agreement, countries commit to Nationally Determined Contributions (NDCs), which outline specific goals to reduce emissions. Recent updates to over 100 NDCs reflect a growing emphasis on integrating gender perspectives into crucial policy areas and industries (UNDP, 2023). This focus aims to ensure that climate action is inclusive and equitable.

2.4.3 ECOWAS Policy for Gender Mainstreaming in Energy Access

The ECOWAS Policy on Gender Mainstreaming addresses the challenges of energy poverty in the West African region. It provides tools and indicators to help Energy Ministries align their initiatives with gender equality principles while unlocking the potential of women as crucial actors in the energy transition. The policy aims to achieve universal energy access by 2030 across the ECOWAS region, as detailed in its implementation plan.

2.4.4 The Gender and Energy Compact

The Gender and Energy Compact is a coalition of public and private sector stakeholders developed in partnership with ENERGIA, GWNET, and the UN Industrial Development Organization. Its signatories include governments from Canada, Kenya, Sweden, and other nations, alongside USAID/Power Africa and numerous civil society groups and academic institutions. The compact prioritizes gender equality in the energy transition and aligns with SDG 7 (Affordable and Clean Energy). Its objectives include:

- We are eradicating energy and time poverty and reducing burdens faced by women.
- We are improving access to productive resources for women-owned businesses.
- We are creating decent work opportunities and promoting career advancement for women.
- We provide knowledge, tools, and gender-specific data to support gender-responsive energy policies.

2.4.5 Agenda 2063: The Africa We Want

Agenda 2063 is Africa's blueprint for achieving inclusive and sustainable development. It embodies the principles of Pan-Africanism, unity, and shared prosperity, with a vision to position Africa as a global leader in the future. African leaders recognized the need to redefine priorities beyond independence struggles, focusing instead on social and economic development, regional integration, governance, and peace. Energy transition plays a crucial role in this vision, ensuring sustainable growth that benefits all communities equitably.

3 Results and Discussion

3.1 Gender Dynamics in the Green Hydrogen Sector

This section explores gender dynamics within the renewable energy sector, focusing on the green hydrogen economy. While the sector shows progress toward gender inclusion, persistent barriers remain. According to the International Labour Organization (ILO, 2023), women in green hydrogen face challenges such as limited access to training, gender biases, and a lack of role models. These barriers highlight the need for targeted policies to ensure women benefit equally from the energy transition. The green hydrogen economy represents a key component in global strategies to combat climate change and achieve carbon neutrality by 2050. Green hydrogen is produced through water electrolysis using renewable energy sources like solar, wind, and hydroelectric power. According to the International Energy Agency (IEA, 2021), hydrogen could meet 13% of global energy demand by 2050, making it a cornerstone of future energy systems.

3.1.1 Gender Inequality in the Energy Sector

Despite the growth of the renewable energy industry, gender inequality persists. According to IRENA (2020):

Women constitute 39% of the global labour force.

They represent 22% of the workforce in the traditional energy sector.

Women comprise 32% of the workforce in the renewable energy sector.

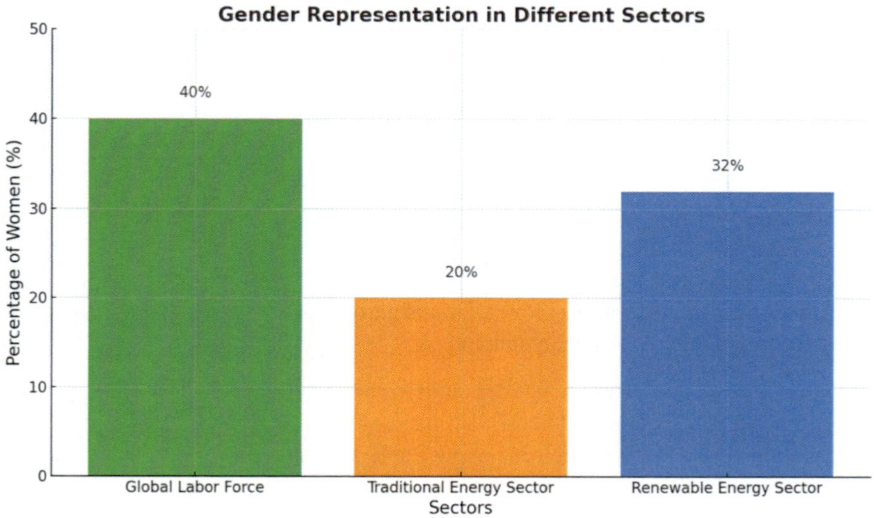

Fig. 1 Gender representation in different sectors of the energy industry

These figures reflect both progress and persistent challenges. The renewable energy sector is relatively more inclusive than traditional energy industries, but achieving gender parity remains challenging. A focused effort is needed to close these gaps, particularly in emerging fields like green hydrogen.

Figure 1 illustrates gender representation across three sectors:

Global labour force: 40% women.

Traditional energy sector: 20% women.

Renewable energy sector: 32% women.

These disparities demonstrate that while women have made progress in the workforce, they remain underrepresented in energy-related fields, particularly in the traditional energy sector. Factors contributing to this gap include:

Cultural norms and stereotypes about gender roles.

Limited access to education and training in engineering and energy-related fields.

Biases in recruitment practices within male-dominated industries.

The higher percentage of women in the renewable energy sector can be attributed to progressive policies promoting diversity and inclusion. Additionally, new roles within renewable energy are less influenced by traditional gender norms, enabling greater participation by women.

3.1.2 Gender and Social Equity in Energy Transitions

Achieving gender and social equity is essential for a just energy transition. Gender-responsive policies can reduce women's socioeconomic vulnerabilities and harness their ability to drive positive change. The shift toward renewable energy offers opportunities to enhance health, safety, and economic well-being for both women and men (Kuriakose et al., 2017). However, the lack of access to clean energy hinders sustainable development, especially in Africa and South Asia.

One-third of the global population lacks access to clean energy, causing nearly four million deaths annually due to household air pollution.

Seven hundred fifty-nine million people live without access to electricity, and 2.6 billion lack clean cooking solutions.

This energy poverty disproportionately affects women and girls, limiting their healthcare, education, and economic opportunities. Addressing these disparities requires comprehensive strategies, including:

- Policy reforms that promote gender equity in the energy sector.
- Gender-sensitive recruitment and training programs.
- Workplace cultures that foster inclusivity and diversity.

3.1.3 Promoting Women's Leadership and Inclusion in Renewable Energy

The renewable energy sector shows potential as a model for gender inclusion, with higher female participation compared to traditional energy industries. This progress reflects the sector's alignment with sustainability goals emphasizing social equity. The relative openness to gender diversity in renewable energy stems from:

- Innovative policies focused on diversity and inclusion.
- A shift toward new roles that challenge traditional gender norms.
- Recognition of the value of diverse perspectives in promoting innovation and sustainability.

Promoting women's leadership, employment, and entrepreneurship in the energy sector is critical to achieving gender equality and ensuring the energy transition's success. Comprehensive strategies are needed to address barriers and promote inclusivity, including:

- Targeted recruitment efforts focused on women.
- Mentorship programs to support career advancement.
- Gender-sensitive policies to create inclusive work environments.

Women in Green Hydrogen (WiGH) represent a more disaggregated representation of women within the energy sectors. Founded in 2020, Women in Green Hydrogen (WiGH) is a network of ambitious women who want to connect, empower, and boost the visibility of women in the sector. As evidenced by this higher percentage of women compared to the traditional energy sector, the renewable energy sector shows promising progress towards gender inclusion. This sector's relative openness to gender diversity may be due to its innovative nature and alignment with global sustainability goals emphasizing social equity. This progress underscores the potential of renewable energy as a model for other sectors seeking to improve gender representation.

However, persistent barriers remain, suggesting that significant gaps remain while renewable sectors are more inclusive. Studies by the International Labour Organization (ILO, 2023) show that women often face additional barriers in sectors like green hydrogen, including lack of access to training, gender biases, and a shortage of role models. As discussed in the background section, the unequal impact of energy production, distribution, and consumption on different genders is widely acknowledged. Women face challenges in accessing and managing resources and services within the energy sector, leading to inequalities. For example, households headed by women bear a disproportionate burden from increasing energy costs, which contributes to higher levels of poverty. Additionally, women are more vulnerable to energy poverty due to rising energy prices, leading to a reliance on biomass for fuel. This unevenly impacts the unpaid care and domestic responsibilities of women and girls, as well as their health and livelihoods.

Another discrepancy is related to women's representation in leadership roles; even though women make up about 32% of the workforce in renewable energy, compared to 22% in the oil and gas sectors, they are underrepresented in senior leadership roles within the sector.

The chart in Fig. 2 shows a gender imbalance in senior leadership positions in renewable energy companies. Men hold 82.0% of these roles, indicating a significant majority, while women only hold 18.0% of senior positions. This suggests that women are underrepresented in senior leadership within this sector. As the energy industry transitions towards more sustainable practices, including developing the green hydrogen economy, it is crucial to integrate gender considerations from the outset. The renewable energy sector's experience demonstrates that proactive measures can significantly enhance gender representation. Applying these lessons to the green hydrogen sector can help avoid the pitfalls experienced in traditional energy and foster a more inclusive, equitable, and innovative industry.

The glass ceiling for women is most challenging to break at the highest echelon of senior management. Women hold less than 20% of board seats and leadership positions in the renewable sector, similar to trends across the broader energy industry(IEA)(Sustainable Energy for All | SEforALL). This disparity highlights the need for advocacy and policy support to address the underrepresentation of women in senior leadership positions in renewable energy companies. The chart illustrates a significant gender gap in senior leadership positions within renewable energy companies, emphasizing the ongoing need for advocacy and policy support to shatter the glass ceiling.

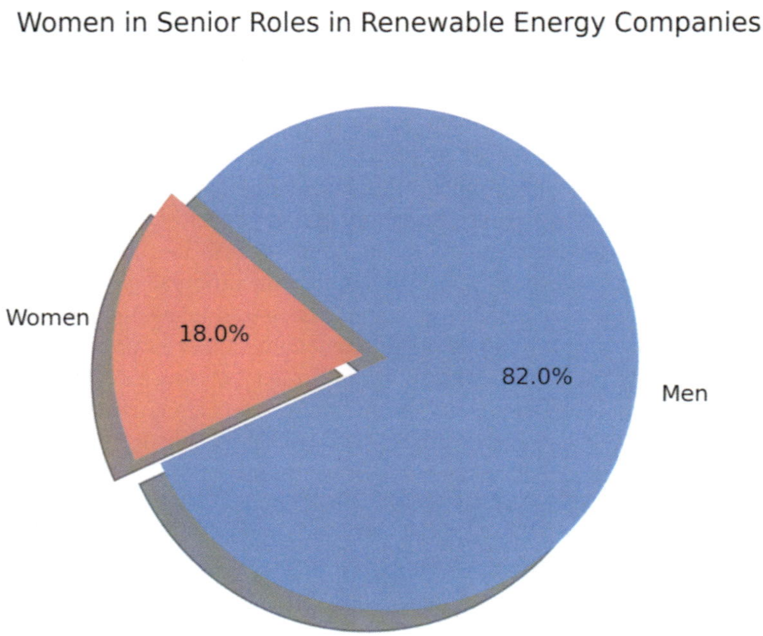

Fig. 2 Women in senior roles in renewable energy companies

3.2 Barriers and Challenges for Gender Mainstreaming in the Energy Transition, with a Focus on Green Hydrogen

This section examines the barriers hindering women's participation and advancement in the energy transition, particularly within the green hydrogen economy. Despite mandates such as the Paris Agreement emphasizing gender equality in just transition policies, gender gaps persist across sectors. Although gender mainstreaming is gaining recognition, implementation remains limited, especially in regions like Asia and the Pacific, due to legal, social, and financial constraints. Women face barriers in accessing resources, financial services, and infrastructure, restricting their ability to benefit from energy initiatives. Legal constraints on land ownership, credit access, and economic decision-making limit women's opportunities to acquire equipment and improve productivity. Addressing these regulatory hurdles is essential to ensure that energy services are inclusive and tailored to the needs of both genders. Raising awareness and collecting gender-disaggregated data will be crucial for tracking progress toward energy equity.

3.2.1 Unpaid Care Labour

During the 68th session of the Commission on the Status of Women, it was emphasized that the disproportionate burden of unpaid care work, places women at an economic disadvantage. Women often serve as household primary caregivers, restricting their workforce participation and exacerbating economic vulnerability. This imbalance affects women's economic empowerment and has broader societal implications. Balancing caregiving responsibilities across genders is essential for promoting career advancement and reducing financial strain. Shifting these societal norms is critical for creating a more equitable and sustainable future.

3.2.2 Limited Availability of Gender-Disaggregated Data

The absence of gender-disaggregated data poses a significant challenge, particularly in developing countries. With such data, tracking progress or designing effective policies for energy equity becomes easier. Mandatory gender impact assessments should accompany policy proposals to ensure accountability. Systematic collection and reporting of gender-specific data across sectors will facilitate fair interventions and empower women in energy-related industries.

3.2.3 Inadequate Understanding of Gender Mainstreaming

Gender mainstreaming is often poorly understood and inconsistently applied across ministries. Budget cuts frequently target gender programs, further undermining progress. Simply increasing women's representation in decision-making without providing capacity-building opportunities is insufficient. Structural barriers in policymaking often ignore gender-specific needs, perpetuating discriminatory practices. Compulsory capacity-building and training programs are essential to ensure meaningful participation.

3.2.4 Gendered Access and Benefits from Renewable Energy Systems

Studies show that energy transitions affect men and women differently. For instance, in rural Peru, renewable electrification provided men with more leisure time, while women used their time to engage in income-generating activities or household duties (Fernández-Baldor et al., 2015). Although new energy systems can create opportunities, social norms often constrain women's participation beyond traditional roles. Energy transitions must address these underlying norms to prevent inequalities from shifting into new areas. For example, while solar and biofuel energy has reduced the burden of activities like firewood collection, women have taken on new responsibilities, such as tending to livestock (Fernández-Baldor et al., 2015).

3.2.5 Land Grabbing and Displacement

Introducing clean energy projects such as hydropower, solar farms, and biofuel plantations can lead to land grabbing and displacement of rural communities. These projects often prioritize environmental benefits over the well-being of local populations. Gendered impacts arise during displacement, as men tend to secure new jobs more quickly than women, increasing women's dependency on male relatives.

3.2.6 Social and Cultural Norms

Gender and social norms shape women's ability to express their energy needs and participate in decision-making. These norms are often overlooked in corporate energy policies. However, research shows that companies with greater gender diversity perform better and empower women more effectively (ENERGIA, 2019a). Overcoming historical, social, and cultural barriers is essential for enhancing women's participation in the energy sector.

Female entrepreneurs need additional challenges, including limited access to information on energy technologies and difficulty obtaining credit. Gender biases often steer women toward less lucrative micro-enterprises, reinforcing inequalities. Mentorship programs and equal access to training opportunities are essential for breaking these barriers. Microcredit and other financial services will enable women to scale their businesses from informal ventures to formal enterprises in energy-related fields.

3.2.7 Gender Technological and Pay Gaps

According to the UN Women Gender Snapshot (2023), women represent 22% of the energy sector workforce and 32% of the renewable energy workforce. However, women still need to be represented in technical and leadership roles. In 2022, only 17% of inventors in international patents were women (DESA, 2023). While the renewable energy sector offers opportunities to close the gender gap, pay disparities persist. Most women in the sector hold administrative roles, while men dominate technical positions. Facilitating women's transition to formal sector employment requires understanding labour market dynamics and targeted efforts to address structural barriers.

3.3 *Strategies for Enhancing Gender Mainstreaming Within the Energy Transition*

This section focuses on strategies to promote gender mainstreaming within the energy transition, with insights from the Integrated Water Resources Management Program Phase II (PGIRE II, 2014–2023.

3.3.1 Case Study: PGIRE II—Integrated Water Resources Management in the Senegalese River Basin

PGIRE II is a multi-sectoral initiative led by the Organization for the Development of the Senegal River (OMVS), involving Guinea, Mali, Mauritania, and Senegal. The project sought to alleviate poverty and promote social inclusion by providing reliable renewable energy in the region. Many development projects adopt a gender-neutral stance, integrating gender considerations only in response to donor requirements (e.g., from the Global Environment Facility, GEF). In contrast, PGIRE II embedded gender mainstreaming throughout the project lifecycle to ensure equity and inclusion.

The strategy involved:

- Participatory design processes with community stakeholders.
- Identifying gender-sensitive objectives, indicators, and activities.
- Establishing monitoring and evaluation systems aligned with Nationally Determined Contributions (NDCs).

Due to funding limitations, the project could not conduct a baseline survey. However, periodic household surveys and monitoring frameworks ensured that gender considerations were integrated across components. The project emphasized liquid biofuels, pico-hydro technologies, and other renewable solutions, targeting the needs of women, men, and vulnerable groups.

3.3.1.1 Outcomes from PGIRE II: Linking Gender Equality and Environmental Sustainability

The PGIRE II project demonstrates how integrating gender-sensitive strategies can promote social inclusion and improve the sustainability of energy initiatives.

Key outcomes include:

- Promoting Climate Adaptation:
- Distributed improved cookstoves to reduce biomass dependence and alleviate women's workload
- Advancing Reforestation and Agriculture:
- Engaged women in agro-sylva-pastoral practices and sustainable gardening to prevent soil erosion and access carbon funds
- Supporting Microcredit Programs:

- Trained women in the carbonization of *Typha australis* using the "3 barrels" technology to foster sustainable energy production
- Enhancing Land Restoration Efforts:
- Trained women in composting techniques to restore degraded land and support reforestation efforts
- Building Community Resilience:
- Facilitated community demonstrations and action research to increase resilience against climate change impacts

These initiatives illustrate how gender-sensitive approaches lead to economic empowerment, improved livelihoods, and enhanced community well-being.

3.3.1.2 Challenges in Implementing Gender Mainstreaming in Energy Projects

Despite progress, barriers to gender mainstreaming persist. Key challenges identified include:

- Fragmented Understanding of Gender Mainstreaming:
- Gender policies are often disconnected across ministries and vulnerable to budget cuts
- More than simply increasing women's representation without capacity-building is required for meaningful progress
- The Burden of Unpaid Care Work:
- Women's disproportionate burden of unpaid care labour limits their economic participation and deepens their vulnerability
- Balancing caregiving responsibilities is essential to narrow gender gaps and promote career advancement
- Limited Availability of Gender-Disaggregated Data:
- The need for more sex-disaggregated data obstructs progress tracking and policymaking efforts
- Mandatory gender impact assessments and regular data collection are essential to ensure accountability.

The PGIRE II case study demonstrates that gender-sensitive approaches embedded within energy projects promote economic empowerment and social inclusion. By addressing the specific needs of women and marginalized groups, energy initiatives become more effective, sustainable, and aligned with climate adaptation goals.

3.4 Lessons Learned and Best Practices for Gender Inclusion in the Green Hydrogen Sector

This section explores the lessons learned from existing gender-inclusive initiatives and provides best practices applicable to the green hydrogen sector. Drawing from the *Intersectionality and Energy Transitions Report* and initiatives such as Solar Sister and ENERGIA it demonstrates how gender-sensitive approaches can empower women, promote social equity, and enhance sustainability within energy transitions (Johnson et al., 2020).

3.4.1 Findings from the Intersectionality and Energy Transitions Report

The report "*Intersectionality and Energy Transitions: A Review of Gender, Social Equity, and Low-Carbon Energy*" offers an in-depth review of gender and social equity in clean energy systems. It underscores the intersection of gender, class, race, and geography in shaping access to energy and participation in renewable sectors. Historically, biomass fuels have dominated energy consumption in low-income countries across Africa and Asia, with women managing the bulk of household energy use. As much as 90% of household energy has come from traditional sources like firewood, emphasizing women's significant, though often overlooked, role in energy systems. The report also highlights the transition from reproductive roles—where women use energy for household tasks—toward productive roles, including entrepreneurship in renewable energy markets (Johnson et al., 2020).

These findings emphasize that gender-inclusive policies can foster women's participation in emerging sectors like green hydrogen, promoting economic growth and social equity.

3.4.1.1 Case Study: Solar Sister and ENERGIA—Models of Gender Inclusion

The Solar Sister initiative and the ENERGIA network exemplify how gender-sensitive strategies can enhance social inclusion and economic empowerment in the energy sector.

3.4.2 Solar Sister—Empowering Women Entrepreneurs in East Africa

Solar Sister, operating across East Africa, empowers women through clean energy entrepreneurship. It provides training in sales, distribution, and maintenance of solar products and clean cookstoves, targeting women from rural communities with limited economic opportunities. Solar Sister equips women to become entrepreneurs and role models within their communities through mentorship and skills development, driving energy access and economic empowerment. Solar Sister's initiative improves access to sustainable energy and reduces health risks by lowering dependence on polluting fuels such as kerosene and wood. This dual economic and environmental impact offers a replicable model for the green hydrogen economy.

3.4.3 ENERGIA—Pioneering Gender Integration in Energy

ENERGIA, a global network focused on gender and sustainable energy, has promoted gender equality across the renewable energy sector. Its initiatives include:

- Smoke-free cookstoves fuelled by sustainably harvested biomass.
- Solar home systems are linked to efficient appliances that reduce indoor pollution and enhance energy access.

Solar Sister and ENERGIA demonstrate that gender-inclusive approaches can address energy poverty while fostering entrepreneurship and leadership among women.

3.4.4 Global Initiatives Supporting Gender Diversity in the Green Hydrogen Sector

Several international initiatives provide valuable lessons on gender integration within the green hydrogen economy:

- WiGH Network—Global
  The Women in Green Hydrogen (WiGH) network promotes women's visibility and participation in the hydrogen sector through webinars, workshops, and mentorship programs.
- Port of Rotterdam Hydrogen Initiative—Netherlands
  This initiative emphasizes gender diversity in project leadership, illustrating the value of inclusive decision-making in advancing hydrogen infrastructure.
- ITM Power—United Kingdom
  ITM Power supports gender equality through diverse leadership teams and targeted recruitment strategies, fostering an inclusive environment for women in STEM roles.
- ARENA—Australia
  ARENA funds hydrogen projects focusing on gender inclusion, providing scholarships, internships, and mentorship programs to support women in the renewable energy sector.
- H2 MOBILITY Deutschland—Germany
  This joint venture emphasizes diversity in developing hydrogen refuelling stations, showcasing the importance of gender-balanced teams.
- Hydrogen Council—Global
  The Hydrogen Council encourages women's participation through mentoring programs and leadership development initiatives.

- Air Liquide—France
  Air Liquide integrates gender considerations into its hydrogen research and development efforts, promoting diversity in technical and leadership roles.

3.4.5 Strategies for the Green Hydrogen Sector: Applying Lessons Learned

Building on insights from Solar Sister, ENERGIA, and the *Intersectionality and Energy Transitions Report*, the following strategies can enhance gender inclusion within the green hydrogen sector:

- Tailored Training and Capacity Building:
  Provide technical training and mentorship programs targeting women in hydrogen production and distribution roles.
- Creating Market Opportunities for Women:
  Support women-led enterprises by ensuring access to procurement and supply chains.
- Promoting Women's Leadership:
  Engage women as ambassadors and educators to foster community acceptance and adoption of hydrogen technologies.
- Inclusive Policy Development:
  Involve women in decision-making processes to align projects with the needs of all community members.

3.4.6 Challenges to Overcome

Despite progress, significant challenges remain in achieving gender equity in the green hydrogen sector:

- Lack of Gender-Disaggregated Data:
  Systematic collection of sex-disaggregated data is essential for tracking progress and informing policy decisions.
- Social and Cultural Norms:
  Deep-rooted gender biases limit women's participation in technical roles. Ongoing awareness campaigns are crucial to shifting these norms.
- Limited Access to Finance and Resources:
  Strengthening financial inclusion through microcredit programs can support women entrepreneurs and foster innovation.

The experiences from Solar Sister, ENERGIA, and global initiatives demonstrate that gender-inclusive policies in the green hydrogen sector can drive economic growth, social equity, and environmental sustainability. By applying these best practices, the sector can promote a just transition that empowers all individuals, ensuring no one is left behind.

4 Conclusion

Promoting gender equality within the green hydrogen economy is essential for building a sustainable, inclusive, and transformative energy future. As green hydrogen emerges as a critical solution to combat climate change, drive economic growth, and reshape the energy landscape, it offers a unique opportunity for Africa to achieve a sustainable future. However, for this transition to succeed, it must actively address existing gender disparities and ensure that all individuals, especially marginalized groups, benefit equitably from the sector's opportunities. Gender equality and sustainable development are inextricably connected. Integrating gender and social equity is not an afterthought but a cornerstone for achieving fair energy transitions. Ensuring equal access to jobs, leadership roles, and entrepreneurial opportunities within the green hydrogen economy is fundamental for empowering women and men. Recognizing and addressing energy transitions' different social and economic impacts is crucial for fostering innovation, economic progress, and community resilience.

A holistic approach is required to tackle structural power imbalances and drive meaningful change. This includes designing and implementing gender-responsive policies that promote fairness, inclusivity, and participation across all stages of the energy transition—from policy development to project implementation. Integrating gender equality into these frameworks enhances climate governance, strengthens social frameworks, and unlocks the full potential of green hydrogen technologies. Creating an inclusive energy transition demands the active involvement of diverse stakeholders—including women, men, youth, the elderly, displaced persons, and individuals with disabilities. Policies and programs that reflect the needs and aspirations of these groups foster more robust community support and ensure that everyone can benefit from the emerging green economy. Cooperation among governments, development organizations, and civil society is essential to overcome barriers—such as legal and financial hurdles—that limit women's access to credit, essential equipment, and energy services.

Achieving gender equality in the energy sector has a far-reaching impact beyond economic growth. It contributes to eradicating poverty and hunger, promoting peace and security, and advancing the broader sustainable development goals. A just and inclusive energy transition strengthens societal frameworks by empowering women, fostering equality, and ensuring that the benefits of renewable energy are shared equitably. This comprehensive approach is essential for the success of the green hydrogen sector and the realization of the 2030 Agenda for Sustainable Development. At the heart of this agenda lies SDG 5, which focuses on achieving gender equality and empowering all women and girls, and SDG 7, which aims for universal access to affordable, reliable, and sustainable energy. Addressing gender disparities within the green hydrogen economy aligns directly with these goals while supporting climate resilience and environmental sustainability (SDG 13 and 15).

However, many of the new job opportunities emerging within the green economy—up to 80%—are concentrated in traditionally male-dominated sectors

(ILO, 2023). Therefore, the energy transition must include proactive efforts to integrate women and marginalized groups into these fields, ensuring they benefit equitably from new economic opportunities. Gender-sensitive recruitment, inclusive workforce strategies, and supportive workplace policies will be essential in bridging gender gaps and fostering a diverse and resilient energy workforce. Achieving gender equality in the green hydrogen sector also requires addressing the systemic barriers women face, including limited access to finance, credit, and technical training. Collaborative efforts between stakeholders, governments, and civil society are critical to ensure that policies promote equity and social inclusion, enabling women to thrive as entrepreneurs, innovators, and leaders in the energy transition.

Ultimately, gender equality is not optional but fundamental to the success of the green hydrogen transition and achieving sustainable development. By placing gender equality at the centre of this shift, we can ensure that the benefits of clean energy extend to all segments of society, promoting economic progress, environmental sustainability, and social equity. Ignoring the importance of gender equity risks undermining the potential of the green hydrogen economy and the broader development agenda. The way forward requires collaborative efforts, gender-sensitive policies, and an unwavering commitment to inclusion and fairness. Together, we can build a more just and sustainable future where everyone has the opportunity to thrive in a world powered by renewable energy.

References

Afridi, F., Mahajan, K., & Sangwan, N. (2022). The gendered effects of droughts: Production shocks and labor response in agriculture, Labour Economics, Volume 78, 2102227, ISSN 0927-5371. https://doi.org/10.1016/j.labeco.2022.102227.

Air Liquide. (n.d.). Gender diversity in hydrogen research and development. Retrieved from [Air Liquide website].

Australian Renewable Energy Agency (ARENA). (n.d.). *Gender diversity in renewable energy*. They were retrieved from [ARENA website].

Ayeb-Karlsson, S. (2020). When the disaster strikes: Gendered (im)mobility in Bangladesh, Climate Risk Management, Volume 29, 100237, ISSN 2212-0963. https://doi.org/10.1016/j.crm.2020.100237.

Cecelski, E., & Oparaocha, S. (2023). The lack of gender targets for clean energy is harming women and girls, SDG action. Retrieved October 25, 2024, from https://sdg-action.org/the-lack-of-gender-targets-for-cleanenergy-is-harming-women-and-girls.

DESA. (2023). Progress on the sustainable development goals, the gender snapshot 2023, UN Women Headquarters Office, ISBN 9789210029063.

Fernández-Baldor, Á., Lillo, P., & Boni, A. 2015. Gender, energy, and inequalities: A capabilities approach analysis of renewable electrification projects in Peru. In: Hostettler, S., Gadgil, A., Hazboun, E. (eds) Sustainable Access to Energy in the Global South. Springer, Cham. https:/doi.org/10.1007/978-3-319-20209-9_1.

GGKP. (2022). Sohna Ngum, Luisa Kim, Working paper: Powering a gender-just energy transition. GGKP Expert Group on Gender.

GGKP. (2023). Sohna Ngum, Luisa Kim, Working paper: Powering a gender-just energy transition. GGKP Expert Group on Gender.

H2 MOBILITY Deutschland. (n.d.). *Hydrogen refuelling network and gender equality*. Retrieved from [H2 MOBILITY website].

Huyer, S. (2016). *Closing the gender gap in agriculture..*

Hydrogen Council. (n.d.). *Promoting gender equality in the hydrogen sector*. Retrieved from [Hydrogen Council website].

IEA. (2021). Global Hydrogen Review, IEA, Paris https://www.iea.org/reports/global-hydrogen-review-2021, (last accessed: 06.06.2025).

ILO. (2017). s.f. World social protection report 2017–19: Universal social protection to achieve the sustainable.

ILO. (2023). s.f. *World employment and social outlook*: Trends 2023.

IPCC. (2014). *Climate change 2014: Synthesis report.*

IRENA and ILO. (2022). *Renewable energy and jobs: Annual review 2022, International renewable energy agency*. Geneva: Abu Dhabi and International Labour Organization.

IRENA. (2020). *Renewable capacity statistics 2020 international renewable energy agency.*

ITM Power. (n.d.). *Diversity and inclusion initiatives*. Retrieved from [ITM Power website].

Jenkins, K., McCauley, D., & Heffron, R. (2016). Energy justice: A conceptual review. *Volume 11, January 2016, Pages 174–182* (Energy Research & Social Science) Volume 11 (January, 174–182): Volume 11, January 2016, Pages 174–182.

Johnson, O. W., Han J. Y-C., Knight, A-L., Mortensen, S., Aung, M. T., Boyland, M., & Resurrecci6n, B. P. (2020). Intersectionality and energy transitions: a review of gender, social equity and low-carbon energy. *Energy Research & Social Science. 70*.https://doi.org/10.1016/j.erss.2020.101774.

Kuriakose, S., & y Nelson, A. T. (2017). *Gender and renewable energy: Entry points for women's livelihoods and employment*. CLIMATE INVESTMENT FUND.

Müller, F., Neuman, M. (2021). Assessing African energy transitions: Renewable energy policies, energy. *Politics and Governance, 9*(1), 119–130.

OECD. (2021). Green economy and energy transition in emerging markets.

Otanicar, T. (2023). A review of solar hybrid photovoltaic-thermal (PV-T) collectors and systems.

Port of Rotterdam. (n.d.). Hydrogen projects and gender diversity. Retrieved from [Port of Rotterdam website].

Tucho. (2019). Gudina terefe, the impacts of policy on energy justice in developing countries. (Energy et al.) (19 October).

UN, (2021). s.f. *The sustainable development goals report 2021.*

UNDP. (2016). *UNDP support to the integration of gender equality across the sdgs including goal 5.*

UNDP. (2017). A review of energy policies in East and Southern Africa.

UNWomen. (1979). Convention on the Elimination af All Forms of Discrimination against Women, available online: https://www.un.org/womenwatch/daw/cedaw/text/econvention.htm (last accessed: 06.06.2025)

Women in Green Hydrogen (WiGH) Network. (n.d.). Retrieved from [WiGH website].

Systematic Literature Review of Hydrogen Production Technologies, Feedstocks and Pathways, and Its Prospects in Sub-Saharan Africa

Robinson J. Tanyi, Lena D. Mensah, Augustine Ntiamoah, David A. Quansah, and Muyiwa S. Adaramola

Abstract This study aims to address the data gaps hindering the development of a hydrogen market in sub-Saharan Africa (SSA) by conducting a systematic literature review on hydrogen production technologies, feedstocks, and pathways. Through analysis, practical avenues for further research and development of hydrogen production in the region are identified, considering financial and environmental factors. After systematically screening four (4) databases (Google Scholar, Scopus, Science Direct and Springer) following the PRISMA Protocol, 62 articles meeting all inclusion criteria underwent full review and quantitative analysis. Results indicate that 80% of articles identified steam methane reforming (SMR) of natural gas as economically viable but environmentally polluting. Meanwhile, 60% favoured water electrolysis with renewable energy and 13% preferred biomass-derived hydrogen as the most environmentally friendly hydrogen production method. Given SSA's abundant solar and biomass resources, and taking environmental friendliness as a key parameter, the authors recommend exploring solar-powered electrolysis and biomass gasification for hydrogen production in the region. Additionally, developing regional and national hydrogen development strategies is crucial to outlining a roadmap, setting deployment targets, and engaging key stakeholders. These findings provide policymakers

R. J. Tanyi (✉) · L. D. Mensah · D. A. Quansah · M. S. Adaramola
Department of Mechanical Engineering, Kwame Nkrumah University of Science and Technology, Kumasi, Ghana
e-mail: robinsontanyi4@gmail.com

A. Ntiamoah
Department of Chemical Engineering, Kwame Nkrumah University of Science and Technology, Kumasi, Ghana

L. D. Mensah · A. Ntiamoah · D. A. Quansah
The Brew-Hammond Energy Centre, Kwame Nkrumah University of Science and Technology, Kumasi, Ghana

M. S. Adaramola
Faculty of Environmental Sciences and Natural Resource Management, Norwegian University of Life Sciences (NMBU), Ås, Norway

M.-M. Narra and S. Narra (eds.), *African Green Transition Through Innovative Pathways*, World Sustainability Series, https://doi.org/10.1007/978-3-031-87043-9_12

and investors with updated data on hydrogen technology, guiding future hydrogen initiatives in sub-Saharan Africa.

Keywords Hydrogen · Production · PRISMA · Technology · Sub-Saharan Africa · Clean energy

1 Introduction

Due to the growing global population, economic development, and severe climate change impacts, there's an urgent need to transition from depleting fossil fuels to clean, renewable energy sources. To mitigate anthropogenic climate change, decarbonizing high-emission sectors like transportation, industry, and power production must be accelerated (Fawzy et al., 2020; Osman et al., 2021). Research is increasingly focusing on hydrogen (H_2) as a sustainable alternative to fossil fuels in these sectors.

Hydrogen is particularly attractive for its potential use in transportation, energy, and industry, and its production can utilize various renewable sources. However, the inconsistent nature of renewable energy poses challenges for a hydrogen-based economy. More research is needed on cost-effective production methods, supportive policies, and development to facilitate this transition (Osman et al., 2022a). Additionally, hydrogen's tendency to cause embrittlement, leading to mechanical degradation and pipeline leakage, has hindered its supply chain development (Dawood et al., 2020). As a result, researchers are exploring ammonia and methanol as safer, easier-to-transport hydrogen carriers (Andersson et al., 2020; Chatterjee et al., 2021; Kim et al., 2022).

To achieve energy security without impacting food, water, and land resources, research has focused on hydrogen production from sources like crop residues, seawater, algae, and other renewable sources like solar and wind ("Green hydrogen"). For instance, Salimi et al. (2016) reviewed hydrogen production from various agricultural residues via gasification and found that barley straw had the highest hydrogen content of 6.5 wt% and produced the most hydrogen (8.38 mmol/g) while canola stalk with the highest carbon/hydrogen ratio, yielded the most total gas (25.3 mmol/g). Li et al. (2021) explored methane pyrolysis for low-carbon hydrogen production, analysing the feasibility of capturing co-produced solid carbon in concrete. They discovered that replacing up to 10% of Portland cement with carbon could increase compressive strength by 20% (for 5% fibrillar carbon) and 16% (for 10% fibrillar carbon), while also reducing cement industry GHG emissions by 10%.

So far, there has been limited research on hydrogen technology development in sub-Saharan Africa. This study aims to stimulate further research in this sector, which currently lacks substantial data or literature to support hydrogen projects and stakeholders. Although some efforts have been made to explore hydrogen production in Sub-Saharan Africa, most advancements remain theoretical or numerical.

Some studies, such as Koumi Ngoh and Njomo (2012), reviewed hydrogen production from biomass, the most abundant renewable energy source in Sub-Saharan Africa. However, they found it impractical due to the simultaneous production of other compounds like carbon monoxide, which must be further converted to carbon dioxide. Koumi Ngoh et al. (2014) simulated hydrogen production via steam electrolysis in tropical conditions, requiring 0.2 hectares of land to produce about 2 tonnes of hydrogen per day. In Ghana, Topriska et al. (2015) examined hydrogen development for green cooking stoves to improve women's lives, as household cooking accounts for nearly 95% of energy use. Their dynamic simulation assessed solar hydrogen's potential to meet Ghana's cooking demand, estimated at 2.5 kWh/day. They found a stove efficiency of 60%, corresponding to 4.2 kWh/day or 0.107 kg of H_2/day, which largely met the demand.

2 Methodology

2.1 Research Scope

This study aims to address data gaps hindering the development of a hydrogen market in sub-Saharan Africa. A systematic literature review was conducted on hydrogen production feedstocks, technologies, and pathways to identify the most environmentally and economically feasible options for SSA. The review followed the PRISMA guidelines for systematic reviews and meta-analyses.

2.1.1 Literature Extraction

To begin the review, inclusion and exclusion criteria were established to screen literature from four databases: SCOPUS, ScienceDirect, Google Scholar and Springer. The search in all four databases was conducted on the 20th of January 2023. Two primary eligibility criteria were set: only literature from the past ten years (2012–2023) was included to focus on recent developments, and only English articles were considered. All types of literature, including published and grey literature from various academic fields, were screened by title and abstract.

A search strategy included keywords: *Hydrogen, energy, feedstock, technology, pathways and production.* In Scopus, an advanced search was performed using the Boolean operator 'AND' with the following query: *TITLE-ABS-KEY (hydrogen AND energy AND feedstock AND technolog* AND pathways AND produc*) AND PUBYEAR > 2011 AND PUBYEAR < 2024 (2012–2023).* The asterisks in *technolog** and *produc** captured variations like technologies and production. Similar advanced search methods were used in Springer and ScienceDirect.

For the Google Scholar search, the software Publish or Perish version 8 was used in searching the Google Scholar site. The search results were filtered using the software's automated tools to display the first 100 articles in order of their citation ranking. For Google Scholar, the Publish or Perish software (version 8) was used to filter the top 100 articles by citation ranking. An additional search focused on the status and trends of hydrogen production in Sub-Saharan Africa, using Google to find scholarly articles and reputable organizational data related to hydrogen production technology.

2.1.2 Literature Screening

The selection process began by exporting search results from the databases into an Excel CSV file, which included information such as titles, abstracts, publication dates, authors, and citations. The initial screening involved reading the titles and abstracts to filter articles based on predefined inclusion and exclusion criteria. Authors independently reviewed the articles, excluding those unrelated to hydrogen technology or not directly connected to hydrogen production pathways and energy uses. If the relevance of an article could not be determined from its title or abstract, the content was reviewed further, and a collective decision was made. Key findings, hydrogen production methods, and economic and sustainability analyses were noted. Ultimately, 62 articles met the inclusion criteria for detailed reading and quantitative analysis, while 34 additional studies identified via other methods (Google and websites) were included in the qualitative analysis, as shown in Fig. 1.

3 Results and Discussion

The results are presented in two main sections: a descriptive analysis of selected articles and a literature classification of these articles. The discussion section examines recent trends in global hydrogen production and demand, the development status in sub-Saharan Africa, and the challenges and opportunities for hydrogen development in the region.

3.1 Descriptive Analysis

3.1.1 Article Distribution Over Databases

Figure 2 illustrates the article sources, with Google Scholar providing the majority, accounting for 73% (45 articles) of the 62 selected articles. SCOPUS followed, contributing 21%, while Springer and ScienceDirect each shared the remaining 6% of the selected articles evenly.

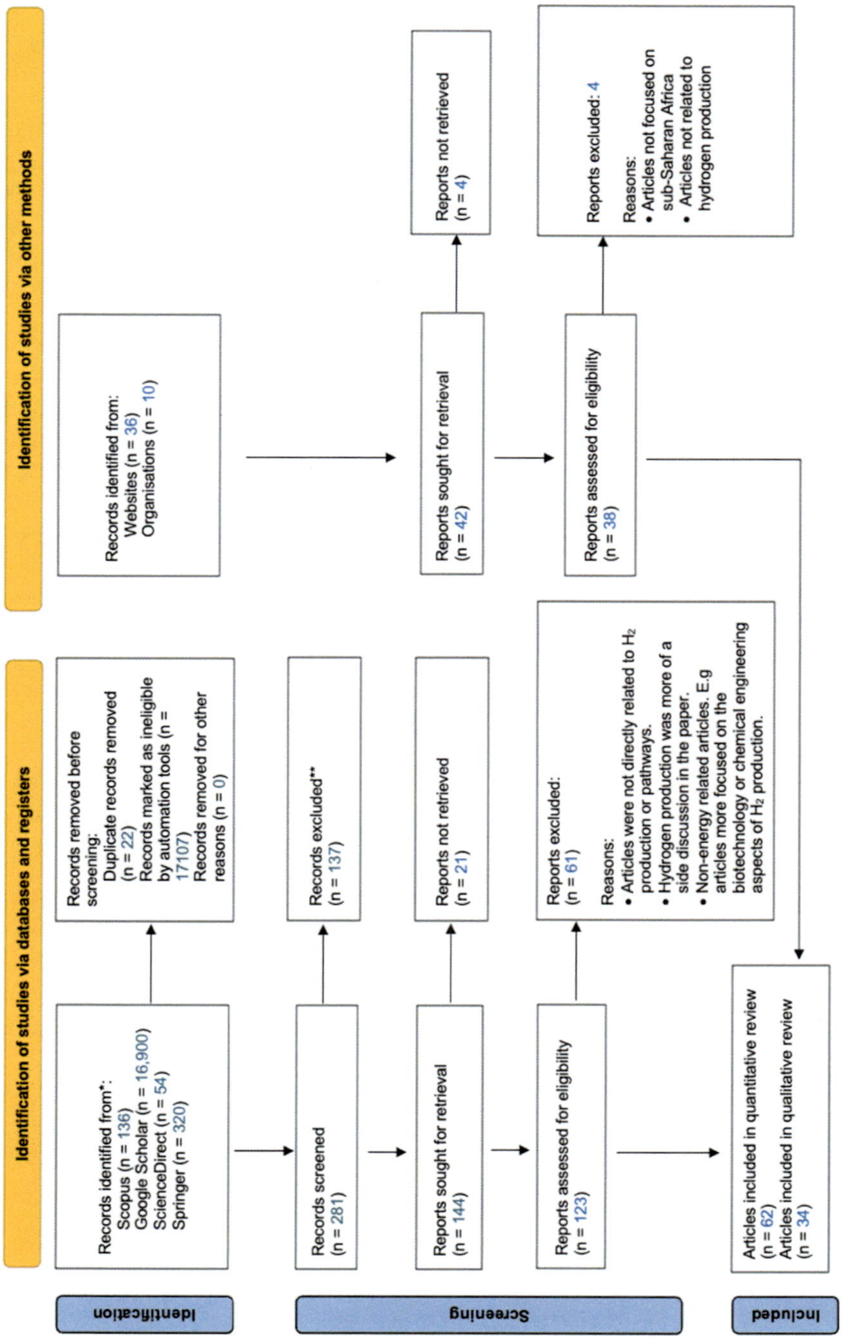

Fig. 1 PRISMA (Preferred Reporting Items for Systematic Reviews and Meta-analysis) flow diagram for searching and extracting data. *Adapted from* PRISMA 2020 statement: an updated guideline for reporting systematic reviews. For more information, visit: http://www.prisma-statement.org/

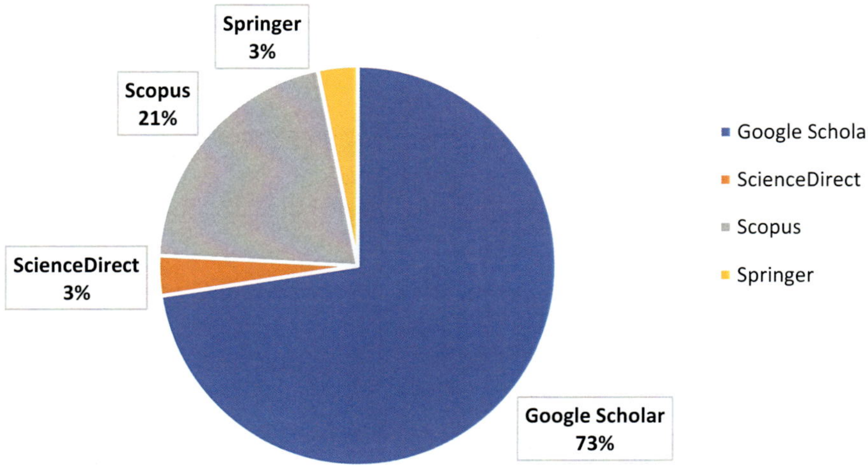

Fig. 2 Distribution of selected articles over various database

3.1.2 Publication Year of Selected Articles

Figure 3 illustrates the growing interest in hydrogen, particularly in the past five years. This surge is fuelled by the search for alternative, cleaner fuel sources to replace costly and polluting fossil fuels. Technological advancements in recent years have further propelled research and attention toward hydrogen as societies seek more eco-friendly options for transportation, electricity generation, heating, and more.

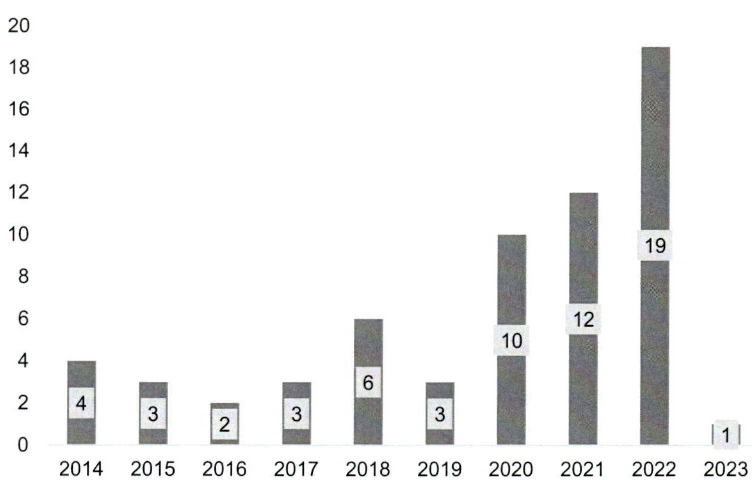

Fig. 3 Publication year of selected articles

Table 1 Technology overview of hydrogen generation from fossil fuels

Technology	Feedstock	Operating conditions	Maturity
Steam reforming	Light hydrocarbons (e.g. natural gas)	800–1000 °C	Commercial
Partial oxidation	Hydrocarbons, heavy fuel oil and coal	>1000 °C	Commercial
Autothermal reforming	Light hydrocarbons	>1000 °C	Early commercial
Pyrolysis	Hydrocarbons	500–800 °C in the absence of oxygen	Commercial
Gasification	Coal	700–1200 °C	Commercial

Source (Megia et al., 2021)

3.2 Literature Classification

3.2.1 Hydrogen Production Methods

Most articles (27) gave a general overview of different hydrogen production methods from non-renewable (fossil fuel) and renewable sources. A summary of the production methods from both sources is presented in Tables 1 and 2 respectively. Other unconventional or novel hydrogen production techniques were identified in the articles and are summarized in Table 3.

3.2.2 Hydrogen Production from Fossil Fuels

Hydrogen production from fossil fuels is dominated by hydrocarbon reforming and pyrolysis (Megia et al., 2021). Table 1 outlines the different fossil fuel technologies, including their feedstock, operational conditions, and technology maturity.

3.2.3 Hydrogen Production from Renewable Sources (Green Hydrogen)

Water and biomass are the most abundant renewable resources for making green hydrogen, and they both produce hydrogen via a variety of different processing techniques, including electrolysis, thermochemical breakdown, and biological approaches. Table 2 shows a summary and general comparison of hydrogen produced from renewable sources.

Table 2 General comparison of hydrogen production from renewable energy sources

Feedstock	Technology	Process summary	Advantages	Disadvantages	Energy efficiency (%)	Cost ($/kg H_2)	References
Water	Electrolysis	– Water + electrical energy passing through two electrodes	– Low temperature – No carbon emissions – Oxygen as a by-product	– Requires high pressure – Low efficiency – High capital cost	55–80	4.15–10.30	a, b, c, i
	Thermolysis	– Water + thermal energy	– Clean and sustainable – Zero carbon emissions – Oxygen as a by-product	– Separation step is required to avoid recombination in an explosive mixture – High capital cost	20–50	7.98–8.40	a, b, c
	Photoelectrolysis	– Water + photonic energy	– Low operating temperature and pressure – Photonic and electrical energy can be converted to chemical energy – Contributes to the sustainability of the energy supply	– Low efficiency – Requires a significant surface – Photocatalytic material required	0.06–14	4.98–10.36	a, b, c, d

(continued)

Table 2 (continued)

Feedstock	Technology	Process summary	Advantages	Disadvantages	Energy efficiency (%)	Cost ($/kg H_2)	References
	Biophotolysis	– Water + microorganisms	– Can produce hydrogen at ambient conditions – Consumes carbon dioxide	– Significant area required for sunlight collection – Larger reactor volume – Difficulty in controlling the different bacteria	10–15	1.42–2.13	a, b, c
Biomass	Dark fermentation	– Anaerobic bacteria on carbohydrate-rich substrates in the absence of light to produce hydrogen, organic acids and CO_2	– Can produce H_2 at any time even in the absence of light and in ambient conditions – Simple reactor design – Waste recycling	– H_2 is metabolically restricted – High by-product generation – Low efficiency – Large reactor volume	60–80	1.68–2.57	a, c, e,
	Photofermentation	– Photosynthetic bacteria use sunlight as an energy source to assimilate small organic molecules present in the biomass, obtaining hydrogen and CO_2 as byproducts	– Allows H_2 production from a wide range of substrates including waste streams – Waste recycling – High removal efficiency of the oxygen demand	– Strict control of environmental conditions is required – Significant area is required to collect sunlight – Large reactor volume	0.1–12	2.57–2.83	a, b, c, e, j, k

(continued)

Table 2 (continued)

Feedstock	Technology	Process summary	Advantages	Disadvantages	Energy efficiency (%)	Cost ($/kg H_2)	References
	Pyrolysis	– Biomass + heat (300 - 1000 °C) → $H_2 + CO + CH_4 +$ other products	– Mature technology – Abundant and cheap feedstock – Solid, liquid and gas product streams – Carbon–neutral emissions	– Hydrogen yield depends upon the feedstock tar formation	35–50	1.59–2.20	a, c, e, l
	Gasification	– Use of oxidation agents like air, oxygen or steam to convert biomass into CO and H_2 with the production of water, CO_2 and CH_4 alongside	– Abundant and cheap feedstock – Carbon–neutral emissions	– Requires oxidation agents – Hydrogen yield depends upon the feedstock tar formation	30–60	1.77–2.05	b, c, e, m, n

(continued)

Table 2 (continued)

Feedstock	Technology	Process summary	Advantages	Disadvantages	Energy efficiency (%)	Cost ($/kg H_2)	References
	Hydrothermal liquefaction	– Hydrothermal liquefaction of wet biomass at moderate temperatures (250–370 °C) and high pressure (4–22 MPa) to yield liquid bio-crude, gaseous steam, and aqueous phase and solid residue byproduct	– Abundant and cheap feedstock – Drying step is not required – High energy efficiency	– Hydrogen yield depends upon the feedstock presence of nitrogenated compounds	85–90	0.54–1.26	f, g, h
	Steam reforming	– Pyrolysis of solid biomass between 500–600 °C to yield bio-oil which is fed into a fluidized bed reactor for steam reforming	– Matured technology – Costly upgrading of bio-oil is avoided	– Produces carbon co-products	74–85	1.83–2.35	c, e

Source Adapted from (Megia et al., 2021)
a = (Nikolaidis & Poullikkas, 2017); b = (Dincer & Acar, 2014); c = (Shiva Kumar & Himabindu, 2019); d = (Parthasarathy & Narayanan, 2014); e = (Lepage et al., 2021); f = (Gollakota et al., 2018); g = (Zhu et al., 2014); h = (Cascioli & Baratieri, 2021); i = (Bolt et al., 2021); j = (Eroglu & Melis, 2011; k = (Mishra et al., 2019); l = (Bolt et al., 2021); m = (Ong et al., 2019); n = (Y. Zhang et al., 2020).

Table 3 Novel or unconventional production technologies

Solar-Driven Advanced Biomass Indirect-Gasification (SABI-Hydrogen)
Syngas chemical looping (SCL) and chemical looping reforming (CLR)
Biogas autothermal reforming and biogas steam reforming
Iron-based chemical looping
Steam reforming of alcoholic waste and steam reforming of glycerol
The AG2S process
Catalytic solar thermochemical splitting of the H_2O molecule
Aquamarine H_2 is produced from solar thermochemical CH_4 pyrolysis with a carbon catalyst
Hydrogen from oil sands (natural bitumen) and oil fields
In-situ combustion of hydrocarbons in reservoirs
Plasmolysis
Reverse electrodialysis, super critical water gasification, microbial electrolysis cell, microbial photoelectrochemical cell, photoelectrochemical cell
Steam cracking
Supercritical water gasification (SCWG)
Aluminium oxidation

3.2.4 Novel or Unconventional Hydrogen Production Technologies

The following novel or unconventional hydrogen production technologies were identified in the selected articles as presented in Table 3.

3.2.5 Production Method Economic and Environmental Analysis

In the product method analysis, articles were assessed based on their examination of hydrogen production pathways concerning:

The most economically feasible technology, referring to pathways with the lowest cost in terms of USD/kg of H_2.

The most environmentally favourable technology, considering pathways with the lowest greenhouse gas (GHG) footprint and other environmental impacts like acidification, eutrophication, and human toxicity potential.

According to Fig. 4, steam methane reforming (SMR) emerged as the most viable hydrogen production method in 80% of the articles. For instance, Moreno et al. (2022) noted that natural gas (71.27%), coal (27.27%), oil and gas (0.7%), and water electrolysis (0.7%) are the primary sources of global hydrogen. Additionally, most articles (91%) also labelled renewable-energy-powered electrolysis as the most expensive production technique due to the high cost of renewable energy and low efficiency, particularly from solar PV. Nonetheless, it is anticipated that this cost will decrease in the future as renewable energy prices decline. For instance, Moreno et al. (2022) reported that while electrolysis with renewable energy has a lower

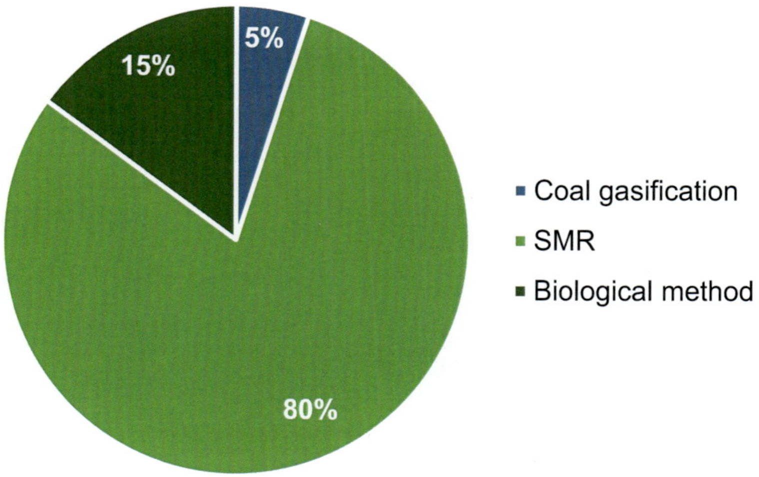

Fig. 4 Most economically viable production techniques

environmental impact compared to carbon-based sources, the resulting H_2 cost is notably higher.

For environmental friendliness, 60% of articles categorized water electrolysis with renewable energy sources like hydro, solar, wind, or biomass as the method with the lowest greenhouse gas (GHG) emissions and environmental impact as shown in Fig. 5. On the other hand, though hydrogen production is predominantly reliant on fossil fuels, it contributes significantly to carbon emissions with a reported emission of 500 million tCO_2 in 2015 and 830 million tCO_2 in 2019 (Futur. Hydrog., 2019). The primary source and technology play a critical role in determining carbon emissions in hydrogen supply (Zacharias et al., 2019). By employing suitable primary sources and carbon capture and storage (CCS) methods, carbon emissions during H_2 generation can be reduced compared to conventional processes (Bakenne et al., 2016; Zacharias et al., 2019). Additionally, if renewable energy sources like wind, solar, hydraulic, biomass and geothermal energy are used as primary energy sources for electricity, electrolysis-based H_2 production might help to lower the global carbon footprint. Nazir et al. (2020) demonstrated that hydrogen production from carbon feedstocks results in a carbon footprint exceeding 6.0 kg CO_2-eq/kg H_2, while renewable sources yield approximately 3.0 kg CO_2-eq/kg H_2. This aligns with findings from Borole and Greig (2019), estimating a minimum carbon footprint of 4.0 kg CO_2-eq/kg H_2 from carbon primary sources, compared to 0.97–1.12 kg CO_2-eq/kg H_2 from wind electrolysis or solar splitting, respectively.

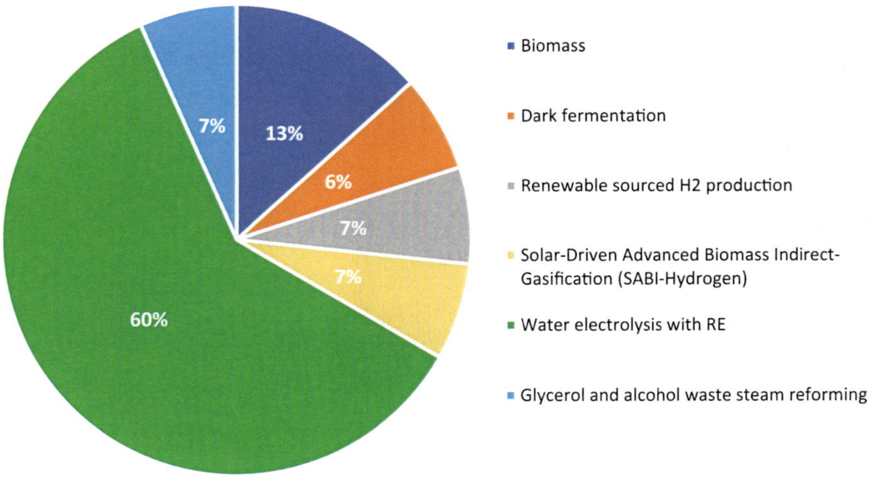

Fig. 5 Most environmentally suitable routes as per articles

4 Discussion

4.1 State of Hydrogen Technology in Sub-Saharan Africa

While numerous studies have explored hydrogen production technologies, feedstocks, and evaluations globally, few have focused on their development in sub-Saharan Africa (SSA).

According to Fopah-Lele (2022), only 17% of SSA countries are actively engaged in hydrogen technology development, with most advancements being theoretical or numerical rather than practical implementations. Moreover, there's a notable absence of norms and standards for hydrogen technology installation in SSA, except for South Africa, which adopted the American National Standard. According to the International Energy Agency's 2022 Global Hydrogen Review, no African country participates in hydrogen technology safety development as shown in Fig. 6.

Despite the absence of clear hydrogen development strategies on the continent, some studies and assessments have been made to investigate the potential of hydrogen technology in different parts of SSA. Koumi Ngoh and Njomo (2012) reviewed methods of hydrogen production from biomass which is the most available and largest renewable energy feedstock for hydrogen production in Sub-Saharan Africa. They however found it inconvenient due to the simultaneous production of other compounds such as carbon monoxide (CO) that need to be converted further to carbon dioxide.

Koumi Ngoh et al. (2014) simulated hydrogen production through steam electrolysis in tropical weather conditions. They found that the electrolyser installation needed 0.2 hectares of land and could create roughly 2 tons of hydrogen per day.

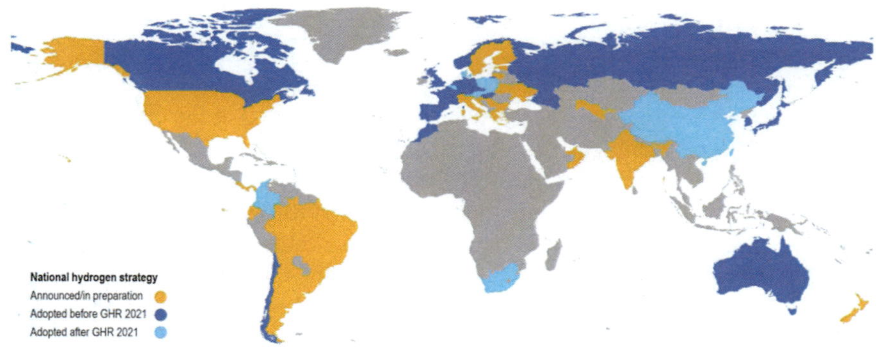

Fig. 6 Countries with a national hydrogen strategy in place or under development. *Source* IEA (2022), Global Hydrogen Review; Licence: CC BY 4.0

For Ghana, Topriska et al. (2015) looked at hydrogen development for the provision of green cooking stoves to improve women's life as energy for household cooking accounted for almost 95% of the household energy use in the country. They ran a dynamic simulation to evaluate the potential of solar hydrogen to cover Ghana's cooking demand estimated at 2.5 kWh/day (Topriska et al., 2016). Their results exhibited a stove efficiency of 60% which corresponded to 4.2 kWh/day or 0.107 kg H_2/day which largely met demand. Ofualagba et al. (2012) sought to find an appropriate fuel cell technology that fed into the economy and energy market of Nigeria while Amoo and Fagbenle (2014a, b) did an integrated assessment for the transportation and power sector to determine the suitability of incorporation of hydrogen energy system on the economy. Their findings showed that hydrogen-powered vehicles did not have any major impact on the transport sector, but that hydrogen played a useful role in backing up residential energy demand. In addition, Douglas (2016) carried out dynamic modelling of a solar PV—battery/fuel cell hybrid system by simulating a residential load based on 4 h of solar radiation on 500 W modules with a battery capacity of 42 Ah and fuel cell power of 1 kW. The operation cycle consisted mostly of the charging and discharging of the solar module's battery. When the batteries were fully charged, the excess energy from the modules was used to generate and store hydrogen in the fuel cells for later use in the fuel cell to be converted back into electricity during peak load hours or when batteries were not charged due to no sunshine. They found batteries could supply power for a longer time for lower loads (233 s) while fuel cells could support larger loads for a shorter time (55 ms for that case). They recommend hydrogen production and storage as a potential solution to peak load constraints on power grids.

In the village of "Bourakebougou" in northern Mali, a naturally occurring hydrogen field was found where 28 × 106 m^3 of H_2 with a 98% purity level was thought to be present. The first pilot plant, which used natural hydrogen to power a generator to electrify the village, was put into operation in 2013. The natural hydrogen is used right away as generator fuel rather than being stored which lowers the overall operational and supply cost of electricity in the village. This venture makes Mali the first nation in sub-Saharan Africa to use natural hydrogen for electrification as a result (Fopah-Lele, 2022).

For Cameroon, Nkanga et al. (2021) evaluated the potential of hydrogen production by electrolysis from solar photovoltaic and wind energy in the city of Douala. They applied a semi-empirical modelling approach of an alkaline electrolyser linked to the solar panel and wind turbines. Their MATLAB/Simulink simulation results showed that the average hydrogen production potential was estimated at 0.55 Nm^3/h for the PV supply corresponding to an average energy efficiency of 70% and 0.675 Nm^3/h for the wind turbine connection giving an average energy efficiency of 84%.

In Namibia, Hyphen Energy was selected as a preferred bidder for the development of the country's first large-scale green hydrogen project in the Tsau Khaeb National Park vicinity (HHE, 2022). This location was ideal as it has an extensive combination of solar and wind resources and proximity to the ocean and port of Lüderitz for export options. The project, which is being developed in stages, is expected to generate 2 gigawatts (GW) of renewable electricity and 300,000 metric tons (MT) of green hydrogen annually for domestic and international markets by 2026 at an estimated cost of $4.4 billion. With a total estimated cost of $9.4 billion, additional construction phases in the late 2020s will scale up to 3 GW of electrolyzer capacity and 5 GW of combined renewable power generation (Fopah-Lele, 2022).

The Project "Nour" in Mauritania, which has been demonstrated in pre-feasibility studies, has the potential to become the greatest large-scale (10 GW) facility for producing green hydrogen. Over 14,400 km^2 of onshore and offshore space are covered by this project, which uses sun and wind energy to generate electricity for electrolysis that creates green hydrogen (Offshore Energy, 2022).

The German company Emerging Energy Corporation (EEC) and the government of Niger inked a contract in February 2021 to develop the hydrogen market. The investigation and development of green hydrogen projects have received approval from the government. EEC intends to use a clean energy-based electrolysis method. This initiative should aid in decarbonizing refineries and oil field operations. Moreover, Niger will export its green hydrogen to other markets, primarily those in Europe and North America. In comparison to the most sophisticated programs in other African nations, this project is still in its early stages (Afrik21, 2022).

These few initiatives in the region however lack policy backing to develop the implementation of various projects. On a regional level, hydrogen projects are being launched by Sub-Saharan African institutions in collaboration with German institutions. An Atlas of Green Hydrogen Generation Potentials in Africa (https://africa.h2atlas.de/ecowas) has been created by the SADC Centre for Renewable Energy and Energy Efficiency (SACREEE), the Southern African Science Service Centre for Climate Change and Adaptive Land Management (SASSCAL), the ECOWAS Regional Centre for Renewable Energy and Energy Efficiency (ECREEE), and the West African Science Service Centre on Climate Change and Adapted Land Use (WASCAL) (Fopah-Lele, 2022). This project was aimed at mapping the hydrogen economy of the Sub-Saharan Africa region while implementing the climate change–energy nexus. While hydrogen technology is not as widespread or developed in Sub-Saharan Africa as in other regions like Asia and Europe, the region holds significant

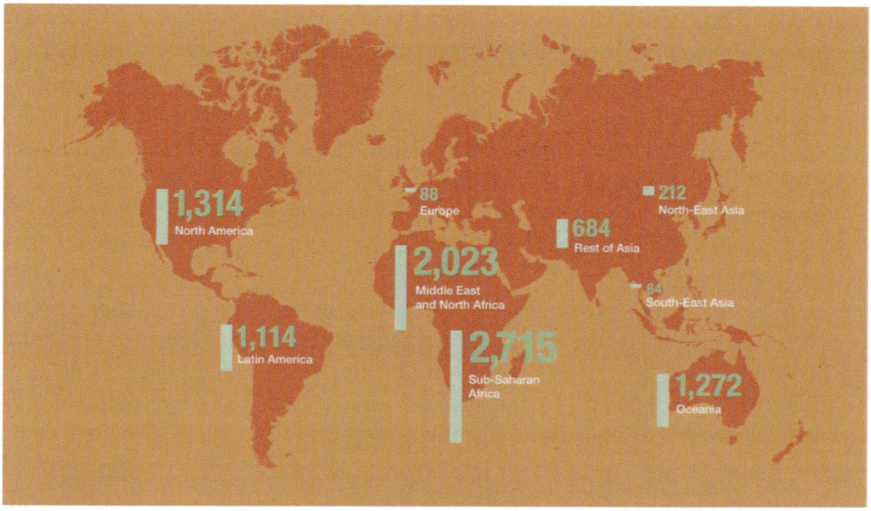

Fig. 7 Technical potential for producing green hydrogen under USD 1.5/kg by 2050, in EJ. *Source* https://hydrogen.revolve.media/2022/features/africa/

potential to produce green hydrogen cost-effectively. The International Renewable Energy Agency (IRENA) forecasts that by 2050, sub-Saharan Africa could lead in producing green hydrogen for less than USD 1.5/kg, surpassing other regions, as highlighted in their 2022 Geopolitics of Energy Transformation Report (see Fig. 7).

4.2 Opportunities and Challenges for Hydrogen Development in SSA

In sub-Saharan Africa, hydrogen's development and utilization as an energy carrier have received minimal attention until recently, with limited progress beyond the research and feasibility stages. Yet, hydrogen presents abundant opportunities to enhance the region's energy security and bolster its renewable energy integration efforts, particularly within the power sector. Sub-Saharan Africa has abundant solar and biomass resources which could be tapped to produce green hydrogen. Katebah et al. (2022) indicate that nearly all countries in the region possess a solar potential nearing 10,000 GW, with approximately 4300 h of annual sunshine and an estimated photovoltaic potential of 6500 TW/yearly. Biomass continues to be a primary energy source in Sub-Saharan Africa (SSA), with projections suggesting it will account for 51 to 57% of total consumption by 2035.

Hydrogen offers a power-to-gas storage solution for intermittent renewables, such as hydro and solar PV, crucial for SSA nations grappling with unreliable electricity supply due to climate-driven hydropower vulnerabilities. By utilizing excess electricity from peak generation periods for water electrolysis and hydrogen production, hydrogen serves as a cleaner and sustainable storage solution. This stored hydrogen can later be converted back to electricity via fuel cells during low water availability, peak demand, or downtime, offering significant energy storage density advantages over pumped hydro storage (PHS) (di Gaeta et al., 2017).

Despite its promising features, hydrogen technology encounters obstacles hindering its large-scale deployment, which vary across national, regional, and international contexts.

Renewable hydrogen production for electricity storage confronts challenges of low round-trip efficiency and higher costs compared to alternative storage technologies like pumped hydro and supercapacitors (Fopah-Lele, 2022). In sub-Saharan Africa, the cost of producing hydrogen from solar PV stands at approximately 13 euros/kg, expected to decrease as renewable energy costs decline (Fopah-Lele et al., 2021).

Transportation of green hydrogen or its conversion for easier transport entails significant energy consumption and losses during both processes, impacting its energy efficiency. Furthermore, specialized pipelines are necessary for transporting pure hydrogen, as existing natural gas pipelines can only accommodate limited hydrogen concentrations (Fopah-Lele, 2022). Recommendations often involve gradually introducing hydrogen into pipelines with concentrations ranging from 5 to 100%.

The lack of reliable or official data on SSA's hydrogen interest or development poses a significant hurdle for stakeholders interested in the region's hydrogen resources. This lack of reliable data peculiar to SSA makes it challenging to guide policymakers or investors in identifying the region's energy gaps that hydrogen development could fill and priority areas of action. Additionally, the absence of a comprehensive regulatory framework outlining technical, operational, environmental, and safety criteria hampers the establishment of uniform usage standards. Overcoming these institutional barriers could foster further research, data management, and policy development for hydrogen in sub-Saharan Africa.

5 Conclusion

This study aimed to address data gaps hindering the development of a hydrogen market in sub-Saharan Africa through a systematic review of hydrogen production feedstocks, technology, and pathways, aiming to identify environmentally and economically viable options for development in the region. The key findings, limitations, and recommendations are summarised as follows:

Quantitative analysis was conducted on 62 articles, while 34 were qualitatively reviewed. Steam methane reforming (SMR) of natural gas emerged as the most economically feasible technology (costing less than \$2/kg H_2) and the most commonly adopted, according to 80% of the reviewed articles. However, criticism arises due to significant CO_2 emissions (exceeding 6 kg CO_2-eq/kg H_2) associated with this method, rendering it environmentally damaging.

Water electrolysis using renewable energy sources was identified as the most environmentally friendly method (resulting in 0.97–1.12 kg CO2-eq/kg H_2 for wind and solar electrolysis, respectively) but also the most expensive (costing between 4.15–10.30 \$/kg H_2) due to the high cost of renewable energy and inefficiencies in technologies like solar PVs. Nevertheless, it is anticipated that costs will decrease over time as renewable costs decline and technology efficiencies improve.

Considering financial and environmental analyses, along with SSA's ample solar, biomass, and water resources, the authors recommend biomass gasification and solar-powered electrolysis as the most practical production methods for further techno-economic analysis and future development in the region.

Due to the limited data on hydrogen production in SSA, this review is limited to hydrogen production data found in related published articles in other regions mostly out of sub-Sahara Africa and the current state of these technologies as reported by reputable institutions like the International Energy Agency (IEA) and the International Renewable Energy Agency (IRENA). The authors recommend additional research and techno-economic analysis focused on biomass gasification and water electrolysis, using data and parameters specific to the region's techno-economic scenarios and policy regulations. Developing regional and national hydrogen strategies outlining a roadmap for hydrogen development should be a priority, enabling deployment targets and engaging key stakeholders. This approach will foster the adoption of clean and alternative energy sources, creating new economic opportunities while meeting climate targets.

Acknowledgements The authors would like to acknowledge the funding provided by the 'Fostering Research And Intra-African Knowledge Transfer Through Mobility And Education (FRAME) scholarship programme and equally appreciate the support from the Brew-Hammond Energy Centre and Department of Mechanical Engineering of the Kwame Nkrumah University of Science and Technology (KNUST), Kumasi, Ghana.

References

Abdin, Z., Zafaranloo, A., Rafiee, A., Mérida, W., Lipiński, W., & Khalilpour, K. R. (2020). Hydrogen as an energy vector. *Renewable and Sustainable Energy Reviews, 120*, 109620. https://doi.org/10.1016/j.rser.2019.109620.

Acar, C., & Dincer, I. (2014). Comparative assessment of hydrogen production methods from renewable and non-renewable sources. *International Journal of Hydrogen Energy, 39*(1), 1–12. https://doi.org/10.1016/j.ijhydene.2013.10.060.

Afrik21. (n.d.). *NIGER: Germany's Emerging Energy to produce and export green hydrogen | Afrik 21.* 2022. Retrieved March 10, 2023, from https://www.afrik21.africa/en/niger-germanys-emerging-energy-to-produce-and-export-green-hydrogen/.

Ahmed, S. F., Mofijur, M., Nuzhat, S., Rafa, N., Musharrat, A., Lam, S. S., & Boretti, A. (2022). Sustainable hydrogen production: Technological advancements and economic analysis. *International Journal of Hydrogen Energy, 47*(88), 37227–37255. https://doi.org/10.1016/j.ijhydene.2021.12.029.

Al-Qahtani, A., Parkinson, B., Hellgardt, K., Shah, N., & Guillen-Gosalbez, G. (2021). Uncovering the true cost of hydrogen production routes using life cycle monetisation. *Applied Energy, 281*, 115958. https://doi.org/10.1016/j.apenergy.2020.115958.

Amiri, T. Y., Ghasemzageh, K., & Iulianelli, A. (2020). Membrane reactors for sustainable hydrogen production through steam reforming of hydrocarbons: A review. *Chemical Engineering and Processing-Process Intensification, 157*, 108148. https://doi.org/10.1016/j.cep.2020.108148.

Amoo, L. M., & Fagbenle, R. L. (2014a). Hydrogen energy's key contributions to the sustainable energy mix of a low-carbon future in Nigeria. http://Ocl.Knust.Edu.Gh:2082/10.1080/14786451.2013.765427, 33(4), 742–765. https://doi.org/10.1080/14786451.2013.765427.

Amoo, L. M., & Fagbenle, R. L. (2014b). An integrated impact assessment of hydrogen as a future energy carrier in Nigeria's transportation, energy and power sectors. *International Journal of Hydrogen Energy, 39*(24), 12409–12433. https://doi.org/10.1016/J.IJHYDENE.2014.06.022.

Andersson, J., Krüger, A., & Grönkvist, S. (2020). Methanol as a carrier of hydrogen and carbon in fossil-free production of direct reduced iron. *Energy Conversion and Management: X, 7*, 100051. https://doi.org/10.1016/j.ecmx.2020.100051.

Arregi, A., Amutio, M., Lopez, G., Bilbao, J., & Olazar, M. (2018). Evaluation of thermochemical routes for hydrogen production from biomass: A review. *Energy Conversion and Management, 165*, 696–719. https://doi.org/10.1016/j.enconman.2018.03.089.

Bakenne, A., Nuttall, W., & Kazantzis, N. (2016). Sankey-Diagram-based insights into the hydrogen economy of today. *International Journal of Hydrogen Energy, 41*(19), 7744–7753. https://doi.org/10.1016/j.ijhydene.2015.12.216.

Barelli, L., Bidini, G., Gallorini, F., & Servili, S. (2008). Hydrogen production through sorption-enhanced steam methane reforming and membrane technology: A review. *Energy, 33*(4), 554–570. https://doi.org/10.1016/j.energy.2007.10.018.

Bolt, A., Dincer, I., & Agelin-Chaab, M. (2021). A Review of unique Aluminum-water based hydrogen production options. *Energy and Fuels, 35*(2), 1024–1040. https://doi.org/10.1021/acs.energyfuels.0c03674.

Borole, A. P., & Greig, A. L. (2019). Life-Cycle assessment and systems analysis of hydrogen production. *Biomass, Biofuels, Biochemicals: Biohydrogen, Second Edition*, 485–512. https://doi.org/10.1016/B978-0-444-64203-5.00020-4.

Brauns, J., & Turek, T. (2020). Alkaline water electrolysis powered by renewable energy: A review. *Processes, 8*(2), 248. https://doi.org/10.3390/pr8020248.

Cao, L., Yu, I. K. M., Xiong, X., Tsang, D. C. W., Zhang, S., Clark, J. H., Hu, C., Ng, Y. H., Shang, J., & Ok, Y. S. (2020). Biorenewable hydrogen production through biomass gasification: A review and future prospects. *Environmental Research, 186*, 109547. https://doi.org/10.1016/j.envres.2020.109547.

Cascioli, A., & Baratieri, M. (2021). Enhanced thermodynamic modelling for hydrothermal liquefaction. *Fuel, 298*, 120796. https://doi.org/10.1016/j.fuel.2021.120796.

Chatterjee, S., Parsapur, R. K., & Huang, K. W. (2021). Limitations of ammonia as a hydrogen energy carrier for the transportation sector. *ACS Energy Letters, 6*(12), 4390–4394. https://doi.org/10.1021/acsenergylett.1c02189.

Chi, J., & Yu, H. (2018). Water electrolysis based on renewable energy for hydrogen production. *Cuihua Xuebao/chinese Journal of Catalysis, 39*(3), 390–394. https://doi.org/10.1016/S1872-2067(17)62949-8.

Cifuentes, B., Gómez, J., Sánchez, N., Proaño, L., Bustamante, F., & Cobo, M. (2021). Bioethanol steam reforming over monoliths washcoated with RhPt/CeO2–SiO2: The use of residual biomass to stably produce syngas. *International Journal of Hydrogen Energy, 46*(5), 4007–4018. https://doi.org/10.1016/j.ijhydene.2020.10.271.

Colella, W., James, B., Moron, J., Saur, G., & Ramsden, T. (2014). Techno-economic analysis of PEM electrolysis for hydrogen production. *Electrolytic hydrogen production workshop, February*, 38 Colella, W., James, B., Moron, J., Saur, G., R. https://www.energy.gov/sites/prod/files/2014/08/f18/fcto_2014_electrolytic_h2_wkshp_colella1.pdf.

Dawood, F., Anda, M., & Shafiullah, G. M. (2020). Hydrogen production for energy: An overview. *International Journal of Hydrogen Energy, 45*(7), 3847–3869. https://doi.org/10.1016/j.ijhydene.2019.12.059.

di Gaeta, A., Reale, F., Chiariello, F., & Massoli, P. (2017). A dynamic model of a 100 kW micro gas turbine fuelled with natural gas and hydrogen blends and its application in a hybrid energy grid. *Energy, 129*, 299–320. https://doi.org/10.1016/j.energy.2017.03.173.

Dincer, I., & Acar, C. (2014). Review and evaluation of hydrogen production methods for better sustainability. *International Journal of Hydrogen Energy, 40*(34), 11094–11111. https://doi.org/10.1016/j.ijhydene.2014.12.035.

Dincer, I., & Zamfirescu, C. (2012). Sustainable hydrogen production options and the role of IAHE. *International Journal of Hydrogen Energy, 37*(21), 16266–16286. https://doi.org/10.1016/j.ijhydene.2012.02.133.

Douglas, T. (2016). Dynamic modelling and simulation of a solar-PV hybrid battery and hydrogen energy storage system. *Journal of Energy Storage, 7*, 104–114. https://doi.org/10.1016/J.EST.2016.06.001.

Dvoynikov, M., Buslaev, G., Kunshin, A., Sidorov, D., Kraslawski, A., & Budovskaya, M. (2021). New concepts of hydrogen production and storage in Arctic region. *Resources, 10*(1), 1–18. https://doi.org/10.3390/resources10010003.

Eckroll, J. (2017). Concepts for large scale hydrogen liquefaction plants. *Master Thesis*, 156. https://ntnuopen.ntnu.no/ntnu-xmlui/handle/11250/2454901.

Eroglu, E., & Melis, A. (2011). Photobiological hydrogen production: Recent advances and state of the art. *Bioresource Technology, 102*(18), 8403–8413. https://doi.org/10.1016/j.biortech.2011.03.026.

Fawzy, S., Osman, A. I., Doran, J., & Rooney, D. W. (2020). Strategies for mitigation of climate change: A review. *Environmental Chemistry Letters, 18*(6), 2069–2094. https://doi.org/10.1007/s10311-020-01059-w.

Fopah-Lele, A., Kabore-Kere, A., Tamba, J. G., & Yaya-Nadjo, I. (2021). Solar electricity storage through green hydrogen production: A case study. *International Journal of Energy Research, 45*(9), 13007–13021. https://doi.org/10.1002/ER.6630.

Fopah-Lele, A. (2022). Hydrogen technology in Sub-Saharan Africa: Prospects for power plants. *E3S Web of Conferences, 354*, 01001. https://doi.org/10.1051/e3sconf/202235401001.

Gollakota, A. R. K., Kishore, N., & Gu, S. (2018). A review on hydrothermal liquefaction of biomass. *Renewable and Sustainable Energy Reviews, 81*, 1378–1392. https://doi.org/10.1016/j.rser.2017.05.178.

Greenbury, T. (2022). The East Africa Hydrogen Corridor | Hydrogen Report 2022. Hydrogen Report. https://hydrogen.revolve.media/2022/features/africa/; CC BY-NC-SA 4.0.

Han, W. B., Kim, I. S., Kim, M. J., Cho, W. C., Kim, S. K., Joo, J. H., Lee, Y. W., Cho, Y., Cho, H. S., & Kim, C. H. (2021). Directly sputtered nickel electrodes for alkaline water electrolysis. *Electrochimica Acta, 386*, 138458. https://doi.org/10.1016/j.electacta.2021.138458.

Hariharan, D., Yang, R., Zhou, Y., Gainey, B., Mamalis, S., Smith, R. E., Lugo-Pimentel, M. A., Castaldi, M. J., Gill, R., Davis, A., Modroukas, D., & Lawler, B. (2019). Catalytic partial oxidation reformation of diesel, gasoline, and natural gas for use in low temperature combustion engines. *Fuel, 246*, 295–307. https://doi.org/10.1016/j.fuel.2019.02.003.

HHE. (2022). *Hyphen Hydrogen Energy*. https://hyphenafrica.com/.

Holladay, J. D., Hu, J., King, D. L., & Wang, Y. (2009). An overview of hydrogen production technologies. *Catalysis Today, 139*(4), 244–260. https://doi.org/10.1016/j.cattod.2008.08.039.

Hosseini, S. E., & Wahid, M. A. (2016). Hydrogen production from renewable and sustainable energy resources: Promising green energy carrier for clean development. *Renewable and Sustainable Energy Reviews, 57*, 850–866. https://doi.org/10.1016/j.rser.2015.12.112.

Hydrogen Tools. (2017). Guide to Safety of Hydrogen and Hydrogen Systems (ANSI/AIAA G-095A-2017). *Guide to Safety of Hydrogen and Hydrogen Systems (ANSI/AIAA G-095A-2017).* https://doi.org/10.2514/4.105197.

IEA. (2022). Global Hydrogen Review 2022, IEA, Paris https://www.iea.org/reports/global-hydrogen-review-2022, Licence: CC BY 4.0.

IRENA. (2022). Geopolitics of the energy transition. *Handbook of Energy Transitions*, 41–59. https://doi.org/10.1201/9781003315353-4.

Khojasteh Salkuyeh, Y., Saville, B. A., & MacLean, H. L. (2017). Techno-economic analysis and life cycle assessment of hydrogen production from natural gas using current and emerging technologies. *International Journal of Hydrogen Energy, 42*(30), 18894–18909. https://doi.org/10.1016/j.ijhydene.2017.05.219.

Kim, J., Huh, C., & Seo, Y. (2022). End-to-end value chain analysis of isolated renewable energy using hydrogen and ammonia energy carrier. *Energy Conversion and Management, 254*, 115247. https://doi.org/10.1016/j.enconman.2022.115247.

King, W. E. H. (1981). Coal gasification. *Fuel.* https://doi.org/10.1016/0016-2361(81)90142-3.

Koumi Ngoh, S., & Njomo, D. (2012). An overview of hydrogen gas production from solar energy. *Renewable and Sustainable Energy Reviews, 16*(9), 6782–6792. https://doi.org/10.1016/J.RSER.2012.07.027.

Koumi Ngoh, S., Ayina Ohandja, L. M., Kemajou, A., & Monkam, L. (2014). Design and simulation of hybrid solar high-temperature hydrogen production system using both solar photovoltaic and thermal energy. *Sustainable Energy Technologies and Assessments, 7*, 279–293. https://doi.org/10.1016/J.SETA.2014.05.002.

Lepage, T., Kammoun, M., Schmetz, Q., & Richel, A. (2021). Biomass-to-hydrogen: A review of main routes production, processes evaluation and techno-economical assessment. *Biomass and Bioenergy, 144*, 105920. https://doi.org/10.1016/j.biombioe.2020.105920.

Li, J., Spanu, L., Heo, J., Zhang, W., Gardner, D. W., Carraro, C., Maboudian, R., & Monteiro, P. J. M. (2021). Sequestration of solid carbon in concrete: A large-scale enabler of lower-carbon intensity hydrogen from natural gas. *MRS Bulletin, 46*(8), 680–686. https://doi.org/10.1557/s43577-021-00118-z.

Luneau, M., Gianotti, E., Meunier, F. C., Mirodatos, C., Puzenat, E., Schuurman, Y., & Guilhaume, N. (2017). Deactivation mechanism of Ni supported on Mg-Al spinel during autothermal reforming of model biogas. *Applied Catalysis B: Environmental, 203*, 289–299. https://doi.org/10.1016/j.apcatb.2016.10.023.

Ma, R., Xu, B., & Zhang, X. (2019). Catalytic partial oxidation (CPOX) of natural gas and renewable hydrocarbons/oxygenated hydrocarbons—A review. *Catalysis Today, 338*, 18–30. https://doi.org/10.1016/j.cattod.2019.06.025.

Maggio, G., Nicita, A., & Squadrito, G. (2019). How the hydrogen production from RES could change energy and fuel markets: A review of recent literature. *International Journal of Hydrogen Energy, 44*(23), 11371–11384. https://doi.org/10.1016/j.ijhydene.2019.03.121.

Mbodji, M., Commenge, J. M., Falk, L., Di Marco, D., Rossignol, F., Prost, L., Valentin, S., Joly, R., & Del-Gallo, P. (2012). Steam methane reforming reaction process intensification by using a millistructured reactor: Experimental setup and model validation for global kinetic reaction rate estimation. *Chemical Engineering Journal, 207–208*, 871–884. https://doi.org/10.1016/j.cej.2012.07.117.

Megia, P. J., Vizcaino, A. J., Calles, J. A., & Carrero, A. (2021). Hydrogen production technologies: From fossil fuels toward renewable sources. A mini review. *Energy and Fuels, 35*(20), 16403–16415. https://doi.org/10.1021/acs.energyfuels.1c02501.

Milani, D., Kiani, A., & McNaughton, R. (2020). Renewable-powered hydrogen economy from Australia's perspective. *International Journal of Hydrogen Energy, 45*(46), 24125–24145. https://doi.org/10.1016/j.ijhydene.2020.06.041.

Mishra, P., Krishnan, S., Rana, S., Singh, L., Sakinah, M., & Ab Wahid, Z. (2019). Outlook of fermentative hydrogen production techniques: An overview of dark, photo and integrated dark-photo fermentative approach to biomass. *Energy Strategy Reviews, 24*, 27–37. https://doi.org/10.1016/j.esr.2019.01.001.

Moreno, J., Cobo, M., Barraza-Botet, C., & Sanchez, N. (2022). Role of low carbon emission H2 in the energy transition of Colombia: Environmental assessment of H2 production pathways for a certification scheme. *Energy Conversion and Management: X, 16*(October), 100312. https://doi.org/10.1016/j.ecmx.2022.100312.

Mularski, J., Pawlak-Kruczek, H., & Modlinski, N. (2020). A review of recent studies of the CFD modelling of coal gasification in entrained flow gasifiers, covering devolatilization, gas-phase reactions, surface reactions, models and kinetics. *Fuel, 271*, 117620. https://doi.org/10.1016/j.fuel.2020.117620.

Muradov, N. Z. (1993). How to produce hydrogen from fossil fuels without CO2 emission. *International Journal of Hydrogen Energy, 18*(3), 211–215. https://doi.org/10.1016/0360-3199(93)90021-2.

Mutsengerere, S., Chihobo, C. H., Musademba, D., & Nhapi, I. (2019). A review of operating parameters affecting bio-oil yield in microwave pyrolysis of lignocellulosic biomass. *Renewable and Sustainable Energy Reviews, 104*, 328–336. https://doi.org/10.1016/j.rser.2019.01.030.

Nazir, H., Louis, C., Jose, S., Prakash, J., Muthuswamy, N., Buan, M. E. M., Flox, C., Chavan, S., Shi, X., Kauranen, P., Kallio, T., Maia, G., Tammeveski, K., Lymperopoulos, N., Carcadea, E., Veziroglu, E., Iranzo, A., & Kannan, A. M. (2020). Is the H2 economy realizable in the foreseeable future? Part I: H2 production methods. *International Journal of Hydrogen Energy, 45*(27), 13777–13788. https://doi.org/10.1016/j.ijhydene.2020.03.092.

NETL. (n.d.). *7.3. Technologies for hydrogen production | netl.doe.gov*. Retrieved March 24, 2023, from https://netl.doe.gov/research/coal/energy-systems/gasification/gasifipedia/technologies-hydrogen.

Nikolaidis, P., & Poullikkas, A. (2017a). A comparative overview of hydrogen production processes. *Renewable and Sustainable Energy Reviews, 67*, 597–611. https://doi.org/10.1016/j.rser.2016.09.044.

Nkanga, E. R. E., Ndoh, D. Z. N., Ntonda, J. N., Nanfak, A., Mekouebe, M. T., & Lontsi, F. (2021). Modeling of hydrogen production in an alkaline electrolyser system connected with a solar photovoltaic panel or a wind turbine: Case study; Douala-Cameroon. *Journal of Power and Energy Engineering, 9*(10), 1–18. https://doi.org/10.4236/JPEE.2021.910001.

Ofualagba, G., Charles, I. K., O. O. (2012). Solar hydrogen fuel cell technology, principle, applications and market. *Journal of Energy Technologies and Policy, 13*(1), 104–116.

Offshore Energy. (2022). *Mauritania's large green hydrogen project in new milestones-offshore energy*. https://www.offshore-energy.biz/mauritanias-large-green-hydrogen-project-in-new-milestones/.

Ong, H. C., Chen, W. H., Farooq, A., Gan, Y. Y., Lee, K. T., & Ashokkumar, V. (2019). Catalytic thermochemical conversion of biomass for biofuel production: A comprehensive review. *Renewable and Sustainable Energy Reviews, 113*, 109266. https://doi.org/10.1016/j.rser.2019.109266.

Osman, A. I., Mehta, N., Elgarahy, A. M., Al-Hinai, A., Al-Muhtaseb, A. H., & Rooney, D. W. (2021). Conversion of biomass to biofuels and life cycle assessment: A review. *Environmental Chemistry Letters, 19*(6), 4075–4118. https://doi.org/10.1007/s10311-021-01273-0.

Osman, A. I., Mehta, N., Elgarahy, A. M., Hefny, M., Al-Hinai, A., Al-Muhtaseb, A. H., & Rooney, D. W. (2022b). Hydrogen production, storage, utilisation and environmental impacts: A review. *Environmental Chemistry Letters, 20*(1), 153–188. https://doi.org/10.1007/s10311-021-01322-8.

Osman, A. I., Deka, T. J., Baruah, D. C., & Rooney, D. W. (2020). Critical challenges in biohydrogen production processes from the organic feedstocks. *Biomass Conversion and Biorefinery*, 1–19.https://doi.org/10.1007/s13399-020-00965-x.

Osman, A. I., Mehta, N., Elgarahy, A. M., Hefny, M., Al-Hinai, A., Al-Muhtaseb, A. H., & Rooney, D. W. (2022a). Hydrogen production, storage, utilisation and environmental impacts: a review.

In *Environmental chemistry letters* (Vol. 20, Issue 1). Springer International Publishing. https://doi.org/10.1007/s10311-021-01322-8.

Parkinson, B., Tabatabaei, M., Upham, D. C., Ballinger, B., Greig, C., Smart, S., & McFarland, E. (2018). Hydrogen production using methane: Techno-economics of decarbonizing fuels and chemicals. *International Journal of Hydrogen Energy, 43*(5), 2540–2555. https://doi.org/10.1016/j.ijhydene.2017.12.081.

Parthasarathy, P., & Narayanan, K. S. (2014). Hydrogen production from steam gasification of biomass: Influence of process parameters on hydrogen yield-a review. *Renewable Energy, 66*, 570–579. https://doi.org/10.1016/j.renene.2013.12.025.

Rowbotham, J. S., Dyer, P. W., Greenwell, H. C., & Theodorou, M. K. (2012). Thermochemical processing of macroalgae: A late bloomer in the development of third-generation biofuels? *Biofuels, 3*(4), 441–461. https://doi.org/10.4155/bfs.12.29.

Safari, F., & Dincer, I. (2020). A review and comparative evaluation of thermochemical water splitting cycles for hydrogen production. *Energy Conversion and Management, 205*, 112182. https://doi.org/10.1016/j.enconman.2019.112182.

Safarian, S., Unnþórsson, R., & Richter, C. (2019). A review of biomass gasification modelling. *Renewable and Sustainable Energy Reviews, 110*, 378–391. https://doi.org/10.1016/j.rser.2019.05.003.

Salameh, Z. (2014). Energy storage. *Renewable Energy System Design*, 201–298. https://doi.org/10.1016/B978-0-12-374991-8.00004-0.

Salimi, M., Safari, F., Tavasoli, A., & Shakeri, A. (2016). Hydrothermal gasification of different agricultural wastes in supercritical water media for hydrogen production: A comparative study. *International Journal of Industrial Chemistry, 7*(3), 277–285. https://doi.org/10.1007/s40090-016-0091-y.

Salkuyeh, Y. K., Saville, B. A., & MacLean, H. L. (2018). Techno-economic analysis and life cycle assessment of hydrogen production from different biomass gasification processes. *International Journal of Hydrogen Energy, 43*(20), 9514–9528. https://doi.org/10.1016/j.ijhydene.2018.04.024.

Sánchez-Bastardo, N., Schlögl, R., & Ruland, H. (2020). Methane pyrolysis for CO2-Free H2 production: A green process to overcome renewable energies unsteadiness. *Chemie Ingenieur Technik, 92*(10), 1596–1609. https://doi.org/10.1002/CITE.202000029.

Sens, L., Piguel, Y., Neuling, U., Timmerberg, S., Wilbrand, K., & Kaltschmitt, M. (2022). Cost minimized hydrogen from solar and wind–production and supply in the European catchment area. *Energy Conversion and Management, 265*, 115742. https://doi.org/10.1016/j.enconman.2022.115742.

Shiva Kumar, S., & Himabindu, V. (2019). Hydrogen production by PEM water electrolysis–A review. *Materials Science for Energy Technologies, 2*(3), 442–454. https://doi.org/10.1016/j.mset.2019.03.002.

Singh Aulakh, D. J., Boulama, K. G., & Pharoah, J. G. (2021). On the reduction of electric energy consumption in electrolysis: A thermodynamic study. *International Journal of Hydrogen Energy, 46*(33), 17084–17096. https://doi.org/10.1016/j.ijhydene.2021.02.161.

Soltani, R., Rosen, M. A., & Dincer, I. (2014). Assessment of CO2 capture options from various points in steam methane reforming for hydrogen production. *International Journal of Hydrogen Energy, 39*(35), 20266–20275. https://doi.org/10.1016/J.IJHYDENE.2014.09.161.

Stiegel, G. J., & Ramezan, M. (2006). Hydrogen from coal gasification: An economical pathway to a sustainable energy future. *International Journal of Coal Geology, 65*(3–4), 173–190. https://doi.org/10.1016/j.coal.2005.05.002.

Suprianto, T., Winarto, W., & W., & Wardana, I. N. G. (2021). Synergistic effect of curcumin and activated carbon catalyst enhancing hydrogen production from biomass pyrolysis. *International Journal of Hydrogen Energy, 46*(10), 7147–7164. https://doi.org/10.1016/j.ijhydene.2020.11.211.

Tariq, R., Maqbool, F., & Abbas, S. Z. (2021). Small-scale production of hydrogen via auto-thermal reforming in an adiabatic packed bed reactor: Parametric study and reactor's optimization through response surface methodology. *Computers and Chemical Engineering, 145*, 107192. https://doi.org/10.1016/j.compchemeng.2020.107192.

The Future of Hydrogen. (2019). The future of hydrogen. https://doi.org/10.1787/1e0514c4-en.

Topriska, E., Kolokotroni, M., Dehouche, Z., & Wilson, E. (2015). Solar hydrogen system for cooking applications: Experimental and numerical study. *Renewable Energy, 83*, 717–728. https://doi.org/10.1016/j.renene.2015.05.011.

Topriska, E., Kolokotroni, M., Dehouche, Z., Novieto, D. T., & Wilson, E. A. (2016). The potential to generate solar hydrogen for cooking applications: Case studies of Ghana, Jamaica and Indonesia. *Renewable Energy, 95*, 495–509. https://doi.org/10.1016/J.RENENE.2016.04.060.

Uddin, M. N., Techato, K., Taweekun, J., Rahman, M. M., Rasul, M. G., Mahlia, T. M. I., & Ashrafur, S. M. (2018). An overview of recent developments in biomass pyrolysis technologies. *Energies, 11*(11), 3115. https://doi.org/10.3390/en11113115.

Wang, S., Dai, G., Yang, H., & Luo, Z. (2017). Lignocellulosic biomass pyrolysis mechanism: A state-of-the-art review. *Progress in Energy and Combustion Science, 62*, 33–86. https://doi.org/10.1016/j.pecs.2017.05.004.

Yu, L., Zhu, Q., Song, S., McElhenny, B., Wang, D., Wu, C., Qin, Z., Bao, J., Yu, Y., Chen, S., & Ren, Z. (2019). Non-noble metal-nitride based electrocatalysts for high-performance alkaline seawater electrolysis. *Nature Communications, 10*(1), 1–10. https://doi.org/10.1038/s41467-019-13092-7.

Zacharias, R., Visentin, S., Bock, S., & Hacker, V. (2019). High-pressure hydrogen production with inherent sequestration of a pure carbon dioxide stream via fixed bed chemical looping. *International Journal of Hydrogen Energy, 44*(16), 7943–7957. https://doi.org/10.1016/j.ijhydene.2019.01.257.

Zhang, C., Kang, X., Liang, N., & Abdullah, A. (2017a). Improvement of biohydrogen production from dark fermentation by cocultures and activated carbon immobilization. *Energy and Fuels, 31*(11), 12217–12222. https://doi.org/10.1021/acs.energyfuels.7b02035.

Zhang, H., Gao, Z., Ao, W., Li, J., Liu, G., Fu, J., Ran, C., Mao, X., Kang, Q., Liu, Y., & Dai, J. (2017b). Microwave-assisted pyrolysis of textile dyeing sludge using different additives. *Journal of Analytical and Applied Pyrolysis, 127*, 140–149. https://doi.org/10.1016/j.jaap.2017.08.014.

Zhang, Y., Wan, L., Guan, J., Xiong, Q., Zhang, S., & Jin, X. (2020). A review on biomass gasification: Effect of main parameters on char generation and reaction. *Energy and Fuels, 34*(11), 13438–13455. https://doi.org/10.1021/acs.energyfuels.0c02900.

Zhu, Y., Biddy, M. J., Jones, S. B., Elliott, D. C., & Schmidt, A. J. (2014). Techno-economic analysis of liquid fuel production from woody biomass via hydrothermal liquefaction (HTL) and upgrading. *Applied Energy, 129*, 384–394. https://doi.org/10.1016/j.apenergy.2014.03.053.

Zulfhazli, Keeley, A. R., Takeda, S., & Managi, S. (2022). A systematic review of the techno-economic assessment of various hydrogen production methods of power generation. *Frontiers in Sustainability, 3*, 117. https://doi.org/10.3389/frsus.2022.943145.

An Evaluation of a Proposed Solar Plant System for Ceramics Production in Ghana

Isaac Mensah, Wendilia Jean Toussaint Garango, and Satyanarayana Narra

Abstract Ceramics are manufactured by firing clay materials at high temperatures in furnaces. High energy and fuel costs are the two main drivers leading to the decline of the ceramic sector of Ghana. Unlike countries such as China, Japan, the United Kingdom, and the United States of America, just to name a few, the cost of these resources is borne by the local manufacturers without any subsidy from the government. This results in high production costs thereby leading to the influx of imported cheap products from China, Japan, and the like. This study seeks to provide a snapshot of a proposed solar photovoltaic (PV) system for ceramics production which is consistent with the Ghanaian government's aim to further develop the sector. The authors consider a solar PV system with battery storage connected on-grid as the most efficient and pragmatic solution for the manufacture of ceramics in Ghana. The solar irradiation levels of Tamale were used to size the solar PV system to meet the energy demand of the electric oven (E-oven). It was estimated that, the cost of electricity saved by supplying 60% of power from the solar PV system was GH¢164,721.60 (€9,336.62) per annum. The implementation of the study findings will lead to the attainment of United Nations Sustainable Development Goals (SDGs); namely: affordable and clean energy (SDG 7), industry, innovation, and infrastructure (SDG 9), responsible consumption and production (SDG 12) and climate action (SDG 13). In conclusion, the proposed design demonstrates the potential of solar energy as a sustainable alternative for the decarbonization of the ceramics industry.

Keywords Ceramics production · Ghana · Sustainability · Firing · Energy transition

I. Mensah (✉) · W. J. T. Garango · S. Narra
Material and Energy Valorisation of Biogenous Residues and Waste and Resource Management, University of Rostock, Rostock, Germany
e-mail: isaac.mensah@uni-rostock.de

M.-M. Narra and S. Narra (eds.), *African Green Transition Through Innovative Pathways*, World Sustainability Series, https://doi.org/10.1007/978-3-031-87043-9_13

1 Introduction

Ceramics production usually involves four basic processing steps; namely: raw materials preparation, forming, drying and firing. Firing is the most crucial stage in the processing of ceramics as the final quality and price are dependent on it (Hay et al., 2024). Electricity, natural gas, liquefied petroleum gas (LPG) and so on are among the energy sources utilized during firing (Agrafiotis & Tsoutsos, 2001). Asante-Kyei et al. (2019) observed that the majority of Ghanaian potters' resort to firewood as a fuel source during the firing of their products. The burning of firewood in kilns releases smoke which contributes to climate change, deforestation, *inter alia.*

It is interesting to note that the production of ceramics was once a profitable industry in Ghana which contributed close to $1.5 billion to the country's economy in 2022. With such gains, the Ghanaian government aimed to capitalize on the industry by establishing academic institutions to train students to acquire the requisite skills needed for its management. Unfortunately, this was not realized due to the several challenges that plagued the industry including outdated equipment and machinery usage, high energy, fuel, water and raw material costs. The influx of cheap products from China, Spain, the United Kingdom, Japan and so on, has collapsed the local market (Asante-kyei, 2019).

In the quest for further development of the dead but promising venture by the Ghanaian government, the main purpose of this study is to target the transitioning of energy in ceramic production and recycling to a more sustainable one. One way that this can be achieved is by integrating solar PV system with battery storage connected on-grid for the production of ceramics in Ghana.

2 Methodology

This section describes the choice of the study area, the solar irradiation levels, the proposed E-oven solar PV system plant configuration, methods for calculating the yearly carbon dioxide (CO_2) emissions decarbonize and theoretical performance evaluation assessment of the E-oven in comparison to firewood and/or gas oven.

2.1 Study Area

Tamale Technical University was selected as the partner institution for the establishment of the proposed plant. The reason being that the University runs a programme in Industrial Art with option in Ceramics and is in the process of erecting a new ceramic studio which can integrate the solar powered E-oven. Tamale is the capital city of the Northern region of Ghana. Figure 1 shows the map of Ghana and the minimum and maximum daily solar irradiation levels of Tamale. The country's solar irradiation

Fig. 1 A map of Ghana showing daily solar irradiation levels of Tamale

levels per day range between 4–6 kWh/m^2 and the sunshine hours per annum are between 1800 to 3000 h. In Tamale, the solar irradiation levels range from 4.80 kWh/m^2/day in August to 5.90 kWh/m^2/day in both April and May. This makes Tamale a suitable place for the installation of the plant (Asumadu-Sarkodie & Owusu, 2016). There are 39 fuel stations in the Tamale metropolis where gas fuel can be sourced for the gas oven (Amoako et al., 2022). Tamale has five clay deposits: Koblimahako with a clay reserve of 9,455,892 tons; Kpaliga with 259,200 tons; Nyankyala with 48,000 tons; Kukuo with 234,582 tons; and Yapei with 38,694 tons which indicates the availability of raw materials for ceramics production (Asamoah et al., 2018).

2.2 *Proposed E-oven Solar PV System Plant Configuration*

The proposed design plans to establish two ceramic ovens. One will be operated using natural gas as the fuel source and the other will be powered by solar PV system connected on-grid. The natural gas oven will be used to simulate the existing

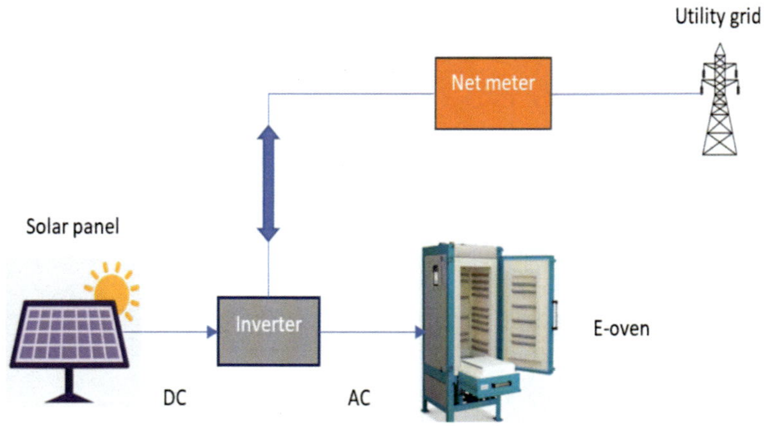

Fig. 2 Process illustration of power generation for the ceramic E-oven

practice in the ceramic industry and the E-oven with 75 kWe capacity will serve as an alternative solution to decarbonize the ceramic industry by comparing the processes and products. To make a clear comparison between the two ovens, the E-oven will be chosen in such a way that it is operated in the same production conditions as the gas oven. Figure 2 shows the arrangement of power generation from the solar PV field to the ceramic oven. The connection includes solar arrays, an inverter, a battery system, a net metering unit, a utility grid and the E-oven. The solar PV field will be connected to a battery energy storage system (1500 VDc Li) to always keep the E-oven in operation during inclement weather conditions. It is planned that the system will also be connected on-grid to keep it working throughout the whole year in the case when the battery storage is used up. The amount of power supplied by the grid will be compensated by the solar PV system when there is a surplus of power production. This will be catered for through the installation of the net metering system.

2.3 CO_2 Emissions Measurement

The yearly CO_2 emissions decarbonize can be estimated using the fuel analysis method from Eq. 1 (US EPA, 2023).

$$X = Y \times HHV \times EF \tag{1}$$

Where: X = mass of CO_2 emitted; Y = mass or volume of fuel combusted in the year; HHV = higher heating value of the fuel in units of energy per mass or volume of fuel; EF = CO_2 emission factor per energy unit.

2.4 *Performance Evaluation Analysis*

To determine the performance evaluation of the two ovens, samples of burnt bricks from the two ovens will be subjected to physical, mechanical and durability tests such as dry density, compressive strength test, splitting tensile strength test, abrasive resistance test and water absorption test following standard protocols as described by Danso and Akwaboah (2021).

3 Results and Discussion

Theoretical solar irradiation levels of some selected cities in Ghana.

Figure 3 shows the monthly solar irradiation levels of some selected cities in the Northern Regions (Wa and Tamale), Middle Belt (Kumasi) and Southern Regions (Accra, Koforidua, Takoradi and Ho) of Ghana. The cities were selected for consideration due to data availability and whether there is a Technical University offering a programme in ceramics or not. The average solar irradiation levels are 5.5 kWh/m^2/day, 5.4 kWh/m^2/day, 4.6 kWh/m^2/day, 5.1 kWh/m^2/day, 4.8 kWh/m^2/day, 5.0 kWh/m^2/day and 5.1 kWh/m^2/day for Wa, Tamale, Kumasi, Accra, Koforidua, Takoradi and Ho, respectively. The higher the solar irradiation levels the greater the power generated by the solar PV system (The Pennsylvania State University, 2023). The relatively higher solar irradiation levels in Tamale, among other factors, to the other cities except Wa in the Upper West region influenced the choice of Tamale Technical University in Tamale as the proposed site for this study. It is worth mentioning that either of the cities could have been chosen due to the higher solar irradiation levels which are greater than 4 kWh/m^2/day (Energypedia, 2018).

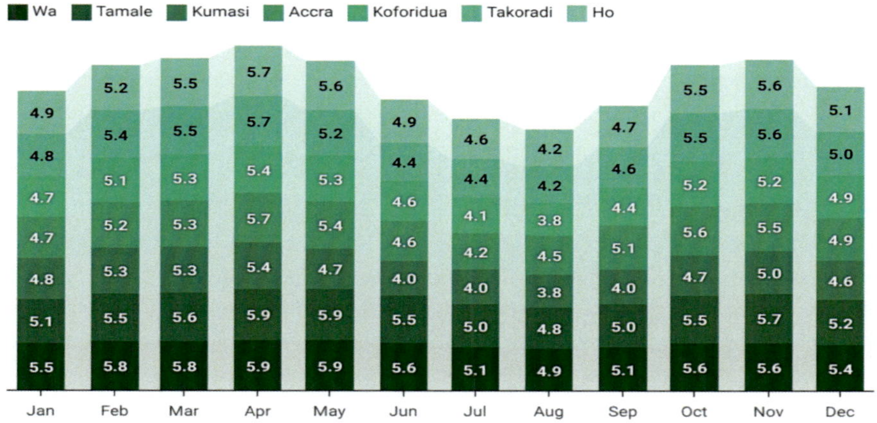

Fig. 3 Monthly solar irradiation levels in kWh/m^2/day of some selected cities in Ghana (Asumadu-Sarkodie & Owusu, 2016)

Table 1 Sizing of the E-oven solar PV system

Parameter	Value
Energy consumption of E-oven	75 kWh
Energy consumption of E-oven operating for 10 h per daily	750 kWh/day
Energy consumption of E-oven per monthly NB: Assume 5 working days	15,000 kWh/month
Energy consumption of E-oven per yearly	180,000 kWh/year
% Power offset with solar energy NB: Assume 60% of power will be supplied by the solar system	108,000 kWh/year
Dividing by annual average PSH NB: Annual average PSH in Tamale 1,971h/year	54.8 kW
Dividing by a derate factor of 86%	63.72 kW ≡ 63,720W
Power of system	63,720 W
Dividing by peak wattage of the solar panels NB: Take 430Wp	148.19
Number of solar panels to be installed	148 panels

3.1 *Theoretical Sizing of the Solar System*

The number of solar PV panels required to meet the energy demand of the ceramic E-oven is given in Table 1. For this study, the derating factor which accounts for the solar system inefficiencies such as wiring losses, temperature, dust, shading, etc. was taken as 0.86 (Dobos, 2014). With this, the system capacity was estimated as 63,720 watts, and the number of solar PV panels required was estimated to be 148 panels (panel size of 430 Wp). Assuming the cost of a 430 Wp panel is 308 € then the estimated cost of the panels is 45,584 € (Ogbomo et al., 2017). According to PURC (2024), the cost of electricity for industrial-class consumers in the year 2024 in Ghana is GH¢1.5252/kWh (€0.086/kWh). With this, the cost of electricity saved by supplying 60% of the power from the solar PV system was estimated as GH¢164,721.6 (€9,336.62) per year.

3.2 *Future Research*

Several attempts have been made to theoretically assess the feasibility of the proposed solar PV system design; however, due to lack of data in the literature, some objectives including performance evaluation of the ovens, GHG emissions measurements, economic and profitability analyses will be addressed in future studies.

4 Conclusion

This study represents a critical step toward achieving a sustainable energy transition in the ceramic industry in Ghana. By integrating solar PV systems with battery storage connected on-grid in ceramic production, the study not only addresses the pressing issue of high energy and production costs but also contributes to significant environmental benefits, including a reduction in greenhouse gas emissions. The proposed solution aligns with global efforts to combat climate change and supports Ghana's commitment to increasing its renewable energy share.

Acknowledgements The authors would like to thank the financial support received from the 7th Energy Research Programme funded by the German Federal Ministry of Education and Research (BMBF).

A special thanks goes to Dr.-Ing. Qahtan Thabit at the Professorship for Material and Energy Valorisation of Biogenous Residues of the University of Rostock for his invaluable contribution towards the write-up of this conference paper.

References

Agrafiotis, C., & Tsoutsos, T. (2001). Energy saving technologies in the European ceramic sector: A systemic review. *Applied Thermal Engineering, 21*, 1231–1249.

Amoako, C., Adamtey, R., & Doe, B. (2022). Fire risks management in emerging Ghanaian cities: Land use planning responses for siting petrol and gas stations in the Tamale Metropolis. *GeoJournal, 87*(4), 3065–3076.

Asamoah, R. B., Nyankson, E., Annan, E., Agyei-Tuffour, B., Efavi, J. K., Kan-Dapaah, K., Apalangya, V. A., Damoah, L. N. W., Dodoo-Arhin, D., Tiburu, E. K., Kwofie, S. K., Onwona-Agyeman, B., & Yaya, A. (2018). Industrial applications of clay materials from Ghana (A review). *Oriental Journal of Chemistry, 34*(4), 1719–1734.

Asante-Kyei, K. (2019). The challenges facing ceramic industries in Ghana: Its impact on ceramic education in Ghana. *Higher Education Research, 4*(2), 30–35.

Asante-Kyei, K., Asiedu, E., & Marfo, S. (2019). Design and fabrication of gas kiln using local materials sources. *Journal of Arts & Humanities, 8*(3), 73–84.

Asumadu-Sarkodie, S., & Owusu, P. A. (2016). A review of Ghana's solar energy potential. *AIMS Energy, 4*(5), 675–696.

Danso, H., & Akwaboah, M. (2021). Assessment of the quality of burnt bricks produced in Ghana: The case of Ashanti Region. *Case Studies in Construction Materials, 15*(2021), 1–13.

Dobos, A. P. (2014). PVWatts version 5 manual (No. NREL/TP-6A20–62641). National Renewable Energy Laboratory (NREL), Golden, CO (United States). Retrieved 2 November, from, https://www.nrel.gov/docs/fy14osti/62641.pdf.

Energypedia. (2018). *Solar irradiation.* Retrieved 2 November, from, https://energypedia.info/wiki/SPIS_Toolbox_-_Solar_Irradiation.

Hay, R., Haider, S., Altantawi, A., & Celik, K. (2024). Color modification of ceramics with controlled firing. *Ceramics International, 50*, 566–574.

Ogbomo, O. O., Amalu, E. H., Ekere, N. N., & Olagbegi, P. O. (2017). A review of photovoltaic module technologies for increased performance in tropical climate. *Renewable and Sustainable Energy Reviews, 75*, 1225–1238.

PURC. (2024). *2024 First quarter tariff review decision for electricity, natural gas and water*. Retrieved 2 November, from, https://www.purc.com.gh/attachment/829688-20240228090210.pdf.

The Pennsylvania State University. (2023). *Irradiance and PV performance optimization*. Retrieved 2 November, from, https://www.e-education.psu.edu/ae868/node/877.

US EPA. (2023). *Direct emissions from stationary combustion sources*. Retrieved 2 September, from, https://www.epa.gov/sites/default/files/2020-12/documents/stationaryemissions.pdf.

Impact of Digestate Formed During Biogas Generation Process on Hydrogen Production

Kalifa Dembele, Kosi Mawuéna Novidzro, and Satyanarayana Narra

Abstract Hydrogen production from renewable sources has become a key objective in the energy transition to a more sustainable, and low-carbon economy. Among the many potential sources for hydrogen production, ammonia generated from anaerobic digestate presents a promising opportunity. Although digestate has its advantages, the presence of toxic ammonia poses environmental problems. Proper control of ammonia volatilization is essential to minimize risks. This study highlights the importance of side stream stripping as a viable method of recovering ammonia from digestate to support hydrogen production. In this study, the ammonium nitrogen (NH_4-N) data of various substrates used for biogas production and its digestate was collected from a brochure on the results of biogas production measurements in Germany published by the Agency for Renewable Resources (FNR: Fachagentur Nachwachsende Rohstoffe). These data were used to estimate the amount of ammonia (NH_3) contained in the digestate of various feedstocks for hydrogen production. Side-stream stripping has been shown to extract ammonia safely, facilitating handling and reducing gaseous and particulate emissions. In addition, the process reduces transport costs by removing excess water from the digestate, while the stripped digestate is returned to the anaerobic digester, ensuring a closed-loop system. The results of this study showed an increase in ammonia content as a function of the ammonium nitrogen content in the digestate. Digestate with a high ammonium nitrogen content produced more hydrogen respectively. The hydrogen potential produced depends on the amount of ammonia available in the digestate. This study explores the potential impact of anaerobic digestate on hydrogen production. Regarding the evaluation of catalysts used in ammonia decomposition. Ru-based catalysts are widely regarded as the best choice for highly efficient ammonia (NH_3) decomposition. However, their

K. Dembele (✉)
WASCAL, University of Lomé, Lomé, Togo
e-mail: dembele.k@edu.wascal.org

K. M. Novidzro
Laboratory of Engineering Process and Natural Resources, University of Lomé, Lomé, Togo

S. Narra
Material and Energy Valorisation of Biogenous Residues and Waste and Resource Management, University of Rostock, Rostock, Germany

M.-M. Narra and S. Narra (eds.), *African Green Transition Through Innovative Pathways*, World Sustainability Series, https://doi.org/10.1007/978-3-031-87043-9_14

considerable cost and limited availability are disadvantages for large-scale applications. In contrast, among non-noble metal catalysts, nickel-based catalysts show the highest activity, making nickel a promising alternative material for ammonia (NH_3) decomposition due to its affordability. At present, challenges remain in improving the efficiency of Ru- and Ni-based catalysts.

Keywords Anaerobic digestate · Side stream stripping · Ammonia decomposition

1 Introduction

Over the past few years, the biogas sector has experienced an important growth in the number of biogas installations all over Europe, and consequently, the quantity of digestate also has had a significant increase. In Europe, biogas production by anaerobic digestion (AD) is a common source of renewable energy and the current number of installations is around 13,000. Together with biogas, digestate is one of the two main by-products resulting from the biogas process. The digested effluent is a liquid product rich in nitrogen (N), phosphorus (P), potassium (K), and micronutrients. Therefore, there is a wide variety of digestate utilization depending on the quality, the origin of the feedstock, the operating conditions of the process as well as the phase of the by-product. The most common end uses are biofertilizer and soil amendment, due to its essential characteristics and when the quality is adequate for agricultural use (Fachagentur Nachwachsende Rohstoffe e.V. (FNR), 2012).

The production of hydrogen from renewable sources has become a key objective in the energy transition toward a more sustainable and low-carbon economy. Among the many potential sources of hydrogen, ammonia from anaerobic digestate presents itself as a promising opportunity. Digestate is a by-product of the degradation of organic waste by anaerobic digestion, and it contains a significant amount of ammonia. The recovery of ammonia contained in the digestate offers a potential route to produce renewable hydrogen.

Ammonia is a promising hydrogen carrier due to its high gravimetric (17.8 wt% H_2) and volumetric (121 kg.m^3 in liquid form) H_2 and 1.4 times greater than that of liquid hydrogen (Aziz et al., 2020; Zhang et al., 2019).

Consequently, transporting and storing ammonia is relatively easy and energy efficient. By breaking down ammonia for hydrogen production, we can fulfil the demand for nearly carbon–neutral hydrogen, leading to an exceptionally low carbon footprint (Liu et al., 2021).

1.1 Problem Statement

Management of digestate involves several topics such as storage, processing, transportation, utilization, economics, and environmental quality. It has many benefits, but also presents several challenges. One of the biggest issues is its ammonia content, which can be toxic to aquatic life and contribute to air pollution. However, due to its hazards and toxicity, precautions are needed to ensure safety (Lee et al., 2019; Makepeace et al., 2019; Zhang et al., 2019).

In addition, digestate can be difficult to handle and transport due to its high-water content.

The main challenge related to the production of hydrogen from ammonia is the efficiency and cost-effectiveness of converting it into hydrogen.

Therefore, one of the major challenges in the production of hydrogen through ammonia decomposition is the need for efficient catalysts that facilitate the conversion process. Additionally, the safe storage and transportation of the hydrogen produced are also significant concerns. Despite the fact that the catalytic synthesis of ammonia is well understood, ammonia decomposition remains a process that requires further comprehension and optimization, especially since there is currently no industrial technology available for its efficient implementation (Lamb et al., 2019).

1.2 Objectives

This research aims to estimate the potential of hydrogen production from digestate of various feedstocks, determine the energy and power potential content in hydrogen production, and compare the values of hydrogen production using different catalysts, such as Ru-based and Ni-based catalysts.

2 Methodology

2.1 Raw Material and Data Collection

The raw material considered in the present work is the digestate from biogas plants. Digestate is a by-product of the degradation of organic waste by anaerobic digestion, and it contains a significant amount of ammonia. The recovery of ammonia contained in the digestate offers a potential route for the production of renewable hydrogen.

In this study, the ammonium nitrogen (NH_4-N) data of different feedstocks used for biogas production and its digestate was collected from a brochure on the results of biogas production measurements in Germany published by the Agency for Renewable Resources (FNR: Fachagentur Nachwachsende Rohstoffe). These data were used to estimate the amount of ammonia (NH_3) contained in the digestate of different feedstocks for hydrogen production.

All these calculations were carried out using Excel software to develop a hydrogen production model. Table 1 shows the list of different feedstocks with their ammonium nitrogen content collected for hydrogen production.

Table 1 Data collection of ammonium nitrogen (NH_4-N) from the digestate and its substrates

Group	Feedstock	NH_4-N (Kg/t) Substrates	NH_4-N (Kg/t) Digestate
A	Pig manure (81.7%), Feed stillage (18.3%)	2.1	2.6
B	Cattle manure (57.5%), Silo corn (20.5%), Chicken manure (17.1%), rapeseed threshing (4%), lawn clippings (0.6%), horse manure (0.3%)	3.2	4
C	Pig manure (86%), Dry chicken droppings (8.8%), Glycerin (2.6%), Coffee residues (1.4%), Sunflower shell (0.8%), Molasses (0.4%)	4.4	5.2
D	Potato peel waste (47.2%), Fat (18.8), Silo maize (15.6), Dry chicken droppings (6.4%), Flour waste (2.3%), Inocilum (2.1%), Apple pomace (2.1%), Whey (1.9%), Cattle manure (1.8%), Grain cleaning (1%), Organic waste (0.5%), Molasses (0.4%)	1.1	2.8
E	Silo maize (50%), Pig manure (23.4%), Grain cleaning (10%), Water (5.2%), Whey (3.6%), Sugar beet leaf (3%), Potatoes (2.4%), Rye (1.2%), Grass silage (1%) Potato pulp (0.2%)	1.7	3.8
F	Silo Maize (27.4%), Cattle manure (14.1%), Pig manure (13.6%), Leftovers (13.2%), Grass silage (7.9%), Whey (5.8%), Potato pulp (5.2%), Stillage (4.4%), Fat (4%), Dog food (3%), Sun lower (0.8), Apple pomace (0.5%), Grain cleanng (0.1%)	1.5	4
G	Pig manure (56.7%), Corn silage (40.4%), Barley grist (1.5%), Dry chicken droppings (1.5%)	2.2	3.2
H	Pig manure (75%), Corn silage (11.8%), Silage (9.1%), Rye grains (6.1%)	3	4
I	Pig manure (49.9%), Corn silage (44.9%), Wheat grains (5.2%)	1.5	2.4
J	Cattle manure (55.6%), Pig manure (5.7%), Turkey manure (2.1%), Corn silage (34.8%), Rye GPS (1.8%)	1.5	3.1
K	Pig manure (49.3%), Turkey manure (7.4%), Corn silage (30.3%), Grass silage (1.2%), Cereal (11.8%)	2.9	5.7
L	Cattle manure (81.4%), Corn silage (15.9%), Wheat grains (2.7%)	2.3	2.7
M	Cattle manure (81.3%), Dry chicken droppings (9.1%), Corn silage (7%), Barley grains (2.5%)	1.7	3.5
N	Corn silage (85.1%), Chicken manure (5.2%), Cattle manure (5.3%), Grass silage (4.4%)	0.8	3.4
O	Corn silage (69.7%), Rye meal (4.2%), Rye ensilage (Full plant) (2.3%), Grass/summer barley (1.1%), Pig manure (8.2%), Cattle manure (14.1%), Grass silage (0.5%)	0.6	2.6
P	Maize silage/ Grain (29.3%), Pig manure (29.1%), Green rye silage (3.1%), Corn silage (38.5%)	1.2	3.2

(continued)

Table 1 (continued)

Group	Feedstock	NH_4-N (Kg/t) Substrates	NH_4-N (Kg/t) Digestate
Q	Cattle/manure (48%), Corn silage (43.5%), Grass silage (1.2%), Wheat grist (4.1%), Rye ensilage (Full plant) (3.3%)	1.1	2.8
R	Cattle manure (31.8%), corn silage (61.6%) Wheat grist (6.6%)	1.2	3.1
S	Cattle manure (30.7%) Corn silage (38.3%), Rye ensilage (Full plant) (17.5%), Cattle manure (1.3%), Grass silage (8.1%), wheat grains (4.1%)	1.1	2.9
T	Pig manure (23.5%), Corn grain silage (36%), Cattle manure (23.6%), wheat (3.6%), Barley ensilage (Full plant) (3.3%), Corn Silage (10%)	1.1	4.2
U	Cattle manure (42,7%), Corn silage (41.2%), Grass silage (14.1%), Wheat grist (2.1%)	1	2.8
V	Pig manure (50%), Turkey manure (7.5%), Corn silage (8.3%), Grass silage (1.8%), Cereal porridge (20.5%) Corn-Cob-Mix (5.6%), Water (4.3%), Wheat grist (2%)	2.7	5.3
W	Cattle manure (30.3%), Corn silage (14.3%), Grass silage (31.2%), Grain meal (3.9%), Oat/rye ensilage (Full plant) (20.2%)	0.8	2.7
X	Corn silage (49.3%), Pig manure (38.1%), grass silage (10.2%), Wheat grist (2.5%)	1	2.5
Y	Corn silage (81%), Rye ensilage (Full plant) (2.6%), Green rye silage (16.3%)	0.5	3.5
Z	Corn silage (89%), Rye ensilage (Full plant) (9%), Millet/sunflower (2%)	0.7	3.2
AB	Turkey manure (5.4%), Corn silage (83.4%), Rye ensilage (Full plant) (6.2%), Corn-Cob-Mix (2.8%), Wheat grains (1.8%), Potatoes (0.3%)	1.2	2.9
CD	Corn silage (96.9%), Corn-Cob-Mix (2.5%), chicken manure (0.6%)	0.7	2.6
EF	Corn silage (98.3%), Grain meal (1.2%), Sugar Beetz (0.5)	0.5	2.4
GH	Corn silage (81.1%), Grass silage (4.8%), Rye ensilage (Full plant) (12.7%), Wheat grist (0.6%), Cattle manure (0.7%)	0.4	2.6
IJ	Turkey manure (29.9%), Corn silage (50.7%), Rye ensilage (6.5%), Sudan grass (9.5%), Green waste (1%), Grass silage (2.3%)	1.8	2.8
KL	Corn silage (70.2%), rye meal (8.3%), Pig manure (17.9%), Rye ensilage (Full plant) (3.5%)	1	4

Source (Fachagentur Nachwachsende Rohstoffe e.V. (FNR), 2009)

2.2 Applied Methods for Digestate Conversion into Ammonia

To convert the measured amount of NH_4-N from the digestate into ammonia (NH_3), NH_4-N is multiplied by the conversion factor of 1.2159. This conversion factor represents the ratio of the molar masses between ammonia (NH_3) and nitrogen (N). This factor converts directly the amount of ammonium nitrogen (NH_4-N) into the corresponding amount of ammonia (NH_3).

The following equation is used to estimate the amount of ammonia (NH_3) produced in the digestate based on the measured amount of ammonium nitrogen (NH_4-N) present.

$$Amount\ of\ (NH_3) = 1.2159 \times Amount\ of\ NH_4 - N \tag{1}$$

Knowing that the molar mass of ammonia (NH_3) is 17.03 g/mole, and the molar mass of nitrogen (N) is 14. 01 g/mole.

2.3 Processes for Ammonia Conversion into Hydrogen

Ammonia can be dissociated to produce hydrogen gas and nitrogen gas. This reaction can be expressed as shown in Eq. (2)

$$2NH_3(g) \leftrightarrows N_2(g) + 3H_2(g) : \Delta H^\circ = 92\ \mathrm{kJ\ mol^{-1}} \tag{2}$$

The decomposition reaction of ammonia is endothermic and reaches 99.99% ammonia conversion at 400 °C and 1 atm according to thermodynamics, considering an inlet flow composed only of ammonia. In practice, the rate of conversion depends on both temperature and the types of catalysts used.

Equation 3 can be used to calculate the hydrogen potential from ammonia:

$$H_2 = \alpha \times NH_3 \tag{3}$$

where:

H_2; is the quantity of hydrogen produced in (moles);

NH_3; is the quantity of ammonia in (moles);

α; is the conversion factor.

In experimental conditions, various factors can affect the conversion factors in a chemical reaction.

Some of these factors include temperature, pressure, the presence of catalysts, and activation energy. Therefore, determining the appropriate value of alpha requires considering these factors and conducting experimental or theoretical studies to establish the conversion factor specific to the given conditions.

The value of alpha can vary depending on the specific reaction system and the particular circumstances under which ammonia is being converted to hydrogen.

Thermal decomposition or catalytic cracking is the most common technique used for the generation of hydrogen from ammonia. It can be carried out with or without the presence of a catalyst, as the presence of a catalyst allows the decrease of the temperature necessary for the decomposition. For this reason, it is important to study the catalysts involved as well as different reactor configurations in order to decrease the supply of energy, in this case in the form of heat, to the system.

The thermal decomposition of ammonia (NH_3) is an endothermic reaction, which means that it requires the input of heat. Without the presence of catalysts, the temperature range for the thermal decomposition of ammonia is generally between 700 and 1,100 °C (Ristig et al., 2022).

Due to the endothermicity of the reaction, ammonia decomposition is favoured by an increase in temperature. Consequently, high efficiency, or even total conversion, in ammonia decomposition requires high temperatures and low pressures for a technical process. Both temperature and pressure must be low for a technical process.

In this work, especially the conversion factor depends on temperature, pressure, catalysts used (Ruthenium-based catalyst and Nickel-based catalyst), activation energy, and conversion rate.

Table 2 summarizes the performance of Ru and Ni-based catalysts used in the decomposition of ammonia. It includes details such as the support material, WHSV, inlet flow, temperature, conversion rate, and activation energy.

Based on the conditions given in Table 2 the conversion factors of different catalysts are presented in Table 3.

Table 2 Catalysts based on Ru, and Ni used to decompose Ammonia and their catalytic performance at 1 atmosphere

Active phase	Support	WHSV (ml/g/h)	%NH_3 inlet flow	Temperature (°C)	Conversion rate (%)	AE (KJ/mole)	Ref
Ru	SiO_2	30,000	100	500	96	41	(Yao et al., 2011)
Ru	CNTs	30,000	100	500	88	69	(Yin et al., 2004)
Ni	Al_2O_3	36,000	100	600	93	123	(Gu et al., 2015)

Table 3 Conversion factors of different catalysts

Catalysts	Conversion factor(α)
Ru/SiO_2	0.72
Ru/CNTs	0.66
Ni/ Al_2O_3	0.69

2.3.1 Determination of the Energy Content from Hydrogen

The calculation of the energy content from hydrogen is made according to the following equation below.

At standard temperature and pressure conditions (STP): 1 mol is equal to 22.4 l.

The following step is to convert the values of hydrogen in liter into volume (m^3) at standard temperature and pressure conditions (STP): 1 L is equal to 0.001 m^3.

$$Energy\ content(Mj) = Mass\ (kg) \times HHV\ of\ hydrogen \tag{4}$$

where: Mass (kg) = Volume of hydrogen × density of hydrogen at (NTP)

Knowing that: The density of hydrogen at normal temperature and pressure (NTP) is equal to 0.08375 (kg/m^3).

The High heating value (HHV) of hydrogen at normal temperature (NTP) is equal to 120 (MJ/kg).

2.4 Determination of Power Potential Based on the Energy Content of Hydrogen

To Calculate the power content in kilowatt-hours (kWh) from the energy content of hydrogen. The following equation is used to convert the energy content of hydrogen from megajoules (MJ) to kilowatt-hours (kWh) by multiplying by the conversion factor of 0.2778 kWh/MJ. After calculating the energy content, the power potential is calculated through this equation:

$$\text{Power content (kWh)} = 0.2778 \times \text{Energy content (MJ)} \tag{5}$$

Knowing that the conversion factor between MJ and kilowatt-hour is: 1MJ is equal to 0.27 (kW/h).

3 Results and Discussion

3.1 Development of Ammonia Recovery Methodology from the Anaerobic Digestion Process for Hydrogen Production

The development of ammonia recovery from the anaerobic digestion is accordingly illustrated by Fig. 1.

The volatilization of ammonia is the primary environmental impact that needs to be effectively controlled and mitigated to minimize the risks of volatilization.

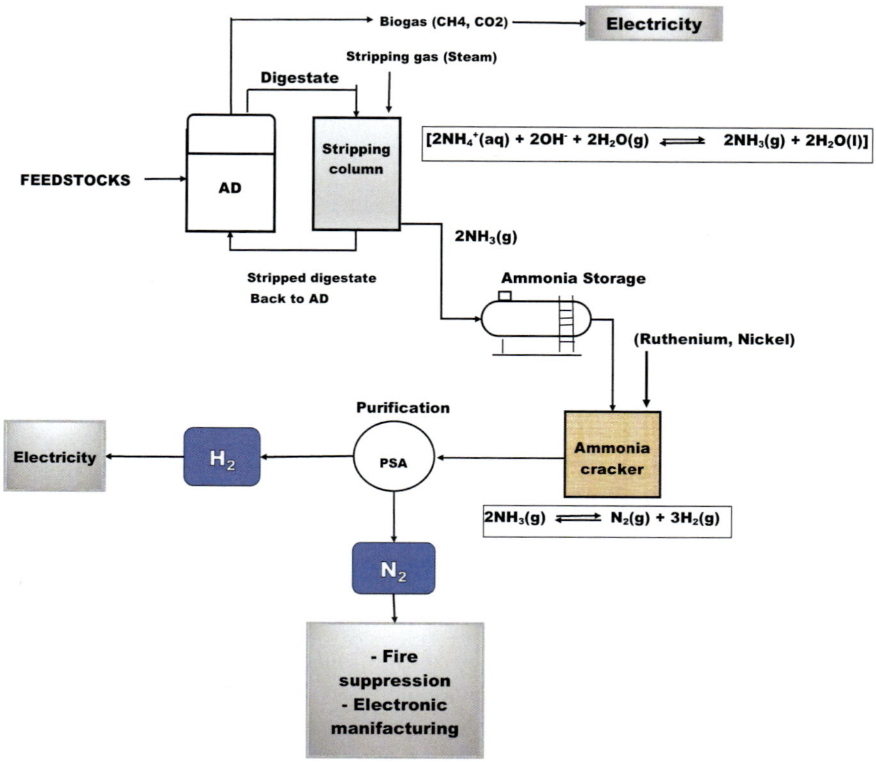

Fig. 1 Overall flow sheets

In addition, digestate can be difficult to handle and transport due to its high water content. To address these challenges, side stream stripping can be used to recover ammonia from the digestate, making it safer to handle and reducing its emission impact. This process can also help reduce transportation costs by removing excess water from the digestate.

Side stream stripping is a process that uses steam to separate ammonia from liquid waste. When the digestate moves from the digester to the stripping column the waste is heated and then passes through a distillation column, where the ammonia is separated from the liquid.

During stream stripping, the reaction between ammonium ions (NH_4^+) and hydroxide ions OH^- with steam results in the formation of gaseous ammonia and water. This reaction can be expressed in Eq. 6:

$$2NH_4^+(aq) + OH^- + H_2O(g) \rightarrow 2NH_3(g) + 2H_2O(l) \quad (6)$$

This process is particularly effective for the recovery of ammonia from anaerobic digestate, as it can remove up to 99% of the ammonia present in the waste and it is also energy efficient, as it uses the heat generated during the digestion process to power the distillation column.

After this step stripped digestate goes back into the digester and the ammonia can then be condensed and collected for use in other applications.

Ammonia gas is stored in a tank and flows into the ammonia cracker. Each ammonia molecule contains one nitrogen and three hydrogen atoms, which are separated as the ammonia comes into contact with the catalyst. The catalyst facilitates the dissociation of ammonia into nitrogen and hydrogen gas. This reaction can be represented as Eq. (7).

$$2NH_3(g) \leftrightarrows N_2(g) + 3H_2(g); \Delta H^\circ = 92 \text{ kJ mol}^{-1} \quad (7)$$

This is an endothermic reaction, requiring heat input to proceed.

After this process the mix of 75% hydrogen and 25% nitrogen pass through the adsorbent materials pressure swing adsorption (PSA) which is commonly used to purify hydrogen. The adsorbent material selectively adsorbs the impurities allowing hydrogen to pass through and separate complete nitrogen from hydrogen to produce electricity.

Nitrogen gas is a byproduct of the purification process which can be used for fire suppression; In electronics manufacturing processes, nitrogen gas is used for cooling cleaning, and as a protective atmosphere during soldering and other operations. It helps prevent oxidation and ensures high-quality production of electronic components; Nitrogen gas is sometimes used to inflate vehicle tires. It offers benefits such as maintaining tire pressure for longer periods, reducing tire wear, and improving fuel efficiency; Nitrogen gas is frequently used to create an inert atmosphere in industrial processes. It displaces oxygen and other reactive gases, preventing oxidation, combustion, or undesirable reactions. This is crucial in industries such as food packaging, pharmaceuticals, electronics manufacturing, and chemical production ("Industrial Uses of Nitrogen Gas," 2021).

3.2 The Level of the Concentration of Ammonium Nitrogen Content from the Digestate and Its Different Substrates

During anaerobic digestion, the decomposition of organic matter by microorganisms releases ammonium nitrogen as a by-product. After substrate digestion, the digestate (the residue left after biogas production often contains higher concentrations of ammonium nitrogen than the original substrate). This is due to the digestion process, which converts organic nitrogen compounds into ammonium.

Figure 2 shows the variation in ammonium nitrogen concentration in kg/t in digestate and its various substrates.

It can be seen from the result of this figure that there is generally an increase in ammonium nitrogen content when comparing the substrate to its digestate.

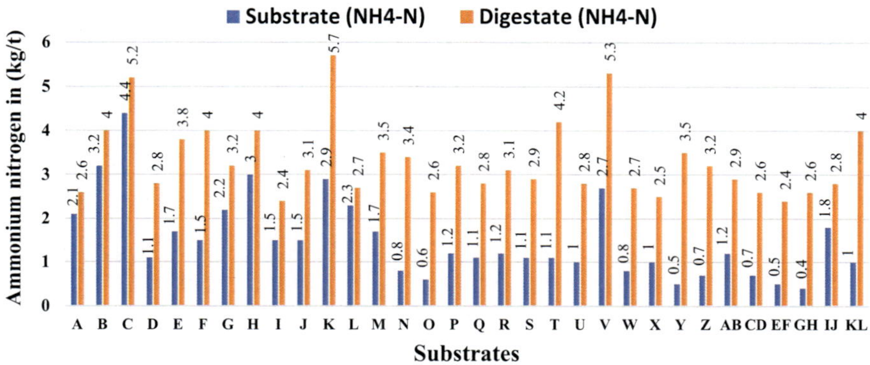

Fig. 2 The level of the concentration of Ammonium nitrogen content in digestate and its different substrates

This can be attributed to several factors, such as organic nitrogen conversion, bound nitrogen release, microbial activity, and ammonium retention.

Therefore, the ammonium nitrogen content in the digestate may not always increase compared to the substrate.

The ammonium nitrogen content in the digestate can be influenced by various factors, including the composition of the substrate, the efficiency of the anaerobic digestion process, and the management practices employed (Peng & Pivato, 2019).

In some cases, the ammonium nitrogen content in the digestate may be lower than that in the substrate due to processes such as microbial assimilation, volatilization, or denitrification. Microorganisms involved in anaerobic digestion can utilize ammonium nitrogen for their own growth and energy needs, resulting in a reduction in its concentration in the digestate. Additionally, under certain conditions, ammonium nitrogen can be transformed into other forms of nitrogen or lost as gaseous ammonia.

It is essential to take into account the specific conditions and characteristics of the anaerobic digestion process, including feedstock composition, operating parameters, and management practices, to accurately understand the variations in ammoniacal nitrogen content between substrate and digestate.

3.3 The Level of Ammonia Content from the Digestate of Different Feedstocks

The ammonia concentration in kg/t as a function of ammonium nitrogen content in digestate from different substrates is shown in Fig. 3.

The results in this figure show that there is an increase in ammonia content as a function of the ammonium nitrogen content in the digestate.

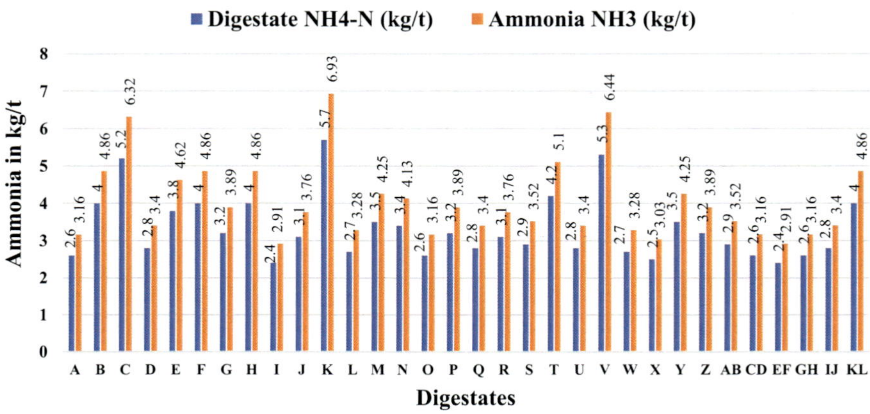

Fig. 3 Variation of ammonia content as a function of the ammonium nitrogen content from the digestate of different feedstocks

The amount of ammonia depends directly on the amount of NH_4-N (ammonium nitrogen) present in the digestate. As the concentration of ammonium nitrogen in the digestate increases, the quantity of ammonia increases.

Looking at the result from the figure, substrate K continues to record the highest concentration of ammonia at about (6.93) kg/t due to its high content of ammonium nitrogen from the digestate at about (5.7) kg/t, followed by substrate V with a concentration of (5.3) kg/t ammonium nitrogen has recorded a concentration at about (6.44) kg/t in ammonia content and substrate C with a concentration of (5.2) kg/t has recorded a concentration at about (6.32) kg/t respectively. The lowest concentration of ammonia was obtained from substrate I with a concentration value of (2.4) kg/t given at about (2.91) kg/t in ammonia and similar to substrate EF with a concentration of (2.4) kg/t recorded (2.91) kg/t respectively.

The potential of hydrogen production from the digestate of different feedstocks.

Figure 4 illustrates the result of the theoretical potential of hydrogen production from the digestate of different feedstocks.

According to the results, of all the digestate, substrate K gave the highest hydrogen potential by volume with approx. (6.84 m^3), followed by digestate from substrate V with approx. (6.36 m^3) and digestate from substrate C with a value of (6.24 m^3) respectively. The lowest hydrogen potential was observed in the digestate from substrates I (approx. 2.88 m^3) and EF (approx. 2.88 m^3) respectively.

Digestate with a high ammonium nitrogen content produced a lot of hydrogen. The hydrogen potential produced depends on the amount of ammonia available in the digestate. As the amount of ammonia increases, more hydrogen is produced, which means that for every 100% of ammonia, 25% of nitrogen and 75% of hydrogen are theoretically produced.

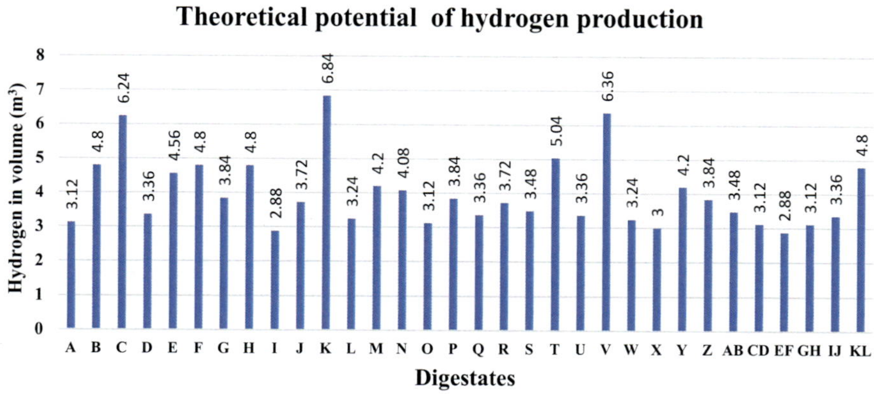

Fig. 4 The theoretical potential of hydrogen production from the digestate of different feedstocks

3.4 *Energy Potential Calculation Based on the Energy Content of Hydrogen Produced*

The estimated energy potentials for the samples studied, including power output in kilowatt-hours (kWh) based on the energy content of hydrogen, are presented in Table 4.

This is calculated by multiplying the energy content of hydrogen produced in MJ by the conversion factor of 0.2778 kWh/MJ. This conversion is beneficial for evaluating the power output with a specific energy content of hydrogen.

3.5 *Evaluation of Hydrogen Production from the Digestate Using Different Catalysts*

In the literature review, numerous catalysts have been investigated for the conversion of ammonia to hydrogen in the catalytic cracking process. Two notable catalyst types that have been extensively studied based on the economic factor, on availability, and found to be scientifically acceptable are ruthenium-based catalysts (Ru/SiO_2; Ru/CNTs) and nickel-based catalysts (Ni/Al_2O_3). These catalysts show promise in facilitating the conversion of ammonia into hydrogen and have been the subject of scientific research and exploration.

The performance of the catalysts can be quantified using the rate of hydrogen production, conversion fraction of ammonia (fraction of ammonia that is converted to hydrogen), and activation energy (Cheddie, 2012).

Figure 5 illustrates a comparison between the applications of Ruthenium supported on silicon dioxide (SiO_2) and Nickel supported on alumina (Al_2O_3) for the conversion of ammonia to hydrogen. With the application of ruthenium supported

Table 4 Estimated energy potentials for the samples studied

Substrates	Digestate (NH4-N)	Hydrogen (m^3)	Energy content (MJ)	Power output (kWh)
A	2.6	3.12	31.39	8.47
B	4.0	4.8	48.30	13.04
C	5.2	6.24	62.79	16.95
D	2.8	3.36	33.81	9.12
E	3.8	4.56	45.88	12.39
F	4	4.8	48.30	13.04
G	3.2	3.84	38.64	10.43
H	4	4.8	48.30	13.04
I	2.4	2.88	28.98	7.82
G	3.1	3.72	37.43	10.10
K	5.7	6.84	68.83	18.58
L	2.7	3.24	32.60	8.80
M	3.5	4.2	42.26	11.41
N	3.4	4.08	41.05	11.08
O	2.6	3.12	31.39	8.47
P	3.2	3.84	38.64	10.43
Q	2.8	3.36	33.81	9.12
R	3.1	3.72	37.43	10.10
S	2.9	3.48	35.02	9.45
T	4.2	5.04	50.71	13.69
U	2.8	3.36	33.81	9.12
V	5.3	6.36	64.00	17.28
W	2.7	3.24	32.60	8.80
X	2.5	3	30.19	8.15
Y	3.5	4.2	42.26	11.41
Z	3.2	3.84	38.64	10.43
AB	2.9	3.48	35.02	9.45
CD	2.6	3.12	31.39	8.47
EF	2.4	2.88	28.98	7.82
GH	2.6	3.12	31.39	8.47
IJ	2.8	3.36	33.81	9.12
KL	4	4.8	48.30	13.04

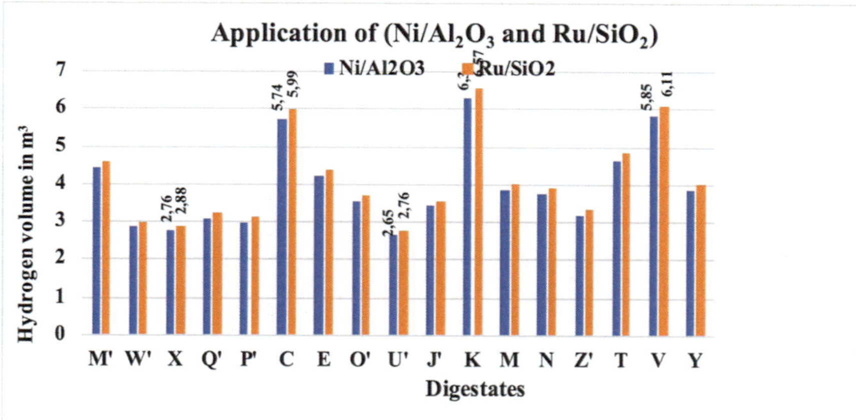

M' = (B, F, H, KL); J' = (J, R); W' = (A, O, CD, GH); Z'= (AB, S); Q'= (D, Q, U, IJ); O'= (G, P, Z); U'= (I, EF); P'(W, L).

Fig. 5 Evaluation of hydrogen production from the digestate using different catalysts (Ru/SiO_2; Ni/Al_2O_3)

on silicon dioxide (SiO_2), according to the results, of all the digestates, substrate K gave the highest hydrogen potential by volume with approx. (6.57 m^3), followed by digestate from substrate V with approx. (6.11 m^3) and digestate from substrate C with a value of (5.99 m^3) respectively. The lowest hydrogen potential was observed in the digestate from substrates U' (I, EF) (approx. 2.76 m^3) and digestate x approximately about (2.88 m^3).

With regard to the application of nickel on alumina, according to the results, of all the digestates, substrate K gave the highest hydrogen potential by volume with approx. (6.3 m^3), followed by digestate from substrate V with approx. (5.85 m^3) and digestate from substrate C with a value of (5.76 m^3) respectively. The lowest hydrogen potential was observed in the digestate from substrates U' (I, EF) (approx. 2.65 m^3) and digestate x approximately about (2.76 m^3).

When comparing the application of Ruthenium supported on silicon dioxide (SiO_2) and Ni/Al_2O_3 in terms of hydrogen by volume, With the application of ruthenium gave the best yield of hydrogen in volume compared to Nickel supported on alumina (Al_2O_3).

It can be seen from the figure an increment with the application of ruthenium supported on (SiO_2) compared to the application of Nickel supported on alumina (Al_2O_3).

With the regards of the evaluation of catalysts in terms of performance, by doing statistical analysis. The result of statistical analysis showed that there is no significant difference between the means of the two catalysts (P-value > 5%) (Fig. 6).

Ru-based catalysts have been documented to exhibit the highest rates of ammonia decomposition, along with the highest conversion rates. It has been reported as the most effective catalyst for low-temperature ammonia decomposition based on its

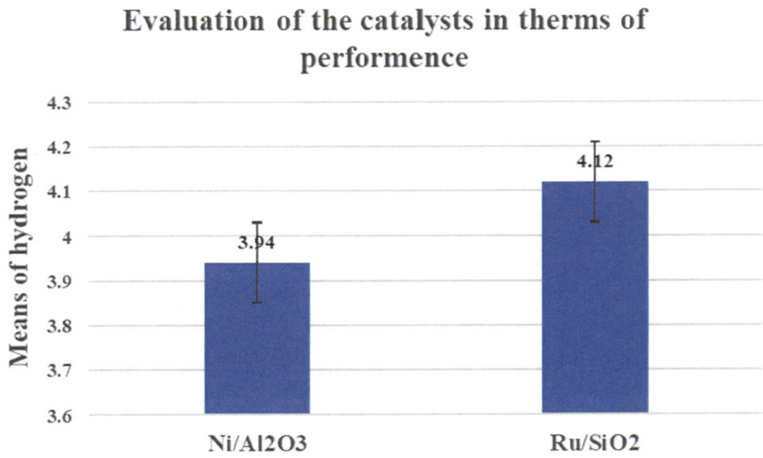

Fig. 6 Evaluation of the catalysts (Ru/SiO_2 and Ni/ Al_2O_3)) in terms of performance

electronic promotion possessing an optimum nitrogen binding energy. The utilization of this catalyst enables a reduction in both temperature and reaction pressure necessary for the decomposition process (Le et al., 2021).

From the discussion, it can be concluded that noble Ru-based catalysts are widely regarded as the optimal choice for achieving highly efficient ammonia (NH_3) decomposition. However, their considerable cost and restricted availability pose disadvantages when it comes to large-scale applications. On the other hand, among catalysts based on non-noble metals, Ni-based catalysts exhibit the highest activity, making nickel a promising alternative material for (NH_3) decomposition due to its affordability. Currently, there are still challenges that need to be addressed to enhance the efficiency of both Ru- and Ni-based catalysts.

4 Conclusion and Perspectives

This study highlights the importance of side stream stripping as a viable method of recovering ammonia from digestate to support hydrogen production. Although digestate has its advantages, the presence of toxic ammonia poses environmental problems. Proper control of ammonia volatilization is essential to minimize risks. Side-stream stripping has proved effective in safely extracting ammonia, facilitating handling, and reducing emissions. In addition, this process reduces transport costs by removing excess water from the digestate, while the stripped digestate is returned to the anaerobic digester, ensuring a closed-loop system.

In conclusion, the analysis of ammonium nitrogen concentration in digestate and in different substrates revealed several key results. Firstly, digestate generally has a higher ammonium nitrogen content than substrates, influenced by factors such as organic nitrogen conversion, microbial activity, and ammonium retention. Furthermore, the ammonia content of the digestate increases in parallel with the ammonium nitrogen content, with substrate K having the highest ammonia concentration. This relationship is reflected in the hydrogen production potential, with digestate with a higher ammonium nitrogen content producing more hydrogen. Substrate K stands out as having the highest hydrogen production potential. In terms of energy production, yield was directly proportional to available hydrogen, with substrate K generating the highest energy yield. Finally, when evaluating different catalysts for hydrogen production, ruthenium on silicon dioxide support proved superior to nickel on alumina support due to its superior catalytic activity in ammonia decomposition.

These results underline the importance of ammonium nitrogen content in digestate and its impact on hydrogen and electricity production potential, highlighting the potential for optimization of anaerobic digestion processes and catalyst technology.

In view of the results obtained, the following are the recommendations:

To elucidate the critical factor influencing ammonia recovery processes from digestate.

Improving the efficiency of Ru- and Ni-based catalysts.

Low-cost catalytic compositions that can exhibit catalytic activity comparable to that of ruthenium, thus facilitating the conversion process.

Acknowledgements This work was supported by WASCAL (IMP-EGH), an initiative funded by the BMBF (Federal Ministry of Education and Research, Germany).

References

Aziz, M., TriWijayanta, A., & Nandiyanto, A. B. D. (2020). Ammonia as effective hydrogen storage: A review on production, storage and utilization. *Energies*, *13*(12). https://doi.org/10.3390/en13123062.

Cheddie, D. (2012). Ammonia as a hydrogen source for fuel cells: A review. In *Hydrogen energy-challenges and perspectives*. InTech. https://doi.org/10.5772/47759.

Fachagentur Nachwachsende Rohstoffe. (2009). *Ergebnisse des biogas-messprogramms*. Agency for renewable resources (FNR). https://www.fnr.de/.

Fachagentur Nachwachsende Rohstoffe (2012). *Guide to biogas from production to use*. Agency For Renewable Resources (FNR). www.fnr.de.

Gu, Y. Q., Jin, Z., Zhang, H., Xu, R. J., Zheng, M. J., Guo, Y. M., Song, Q. S., & Jia, C. J. (2015). Transition metal nanoparticles dispersed in an alumina matrix as active and stable catalysts for COx-free hydrogen production from ammonia. *Journal of Materials Chemistry A, 3*(33), 17172–17180. https://doi.org/10.1039/C5TA04179A.

Industrial uses of nitrogen gas. (2021). COMPRESSED GAS TECHNOLOGIES INC.

Lamb, K. E., Dolan, M. D., & Kennedy, D. F. (2019). Ammonia for hydrogen storage; A review of catalytic ammonia decomposition and hydrogen separation and purification. *International Journal of Hydrogen Energy, 44*(7), 3580–3593. https://doi.org/10.1016/J.IJHYDENE.2018.12.024.

Le, T. A., Do, Q. C., Kim, Y., Kim, T. W., & Chae, H. J. (2021). A review on the recent developments of ruthenium and nickel catalysts for COx-free H2 generation by ammonia decomposition. In

Korean Journal of Chemical Engineering (Vol. 38, Issue 6, pp. 1087–1103). Springer. https://doi.org/10.1007/s11814-021-0767-7.

Lee, B., Park, J., Lee, H., Byun, M., Yoon, C. W., & Lim, H. (2019). Assessment of the economic potential: COx-free hydrogen production from renewables via ammonia decomposition for small-sized H2 refueling stations. *Renewable and Sustainable Energy Reviews, 113*. https://doi.org/10.1016/j.rser.2019.109262.

Liu, Y., Xue, Y., Hui, L., Yu, H., Fang, Y., He, F., & Li, Y. (2021). Porous graphdiyne loading CoOx quantum dots for fixation nitrogen reaction. *Nano Energy, 89*. https://doi.org/10.1016/j.nanoen.2021.106333.

Makepeace, J. W., He, T., Weidenthaler, C., Jensen, T. R., Chang, F., Vegge, T., Ngene, P., Kojima, Y., de Jongh, P. E., Chen, P., & David, W. I. F. (2019). Reversible ammonia-based and liquid organic hydrogen carriers for high-density hydrogen storage: Recent progress. *International Journal of Hydrogen Energy, 44*(15), 7746–7767. https://doi.org/10.1016/J.IJHYDENE.2019.01.144.

Peng, W., & Pivato, A. (2019). Sustainable management of digestate from the organic fraction of municipal solid waste and food waste under the concepts of back to earth alternatives and circular economy. In *Waste and Biomass Valorization* (Vol. 10, Issue 2, pp. 465–481). Springer Netherlands. https://doi.org/10.1007/s12649-017-0071-2.

Ristig, S., Poschmann, M., Folke, J., Gómez-Cápiro, O., Chen, Z., Sanchez-Bastardo, N., Schlögl, R., Heumann, S., & Ruland, H. (2022). Ammonia decomposition in the process chain for a renewable hydrogen supply. In *Chemie-Ingenieur-Technik* (Vol. 94, Issue 10, pp. 1413–1425). John Wiley and Sons Inc. https://doi.org/10.1002/cite.202200003.

Yin, S. F., Zhang, Q. H., Xu, B. Q., Zhu, W. X., Ng, C. F., & Au, C. T. (2004). Investigation on the catalysis of COx-free hydrogen generation from ammonia. *Journal of Catalysis, 224*(2), 384–396. https://doi.org/10.1016/J.JCAT.2004.03.008.

Zhang, Z., Liguori, S., Fuerst, T. F., Way, J. D., & Wolden, C. A. (2019). Efficient ammonia decomposition in a catalytic membrane reactor to enable hydrogen storage and utilization. *ACS Sustainable Chemistry and Engineering, 7*(6), 5975–5985. https://doi.org/10.1021/acssuschemeng.8b06065.

Technological Advances and Data Driven Solutions in Energy Optimization

Performance Parameters of an Off-Grid Photovoltaic System in the Gambia

Sherif Muhamedou Hydara, Halidou Ibrahim, Muhammed Lamin Sanyang, and Mounkaila Moussa Saley

Abstract The Gambia is actively pursuing renewable energy integration, targeting a 40% threshold by 2021 in alignment with SDG7. However, regulatory and logistical obstacles impede this progress. Leveraging the nation's abundant solar resources presents a significant opportunity, particularly in reaching remote rural communities beyond the utility grid through off-grid photovoltaic systems. Monitoring the performance of such a system is essential for sectoral growth and development. Over the course of a year, a study monitored the performance of a typical off-grid photovoltaic system, revealing notable seasonal variations. Optimal performance occurred during months with high temperatures and irradiance, notably in February, March, and April. Key performance indicators included an average reference yield of 190.55 kWh/kWp/month, array yield of 113.94 kWh/kWp/month, final yield of 99.46 kWh/kWp/month, capture loss of 77.36 kWh/kWp/month and system loss of 14.48 kWh/kWp/month. The system exhibited average array, inverter, and system efficiencies of 12.4%, 87.017% and 19.53% respectively, resulting in an average performance ratio (PR) of 52.22%. This study underscores the importance of understanding and optimizing the performance of off-grid photovoltaic systems to enhance renewable energy utilization in The Gambia and similar regions.

Keywords Renewable energy integration · Off-grid photovoltaic system · Performance monitoring · Solar energy · System efficiency · Performance ratio · Rural electrification

---

S. M. Hydara (✉) · M. M. Saley
UWest Africa Science Service Centre on Climate Change and Adopted Land Uses (WASCAL)—Doctoral Research Program on Climate Change and Energy (DRP-CCE), Universite Abdou Moumoui (UAM), Niamey, Niger
e-mail: rigndaara@gmail.com

H. Ibrahim · M. M. Saley
Laboratoire d' Energetique, d'Electronique, d'Automatique et d'Informatique Industrielle (L3EA2I), Universite Abdou Moumini (UAM), Niamey, Niger

M. L. Sanyang
West Africa Coastal Area Resilience Investment Project 2 (WACA ResIP 2)—The Gambia, Kanifing, The Gambia

M.-M. Narra and S. Narra (eds.), *African Green Transition Through Innovative Pathways*, World Sustainability Series, https://doi.org/10.1007/978-3-031-87043-9_15

1 Introduction

The Gambia heavily relies on imported petroleum products for electricity generation, facing challenges of unstable supply and unsatisfied demand (AFREC, 2022). Despite the country's rich solar potential, renewable energy only contributes 19.2% to electricity generation by 2030 (Bass & Lopez-Aguera, 2023).

The integration of solar PV systems, like mini grids, has shown promise in rural electrification projects in Sub-Saharan Africa (SSA), offering reliable and cost-effective electricity solutions (Bass & Lopez-Aguera, 2023; Mayanjo & Justo, 2023). To offset the high dependence on fossil fuels, The Gambia aims to achieve a 40% renewable electricity integration rate by 2021 but faces regulatory and policy hurdles hindering this goal (Obeng-Darko, 2022). Monitoring the performance of existing PV systems is crucial for advancing the PV market in The Gambia and ensuring a sustainable energy future (Samikannu et al., 2022).

Off-grid PV systems play a crucial role in providing electricity to locations without access to the utility grid, offering energy independence to individuals (Yudin et al., 2023). These systems are particularly vital in remote African communities, where connecting to the national grid is challenging. By utilizing solar PV systems, these off-grid setups can efficiently generate electricity, powering lights, appliances, and other electronics devices (Mayanjo & Justo, 2023). The technology involves solar panels converting sunlight into direct current, which is stored in batteries and then converted into alternating current for use by appliances through an inverter (Aljuboury et al., 2022). Off-grid solar power systems are cost-effective, efficient, and customizable to meet specific energy needs, contributing to long-term economic growth and improving living standards in rural areas.

Despite the significant potential of Building-Integrated Photovoltaic (BIPV) systems, their global adoption remains limited, constituting only about 1% of total PV installation worldwide. The technology's underrepresentation in regions like The Gambia underscores the need for continuous research and innovation to drive its growth (desa & Sabry 2023; Abdelrazik et al., 2022). Various types of BIPV installations such as rooftop, curtain and wall façade system offer opportunities to enhance power efficiency and reduce reliance on traditional energy sources like air conditioning (Weerasinghe & Yang, 2022). Additionally, advancements in BIPV technologies have addressed previous technical complexities, paving the way for improved system performances and economic viability (Skandalos et al., 2022). To further promote the widespread adoption of BIPV systems, ongoing research efforts are crucial to optimize design, efficiency, and cost-effectiveness, ultimately contributing to the sustainable development of energy efficient building globally.

The performance ratio (PR) is a crucial metric for evaluating the efficiency of PV systems. Studies have shown a significant improvement in PR over the years, with values increasing from 0.5 to 0.75 in the late 1980s, 0.65 to 0.7 in the 1990s, to over 0.8 currently. However, stand-alone PV systems often exhibit lower PR values, with common values as low as 0.35 (Menga & Owona, 2023; Kim et al., 2023; Ganga Bhavani et al., 2022; Jayaudhaya et al., 2022). Factors such as weather conditions, system design and components capacities play a significant role in determining the PR of a PV system. Continuous research and advancements aim to enhance PR values ensuring better energy efficiency and performance of PV systems across various applications.

The PR is crucial metric in assessing the impact of losses on a photovoltaic system's rated output (Deceglie et al., 2023; Abdoulaye et al., 2023; Livera et al., 2023). While PR does not directly represent the energy produced, it is essential for evaluating system efficiency and performance (Jamil et al., 2022). A system with low PR in a high solar resource location may still produce more energy than a system with a high PR in a low solar resource location (Jamil et al., 2022). Increasing the PR through component or design changes can lead to a corresponding increase in energy yield for a given system, location, and time. However, while PR can signal performance issues, identifying the root cause requires on-site research to point specific problems.

The performance evaluation of a stand-alone PV systems can be complex as poor performance may not always indicate technical issues but could result from mismatches between energy production and consumption (Menga & Owona, 2023; Surendra et al., 2022). The IEA PVPS task II collected operational data from various stand-alone PV systems, revealing a wide range of PR values, typically falling between 0.1 and 0.7. However, most systems exhibited PR values between 0.3 and 0.4, with only a small percentage ranging from 0.6 to 0.7 (Surendra et al., 2022). Oversized stand-alone PV systems often experience array disconnections to prevent battery overcharging, negatively impacting the PR (Tejaswini, 2022). This PV systems ensure optimal performance and efficiency over time (Ibrahim et al., 2022).

The study of a 120 kWp stand-alone PV system in Nyamanarri, Kantora, URR, The Gambia, aims to showcase the system's performance in addressing energy poverty and insecurity in remote rural areas, particularly in less developed African countries (Aljuboury et al., 2022; Bouaziz & Bacha, 2023). Stand-Alone PV systems with battery storage play a crucial role in providing electricity to off-grid communities, enhancing education, living standards and health care (Naqvi et al., 2021). In India, stand-alone solar PV systems have been analyzed for rural electrification, emphasizing the economic viability and optimal design for different load profiles (Nsengimana et al., 2022). Additionally, the design analysis of grid-connected and off-grid PV systems in Nigeria highlights the importance of clean and reliable energy access for sustainable development goals (Makinde et al., 2021). These studies collectively underscore the potential of such systems in combatting climate change, energy poverty and insecurity in remote regions globally.

1.1 System Description

This off-grid system is located at Nyamanarri, Kantora in The Gambia with Latitude 13.20 North and Longitude 13.52 West. The system has a peak capacity of 120 kWp.

The JA SOLAR Crystalline Silicon PV modules used here have a high capacity to frame area ratio, with a conversion efficiency of 19.8%. These PV modules have been integrated to form part of the community base off-grid system. The 310 PV modules occupy a total of 611 square meters at an inclination of 15°.

A SUNNY ISLAND 8.0H inverter is used in this system which converts the DC energy produced by the panels to AC energy needed by the loads. The inverter has a nominal output power of 6 kVA at 25 °C. There are 12 inverters grouped in 4 clusters and each cluster is composed of 3 inverters of one main inverter and 2 supportive.

There is also a SMA DATA MANAGER. This regulator ensures peak performance from the PV panels, by letting them deliver their maximum power at any given instance. The charge controller also protects the battery by controlling its charging current and voltage, thereby provoking the batteries life span.

192 HOPPECKE c/100 1695 AH rated capacity flooded lead acid batteries are used in this system to store energy 576 kWh during excess generation. The batteries are grouped in 4 clusters of 48 batteries. The battery bank is a 48 V system.

There is a portal system used to monitor the system from a distance. The portal measures and records the system performance and shows the meteorological data.

The configuration of the panels, as observed in Fig. 1, highlights the solar panel array installed to harness solar energy efficiently. The accompanying structure houses essential equipment such as inverters, batteries and other components critical to energy storage and distribution. Both pictures were taken by me during fieldwork conducted at the site.

The location of Nyamanrri is shown in Fig. 2. Nyamanarri is situated in the Kantora District within the Upper River Region (URR) of The Gambia. The map highlights the settlement of Nyamanrri and its surrounding areas, emphasizing its geographical positioning.

Fig. 1 A view of the PV infrastructure in Nyamanarri, Kantora, Upper River Region (URR)—The Gambia (Pictures taken by Sherif Muhamedou Hydara)

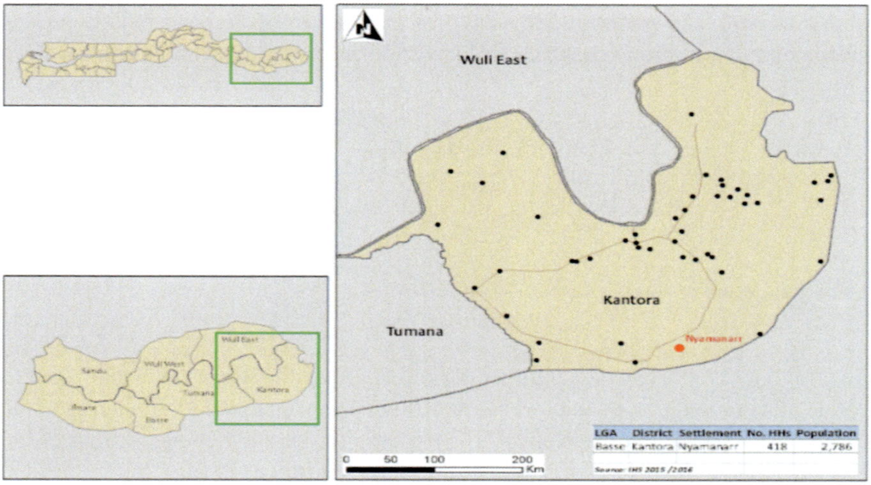

Fig. 2 Map of Nyamanarri, Kantora, Upper River Region (URR)—The Gambia (GBOS, 2015/16)

2 Methodology

The system is supervised by a portal for measuring and recording electrical parameters. The meteorological parameters are shown on the screen but cannot be retrieved. The electrical parameters include the current and voltage flowing from and into each balance of system component. The meteorological parameters include solar irradiance on the plane array, ambient temperature, cell temperature, relative humidity, wind speed and direction. There is a data logger used for measuring the electrical. The electrical data is sampled every 5 min.

Data collected for a period of one year between April 2022 and March 2023. This data was used to evaluate the performance of this off-grid system based on certain normalized system performance parameters (Yibeltal et al., 2023). This was done based on the IS/IEC 61,724 1998 standards (Meyer et al., 2017). These performance parameters are discussed below.

The yields have units of kWh/kW$_p$/day or kWh/kW$_p$/month and indicate the amount of time during which the array would be required to operate at the array's rated output power, P_0 to provide a particular quantity (Meyer et al., 2017; IS/IEC 61,724, 1998).

2.1 *Reference Yield* (Y_R)

The reference yield is the theoretical possible energy output of the PV plant. It is defined as the anticipated output from the same system with nominal efficiency determined under STC of PV modules. In other words, the reference yield is the number of peak sun-hours. It is the ratio of the total in-plane solar radiation (H_{id}) [kWh/m^2], to the reference irradiance (1 kWh/m^2) at standard test conditions as given in Eq. (1).

$$Y_R = H_{id}/G_{STC} \tag{1}$$

2.2 *Array Yield (Y_A)*

The array yield represents the number of hours per day that the array would need to operate at its nominal or peak power to contribute the same quantity of energy to the system as that measured in practice. It is the ratio of the daily energy output of the PV array (E_A) to the peak power of the installed PV array as in Eq. (2).

$$Y_A = E_A/P_0 \tag{2}$$

2.3 *Final Yield (Y_F)*

The final yield is the usable portion (used by the load, E_L) of the energy derived from the PV modules (E_{PV}). In an off-grid system, the energy stored in the battery bank (E_{bat}) is equally taken into consideration. The final yield is obtained from Eq. (3):

$$Y_F = E_{PV}/P_0 \text{ where } E_{Pv} = E_L/1 + (E_{bat}/E_A) \tag{3}$$

2.4 *Performance Ratio (PR)*

The performance ratio is one of the most important parameters to consider when evaluating the performance of a PV system (Meyer et al., 2017; Sark et al., 2012). It is the ratio of the actual energy output to the theoretically possible energy output. It is independent of the orientation of the PV plant and the incident solar radiation of the PV plant. This is why the performance ratio can be used to compare PV plants at different locations all over the world or for performance comparison of a particular system over the years. It is the ratio of utilizable AC electricity to the amount of energy which could be generated in case modules were operated under STC continuously and without any further losses in the system (Meyer et al., 2017; Sark et al., 2012) as shown in Eq. (4).

$$PR = Y_F/Y_R \tag{4}$$

2.5 Losses

The losses in a PV system may be caused by high array temperature, incomplete utilization of the irradiation, improper system sizing and system component inefficiencies or failures (Wu et al., 2015). Energy losses indicate the amount of time during which the array would be required to operate at its nominal power in order to produce power for the losses (Wu et al., 2015). The losses are in two parts: system losses (L_s) and array capture losses (L_c). The capture loss is the difference between the reference yield and the array yield (Jed et al., 2020). It is caused by incomplete utilization of the available radiation. The system loss is the difference between the array yield and the final yield. It mainly comes from the inverter in converting the DC power to AC (Kymakis et al., 2009) and from losses in the other balance of the system components as shown in Eqs. (5) and (6).

$$L_c = Y_R - Y_A \tag{5}$$

$$L_S = Y_A - Y_F \tag{6}$$

2.6 Efficiencies

The efficiencies involved in a PV system are in two major levels: at the level of the array and the whole system in general. The system efficiency embodied the efficiencies of the balance of the system component where losses are incurred is the inverter, the system efficiency is simply the product of the array efficiency and the inverter efficiency as seen in Eqs. (7) and (8).

$$\eta_{array} = E_A/\left(A_{array} \times G_{id}\right) \tag{7}$$

$$\eta_{sys} = \eta_{array} \times \eta_{inv} \tag{8}$$

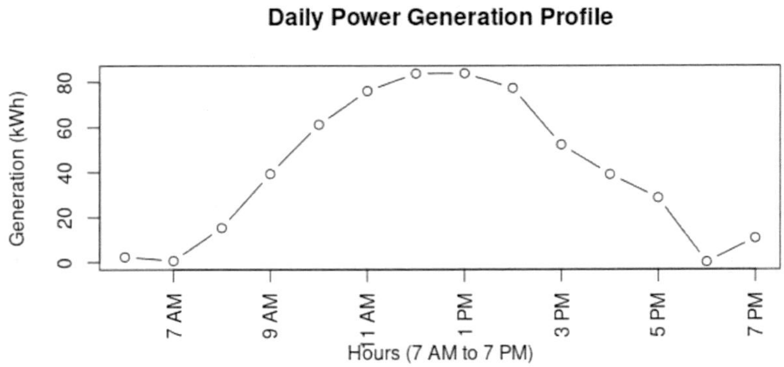

Fig. 3 Daily power generation profile of the Nyamanarri solar mini-grid between February and April 2023

3 Results and Discussions

3.1 Load Profile for Nyamanarri

Between April 2022 and March 2023, Nyamanarri Village off-grid PV system has issued over 300 m. The figure should be on the rise because it is a one-year-old project. Figure 3 shows the daily power generation profile which highlights peak output during midday when sunlight is must intense.

Figure 3 shows the load profile for an average of 2 months that starts from 12th February 2023 to 13th April 2023. An average generation profile for the whole period is also presented.

The generation pattern is a slightly parabola function. On average, morning electricity generation starts at 7:00 AM while evening generation ends at 7:00 PM. The peak generation falls between midday and 2 PM.

At night, the system is supplied by the battery bank.

3.2 Solar Radiation and Temperature Variation

The performance of a PV system is influenced by so many factors amongst which are meteorological factors. The principal meteorological factors affecting the power output and efficiency of the system are ambient temperature and solar radiation. Figure 4 shows that the variation of solar radiation (average number of peak sun hours per day) and ambient temperature for the various months considered for this study.

It can be observed that the highest radiation occurs in March, with an average of 9.4 peak sun hours per day while the least radiation occurs in September, with

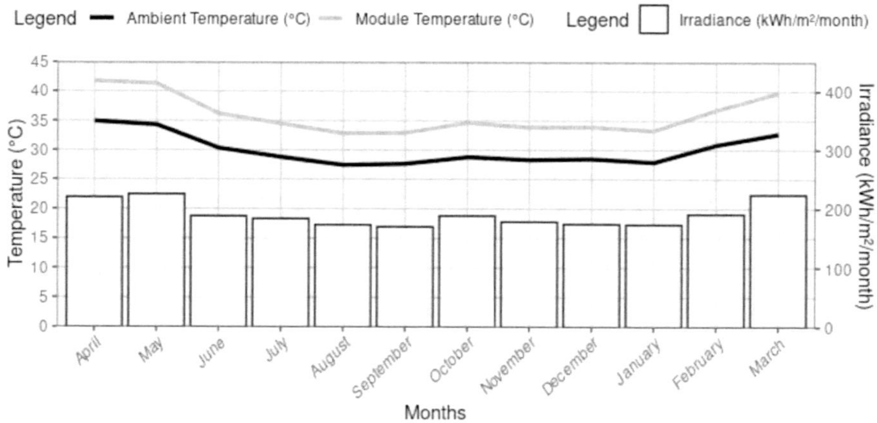

Fig. 4 Average ambient temperature, module temperature and irradiance profiles

an average daily radiation of 2.7 peak sun hours per day. The temperature variation follows a similar pattern.

However, the highest daily temperature is noticed in April (34.89 °C) and while the least temperature is observed in September (°C). The PV system is expected to have its best performance around April, May, and March. It is also seen that the month of October has a high production. The higher the temperature the greater the performance of the system.

3.3 *Yield Factors for the System*

The yields indicate the amount of time during which the array would be required to operate at its rated output power to provide a particular quantity. The figure below shows the relationship between the reference, array, and final yields for this system for the various months considered for the study (Fig. 5).

The reference yield is highest followed by array yield then the final yield. The reference yield depicts the number of sun hours for a particular month. It represents the amount of energy which would be produced if the system operated at STC conditions and at the rated efficiency. This is highest in March with a total number of 9.4 peak sun hours per day. The least number of sun hours per day occurs is September with 2.1 sun peak hours per day.

Nyamanarri PV system is an off grid with a battery back-up and a charge controller. Only the amount of energy needed to charge the battery bank and to feed the loads can be used irrespective of the available energy from the sun. The huge difference between the reference yield and array yield can be explained by the above reason. The low conversion efficiency of the solar radiation to electricity also contributes

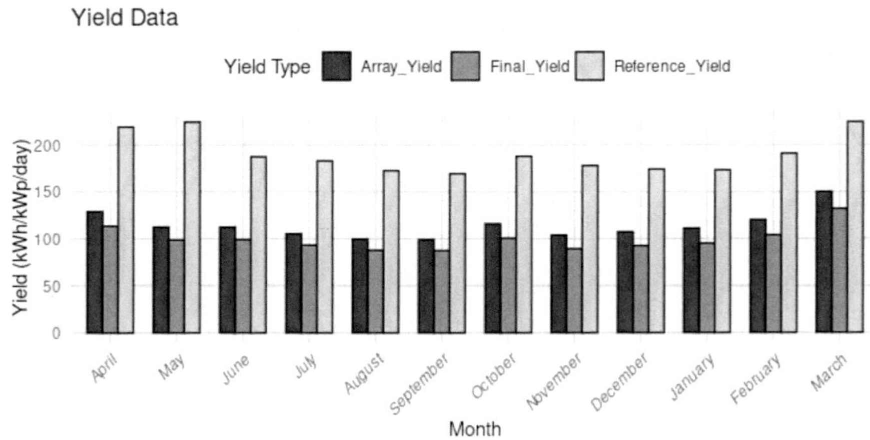

Fig. 5 Monthly yield factors of the system

to this huge difference. The energy difference between the solar radiation and the amount of energy tapped by the MPPT charge controller is called Capture Loss.

The energy tapped by the charge controller to feed the battery bank and the loads, parts of it is lost before it finally gets to the loads. This is a result of energy losses in the various system components such as the charge controller, the battery bank, the inverter, and other wiring losses. These resulted in the gap between the array yield and final yield and is called System Loss (Fig. 6).

The combination of the final yield, system loss and array losses result in the reference yield. The length of each bar represents the amount of energy this system would have produced at standard test conditions and at the arrays rated, in the absence of losses.

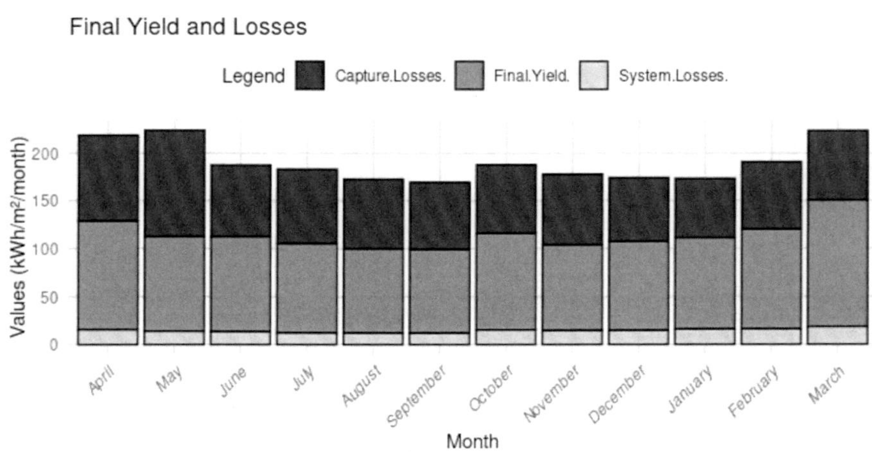

Fig. 6 Monthly two major types of losses and final yield incurred in this system

It can be noticed that the system encounters a lot of capture losses. This system has been in existence for one year, the module is still new. There is no sign of degradation. The high capture loss can be due to high temperature. It is seen that the minimum average monthly temperature is above 27 °C which is higher than the STC temperature. There is a lot of energy not tapped due to the battery banks getting fully charged and the load is demanding very minimal energy when the PV is still at its peak generating capacity.

It can also be observed that most of the capture losses occur in the summer months when temperature is still high. During the winter, the capture loss is reduced due to reduced temperature.

On average, the system recorded capture losses of 919.317 kWh/kW$_p$/month and system losses of 173.7309 kWh/kW$_p$/month.

3.4 *Performance Ratio of the System*

Performance Ratio or PR is a globally accepted parameter used to compare similar PV systems at various locations on the globe. A low value for PR in an off-grid PV system does not necessarily mean poor system performance or technical problems. It is an indication of poor matching between the PV generator and the electrical appliances powered by the system. An oversized system incurs losses because the PV system is not being used at its full potential. Figure 7 illustrates the changes in the performance ratio for this system at various months of the year. Figure 7 reveals that the dependence of the performance ratio on temperature.

This system shows a relatively low Performance Ratio (PR) compared to other stand-alone PV systems in different parts of the world. It, however, falls within the range of the PR values obtained by the IEA PVPS task II experts. They carried out

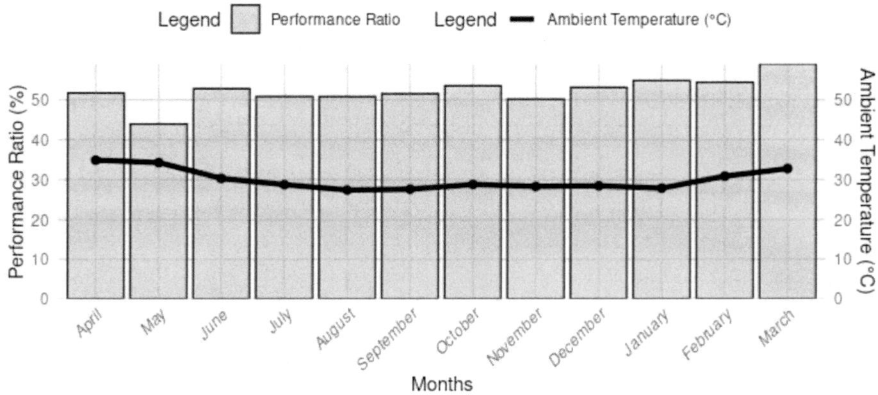

Fig. 7 Variation of performance ratio and ambient temperature for the year

a survey on about thirty off-grid PV systems and discovered that over 80% of the PR values ranged between 0.3 and 0.4 while just 10% ranged between 0.6 and 0.7. The month with the highest average daily performance ratio is March while the PR is least in May (a daily of 45%). However, the highest temperature is in April and the least is in January.

The system recorded an average performance ratio of 55.22%. It is well known that the performance of a PV system is already affected by high temperature. The fact that the month with the highest performance does not exactly correspond to the month the least ambient temperature justifies the fact that the performance ratio of this off-grid system is also affected by other factors, load matching being one of the key factors. The figure below illustrates the correlation between the performance ratio for this system and the energy demand. It shows a linear relationship with a correlation of 0.544. This goes to justify the fact that the performance ratio for an off-grid system is highly dependent on its local consumption.

3.5 System Efficiency

The efficiency of the system is equally an important parameter used in describing the performance of a PV system. It encompasses the efficiencies of the various balance of system components used in the system. The efficiency shown in the figure below is a product of the array efficiency and the inverter efficiency. The inverter efficiency was used because it is one of the major balances of system component where losses are incurred. The array efficiency accounts for the percentage of incident radiation converted to electricity by the modules.

The least array, inverter, and system efficiencies of 9.866%, 85.568% and 17.168% respectively occur in May, January, and September. Similarly, the highest array, inverter, and system efficiencies of 13.218%, 88.466% and 25.942% respectively occur in March and July.

Figure 8 represents the monthly efficiency trends of the array, inverter and the system.

The inverter is one of these components whose efficiency is highly dependent on load demand. This is because being a stand-alone PV system, the input power into the inverter is determined by the demand from the loads. In the absence of any demand, little or no energy is fed to the inverter. The figure above reveals the comparison between the inverter efficiency, array efficiency and system efficiency.

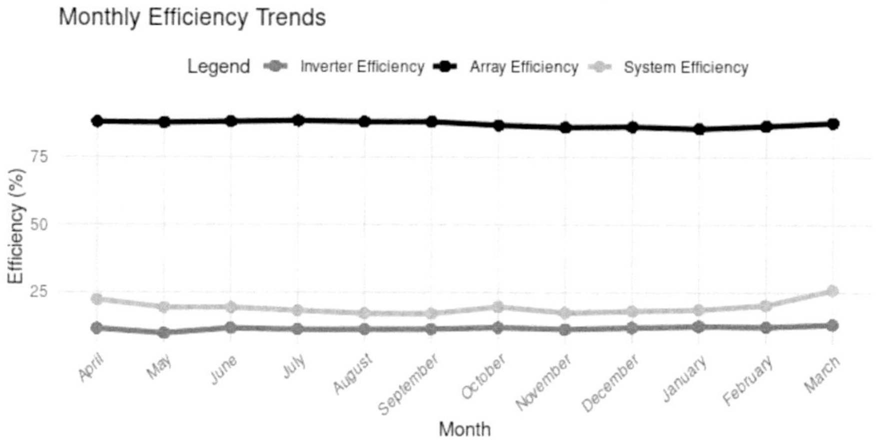

Fig. 8 Variation of array, inverter and system efficiency for the year

4 Conclusion

It is recommended that off-grid PV systems be used only in location inaccessible to the grid. In location where the PV system is installed close to the utility grid, the system should be connected to the utility grid to feed-in the surplus energy that is otherwise wasted due to poor system sizing, which is of course difficult to avoid.

The study also shows that temperature is a very important parameter affecting the performance of a PV system. The system performed better in the spring months with higher temperatures than in the summer months. The highest irradiation values were recorded in the spring months. The performance ratio ranged from 58.9% in March to 43.9% in May with an average of 55.4% annually. Furthermore, the higher values for the array efficiency, inverter efficiency and system efficiency were 13.2% in March, 88.4% in July and 25.9% in March and the lowest values are 9.8% in May, 85.5% in January and 17.2% in August.

The system encounters a lot of array losses principally due to the mismatching between the array capacity and the load consumption.

It is recommended that energy intensity appliance be used at about midday when PV generation is maximum to minimize the waste PV generation potential.

References

Abdelrazik, A. S., Shboul, B., Elwardany, M., Zohny, R. N., & Osama, A. (2022). Recent developments in building integrated photovoltaic/thermal (BIPV/T) systems: A contemporary review. *Renewable and Sustainable Energy Reviews* 112988. https://doi.org/10.1016/j.rser.2022.112988.

Abdoulaye, B. M., Abdourahimoun, D., Harouna, S. D. N., Saidou, M., & Makinta, B. (2023). Performance ratio and loss analysis for a grid-connected solar photovoltaic system: Case of the 7 MW plant in Malbaza, Niger. *Journal of Energy Research and Reviews.* https://doi.org/10.9834/jenrr/2023/v13i1256

Adeleye Makinde, K., Adewuyi, O. B., Amole, A., & Adeaga, O. A. (2021). Technical analysis of net zero energy houses and buildings: a root cause analysis. *Journal of Energy Research and Reviews, 8*(2), 203–219, 2021; Article no.JENRR.68373. https://doi.org/10.9734/JENRR/2021/V8I130203.

African Energy Commission. (2022). *Gambia.* Retrieved February 11, 2025, from https://au-afrec.org/gambia

Aljuboury, A. S., Al-Azzawi, W. K., Shakier, L. M., Al-Amiery, A. A., & Kadhum, A. A. H. (2022). Analysis of off-grid community photovoltaic microgrid. F1000Research, https://doi.org/10.12688/f1000research.128868.1.

Bass, M. S. & Lopez-Aguera, A. (2023). WEAP.FM: Long term optimization model for the Gambia's energy transition. *Renewable Energy and Power Quality Journal.* https://doi.org/10.24084/repqj21.412.

Bouaziz, B., & Bacha, F. (2023). *Building-integrated photovoltaic system: Real-time monitoring system design and implementation.* https://doi.org/10.1109/IC_ASET58101.2023.10151167.

Deceglie, M. G., Anderson, K., Fregosi, D., Hobbs, W. B., Mikofski, M. A., Theristis, M., & Meyers, B. (2023). Outlook on the relative operational loss of photovoltaic systems. *Solar RRL.* https://doi.org/10.1002/solr.202300196.

Desa, M. K. M., & Sabry, A. H. (2023). Future opportunities with building-integrated solar PV systems and the use of bifacial PVs. *Frontiers in Energy Research, 11*, 1164494. https://doi.org/10.3389/fengr.2023.1164494.

Dhian, N. A., Tresna, T., Hermawan, I. M., Mankuto, R. A., Pradipta, J., Jamala, N., Paramita, B., & Atthailah. (2023). Experimental evaluation of daylighting performance and energy output of building integrated photovoltaic (BIPV) panel in Bandung. Indonesia IOP Conference Series, Vol. 1157 (1), pp. 012004. https://doi.org/10.1088/1755-1315/1157/1/012004.

Ganga Bhavani, C. S., Prasad, N. B., & Kishore, D. R. (2022). Analysis of performance of solar photovoltaic system. In: IEEE 2nd International Symposium on Sustainable Energy, Signal Processing and Cyber Security (iSSSC), Gunupur, Odisha, India, pp. 1–6. https://doi.org/10.1109/iSSSC56467.2022.10051250.

Gambia Bureau of Statistics (GBOS) (2015/16). HIS report.

Ibrahim, K. H., Hassan, F. Y., Abdelrazek, A. S., & Saleh, S. M. (2022). Economic analysis of stand-alone PV-battery system based new power assessment configuration in Siwa Oasis–Egypt. *Alexandria Engineering Journal.* https://doi.org/10.1016/j.aej.2022.07.034.

IS/IEC 61724. (1998). Performace monitoring–guidelines for measurement, data exchange and analysis. Indian Standards.

Jamil, I., Lucheng, H., Habib, S., Aurangzeb, M., Ali, A., & Ahmed, E. M. (2022). Performance ratio analysis based on energy production for a large-scale solar plant. *IEEE Access, 10*, 5715–5735. https://doi.org/10.1109/ACCESS.2022.3141755.

Jayaudhayal, J., Kumar, K. R., Selvi, V. T., & Padmavathi, N. (2022). Improved performance analysis of PV array model using the flower pollination algorithm and gray wolf optimization algorithm. *Mathematical Problems in Engineering.* https://doi.org/10.1155/2022/5803771.

Jed, M. E. H., Ihaddadene, R., Ihaddadene, N., ElBanany, C. E. S., & EL Bah, M. (2020). Performance analysis of the 954,809 kWp PV array of Sheikh Zayed solar power plant (Nouakchott, Mauritania). *Renewable Energy Focus, 32*, 45–54. https://doi.org/10.1016/j.ref.2019.11.002.

Kanchikere, J., Ghosh, A. K., & Kumar, K. (2018). Analysis of 80kW grid-connected rooftop solar power plant using SISIFO. *International Journal of Mechanical Engineering and Production Engineering Research and Development (IJMPERD), 8*(6), 33–46. https://doi.org/10.24247/IJMPERDDEC20184.

Kebour, O., & Arab, A. H. (2021). Performance analysis of a stand-alone photovoltaic system in the sub-Saharan environment. https://doi.org/10.48393/IMIST.PRSM/JASERS-V4I1.24645.

Kim, G. G., Hyun, J. H., Choi, J. H., Ahn, S.-H., Bhang, B. G., & Ahn, H.-K. (2023). Quality analysis of photovoltaic systemusing descriptive statistics of power performance index. *IEEE Access, 11*, 28427–28438. https://doi.org/10.1109/ACCESS.2023.3257373.

Kymakis, E., Kalykakis, S., & Papazoglou, T. M. (2009). Performance analysis of a Grid-connected photovoltaic park on the island of Crete. *Journal of Energy Conversion and Management, 50*, 433–436.

Livera, A., Tziolis, G., Theristis, M., Stein, J. S., & Georghiou, G. E. (2023). Estimating the performance loss photovoltaic systems using time series change analysis. *Energies*. https://doi.org/10.3390/en16093724.

Lofti, H., Adar, M., Bennouna, A., Izbaim, D., Oum'bark, F., & Ouacha, E. H. (2022). Silicon photovoltaic systems performance assessment using the principal component technique. *Materials Today: Proceedings, 51*(6), 1966–1974. https://doi.org/10.1016/j.matpr.2021.04.374.

Mayer, D., & Heidenreich, M. (2003). Performance analysis of stand-alone photovoltaic system from a rational use of Energy point of view. *International Energy Agency Photovoltaic Power System Program.*

Mayanjo, S., & Justo, J. J. (2023). Development of solar PV System for mini-grid applications in Tanzania. *Tanzania Journal of Engineering and Technology*. https://doi.org/10.52339/tjet.v42i1.899.

Menga. F. D., & Owona, J. L. (2023). Determination of a new performance indicator for assessment of stand-alone PV system. *Energy Environment and Storage*. https://doi.org/10.52924/qbmk2848.

Meyer, E. L., Buma, C. L., & Taziwa, R. (2017). Performance parameters of an off-grid building-integrated photovoltaic system in South Africa. *33rd European Photovoltaic Solar Energy Conference and Exhibition* (pp. 2450–2455). University of Fort Hare.

Mustafa, K., & Ashraf, I. (2021). Performance Assessments of a stand-alone solar photovoltaic system for different load profiles in rural areas. *Journal of The Institution of Engineers: Series B*. https://doi.org/10.1007/S40031-021-00576-8.

Nagarajan, P. C., Muthuvinayagam, M., Siva, M., Maniraj, R. P., & Kumar, N. S. (2023). *Review of an off-Grid Solar DC System for Rural Houses*. https://doi.org/10.1109/ICICCS56967.2023.10142541.

Naghaviha, D., Nikkhajoei, H., & Karimi, H. (2022). PV system losses and energy yield. *John Wiley and Sons*. https://doi.org/10.1002/9781119736592.ch8.

Naqvi, A. A., Nadeem, T. B., Ahmed, A., & Zaidi, A. A. (2021). Designing of an off-grid photovoltaic system with battery storage for remote location. *Tecciencia, 16*(31), 15–28. http://dx.doi.org/10.18180/tecciencial.2021.31.2

Nsengimana, C., Liu, K., Yuhao, C., & Lingling, L. (2022). *Stand-Alone Photovoltaic and Battery Microgrid Design for Rural Areas*. https://doi.org/10.1177/01445987221102196.

Obeng-Darko, N. A. (2022). *Achieving sustainable electricity through renewable energy development in the Gambia: Regulatory and policy consiceration*. https://doi.org/10.1007/978-3-030-76221-6_66.

Samikannu, R., Oladiran, M. T., Gamariel, G., Makepe, P., Keisang, K., & Ladu, N. S. D. (2022). An assessment and design of a distributed hybrid energy system for rural electrification: The case for Jamataka Village, Botswana. *International Transactions on Electrical Energy Systems, 2022*(4841241), 1–12. https://doi.org/10.1155/2022/4841241

Sark, G. V. W., Reich, N. H., & Muller, B. (2012). A review of PV performance ratio development. *Conference Proceedings of World Renewable Energy Forum*. Vol. 6.

Skandalos, N., Wang, M., Kapsalis, V., D'Agostini, D., Parker, D. S., Bhuvad, S. S., Peng, J., & Karamanis, D. (2022). Building PV Integration according to regional climate conditions: BIPV Regional adaptability extending Koppen-Geiger climate classification against urban and climate-related temperature increase. *Renewable and Sustainable Energy Reviews*. https://doi.org/10.1016/j.rser.2022112950.

Soy, S. K., Rajput, S. K., Prakash, R., & Dheer, D. K. (2022). *Performance and Energy Metrics Analysis of Stand-Alone Rooftop PV Plant*. https://doi.org/10.1109/ICEFEET51821.9848352.

Surendra, H. H., Sudhindra, K. R., Archana, H. R. & Madhusundhan, K. N. (2022). Performance analysis of stand-alone PV system using advisor model-case study. *IEEE International Conference on Current Development in Engineering and Technology (CCET)*, Bhopal, India. https://doi.org/10.1109/CCET56606.2022.10080186.

Tejaswini, V. (2022). Performance analysis of a stand-alone PV system using single voltage sensor based MPPT method. *International Journal for Science Technology and Engineering*. https://doi.org/10.22214/ijraset.2022.444414.

Wassie, Y. T., & Ahlgren, E. O. (2023). Performance analysis and reliability of an off-grid mini grid in rural tropical climate Africa: A case study in southern Ethiopia. *Development Engineering, 8*, 100106. https://doi.org/10.1016/j.deveng.2022.100106.

Weerasinghe, R. J., & Yang, R. (2022). Impact of technological developments on multicriteria performance BIPV in non-domestic Buildings. *IOP Conference Series: Earth and Environmental Sciences*. https://doi.org/10.1088/1755-1315/1101/2/022029.

Wu, X., Liu, Y., Xu, J., Lei, W., Si, X., Du, W., Zhao, C., Zhong, Y., Peng, L., & Lin, J. (2015). Monitoring the performance of the building attached photovoltaic (BAPV) system in Shanghai. *Energy and Buildings, 88*, 174–182. https://doi.org/10.1016/j.enbuild.2014.11.073.

Yudin, A., Keerthy, J., & Revanna, C. (2023). Off-grid solar PV system sizing for a typical East African household. https://doi.org/10.36227/techrxiv.22722511.

Higher Heating Value Analysis for Waste to Energy Intelligence Model Optimization-Prediction

Hasiniaina Rojosoa, Sayon dit Sadio Sidibe, Marie Sawadogo, Francesco Di Maria, and Salifou Koucka Ouiminga

Abstract Physico-chemical characterization using a statistical method is a crucial first step in the process of recovering energy from solid waste. The objective of this study is the creation of the Optimization-Prediction of Higher Heating Value (HHV) Model ($OPH_{hv}M$) to be classified as an innovation in the field. In addition to a more complex model optimization problem solving applied in Food and Combustible Waste (FCW) in Ouagadougou, the capital of Burkina Faso in West Africa, using a structured methodology. This study focuses on FCW generated by households in order to optimize the energy source through the calorific value in electricity produced by different technological processes such as combustion, incineration, gasification and cogeneration. Several methods have been used to obtain a reliable result, one being linear regression through different models, the other being artificial neutral network through Support Vector Machine (SVM) with Support Vector Classifier (SVC) and Support Vector Regression (SVR). The Optimization and Prediction of HHV Models ($OPH_{hv}M$)) are qualified using six performance matrices, such as coefficient of determination (R2), Mean Square Error (MSE), Root Mean Square error (RMSE), mean absolute error (MAE), mean absolute percentage error (MAPE) and standard error of estimate (SEE). The results showed that the linear optimization and prediction model with the combination of volatility material, ash content and cabon fix is the best model, as indicated by the following equation: HHV = 10.982 + 0.1136VM-0.2848AC + 4.91694FC. The results of the $OPH_{hv}M$ model show a difference between the maximum and minimum value of R2[0.91; 0.87], MSE_SVM [0.17; 0.12], MAE_SVM [0.40; 0.27], RMSE_SVM [0.41; 0.35], MAPE_SVM (%) [10.15; 6 .06], SEE_SVM [0.39; 0.34]. These values indicate the significant

H. Rojosoa (✉) · S. dit Sadio Sidibe · M. Sawadogo
Institut International d'Ingénierie de l'Eau et de l'Environnement, Ouagadougou, Burkina Faso
e-mail: hasiniaina.rojosoa@2ie-edu.org; hasiniainarojosoa@gmail.com

F. Di Maria
iiLAR5 Laboratory, Department on Egineering, University of Perugia, Perugia, Italy

S. K. Ouiminga
LPCE, Laboratoire de Physique et Chimie de l'Environnement, Université Joseph Ki-Zerbo, Ouagadougou, Burkina Faso

M.-M. Narra and S. Narra (eds.), *African Green Transition Through Innovative Pathways*, World Sustainability Series, https://doi.org/10.1007/978-3-031-87043-9_16

relationship between the endogenous and exogenous variables, and the $OPH_{hv}M$ can use the best tools to quantify the future value of the energy released through HHV.

Keywords Higher heating value · Optimization model · Prediction-model · Foods waste · Combustible waste · Support vector machine · Ouagadougou

1 Introduction

Energy economics is a key factor in development of the economy in general. Many countries are now turning waste into energy, using direct combustion, incineration or gasification for renewable and clean energy. However, one of the problematic factors is that as Africa's population grows, so does its electricity demand, particularly in sub-Saharan Africa (SSA). In Madagascar, one of the SSA, the energy sector has been developing since September 17, 2024, thanks to the transformation of waste into energy. Two technologies could be combined to optimize the energy available, either through methanization to produce biogas as the main source of electricity, or by incinerating non-recyclable waste and plastic to convert it into heat and electricity (Jameel et al., 2024). The project is still very small and is an initiative financed by the Malagasy government, with the aim of transforming the 700 tonnes of waste produced daily at the Andralanitra site into an estimated 16 and 22 MW of electricity (Eiji Iwasaki, 2022). In Ivory Coast, green waste can be transformed into energy at a rate of 748 MW by 2030, with a planned 15 power plants costing an estimated 230 million euros (Frank Rijsberman, 2022). In Cameroon, waste from papaya peelings, bananas and chicken rejects is set to be transformed into renewable energy to produce biogas as a clean source of energy for cooking.

In 2023, 675 million people in SSA don't have access to electricity (IRENA, 2023). However, the abundance of waste and the inadequacy of electrical energy coverage lead to load shedding and untimely rationing of power. This increase will have a major negative/positive factor in different areas such as socio-economic and environmental which makes people more distant from each other and weakens work performance, so in this promotes the destruction of the ozone layer and makes air quality more polluting. However, the quantity of waste is an enormous resource in SSA, transforming it through the production of heat (Combustion or Incineration, or Gasification) into electricity. A study by Efstratios N. Kalogirou (2018) mentioned a new method called cogeneration which consists of transforming waste from 5 to 7 tons of fuel oil with 70% of combustible waste. This process is classified as an innovative solution in SSA to solve the problem of sustainable development goal 7 "access to energy for all".

A revolution in scientific research has brought a deeper analysis in the application of artificial intelligence (AI) in mathematical problem-solving processes by applying them to the transformation of waste or biomass into energy. The application of machine learning in this field makes it possible to refine and improve the results of experimental studies. In this way, this method could solve the problem of time and materials during the study process in the laboratory. Amen et al. (2020) developed 13 linear models ($R2 > 0.8$) based on the ultimate analysis in the determination of the energy content of solid waste using different direct combustion technologies.

Another study clarified the real value of biomass HHV prediction Root Mean Square Error (RMSE = 3,587), Mean Absolute Deviation (MAD = 2,409), Mean Absolute Percentage Error (MAPE = 21,68%), Coefficient of correlation (R^2 = 0;97) to create a linear model with artificial neural network (Olatunji et al., 2019). This study brought an improvement in the context of proximate analysis of Foods and Combustible Waste (FCW) by substituting analyses through AI model methods. A case study was chosen in the western part of Africa, in the capital Burkina Faso, called Ouagadougou, where enormous quantities of FCW were generated for the future (Rojosoa et al., 2024).

Ouagadougou, the capital of Burkina Faso, has been used as a case study in all analysis, because of its potential waste deposits, which continue to increase gradually and the low level of waste-to-energy conversion, such as gasification/incineration/pyrolysis. The city of Ouagadougou has a population of 1,300,000, a surface area of 219,3 km^2 and an energy access rate of 15%. This rate is classified as low, indicating limited energy production and consumption in each household (Word Bank, 2015). One of the most viable solutions is to exploit the gigantic FCW by transforming it into energy. The general objective of this analysis is to demonstrate the energy recovery of FCW in Ouagadougou using Artificial Neural Network (ANN) optimization and performance assessment for the prediction model of Higher Heating Value (HHV).

To obtain accurate results, the use of AI in the Ouagadougou—Burkina Faso case study in the creation of the model enables it to be classified as an innovation in the field. In addition, a more complex model optimization problem solving applied in FCW distinguishes this study among the different reviews analysed.

To concretize the results, the following four specific objects were identified:

- Determining the energy content and carried out using the calorific value to optimize the energy source in the electricity produced
- Create the model through the different variables in the field, secondary data and experimental study
- Optimize the model using Support Vector Machine with Support Vector Classifier (SVM_svc) and Support Vector Machine (SVM) with Support Vector Regression (SVM_svr) using AI methods
- Select the best model across the six (06) matrix performances.

Most studies in the literature analyse the regression model with the classical method, by developing new regression models that optimise errors using AI-SVM methods of FCW, this paper presents a novel approach to classification. Secondly, one of the challenges the author faces in this work is to create an optimization and prediction of HHV model ($OPH_{hv}M$) using both variables of proximal analysis, statistical analysis and machine learning. Finally, due to the constraints of time, laboratory equipment and pressure of results, this study used new approaches combining chemical and mathematical analysis through the optimization method and results.

1.1 Literature Review

Several analyses have been carried out in the recent journal concerning the HHV Model based in waste or biomass to have released a finished product such as Jiang et al. (2024) for example analysed a study on establishing a generalized model for accurate prediction of higher heating values of substances with large ash fractions in China. This work proposed a novel HHV optimization-prediction model based on its Reduction Degree (RD) and ash content (AshC).

First, ultimate analysis of biomass was applied to establish the calculation method of RD; then, the correlation between RD, AshC, and HHV was analyzed using the Pearson Correlation Coefficient; subsequently, the HHV ¼ f (RD, Cash) model was developed using regression analysis (Jiang et al., 2024).

In addition, the authors optimized the use of waste biomass through the process of the HHV prediction model based on its RD and ashC. This work proposed a new prediction model $HHV = 0.734DRS_{,B} + 0.276ashC + 6.801$ with a value $R^2 = 0.854$, Root Mean Squard Error (RMSE) = 0.900, and Mean Absolute Error (MAE) = 0.773. The analysis was performed with a precision of HHV for biochar, bio-oil and charcoal which take performance matrix R^2 values of 0.989, 0.939 and 0.961 respectively.

Another study was carried out by Özyuğuran and Yaman (2017) on the Prediction of Calorific Value of Biomass from Proximate Analysis in Turkey. The main objective in this work is the prediction of energies released in a linear and nonlinear model between proximate analysis variables. In this context, the biomass of twenty-seven species was analysed, including pulps, stems and husks, fruit kernels, agricultural residues, herbaceous and woody plants, and nut shells.

The results showed that for the different fuel characteristics and structures analysed, the correlation coefficient between the experimental study (0.812) was lower than that of the mathematical analysis (0.837). Another statistical parameter used in this study to determine the accuracy of the HHV prediction was the larger standard deviation, which can be taken as indicative of the model's good performance. In this case, the results mentioned a high value of standard deviation in the mathematical analysis (1.469 and 1.493 MJ/kg), while in the experimental analysis it varied between 1.469 and 1.493 MJ/kg. Besides, satisfactory predictions could be obtained when the calorific value of biomass is bigger than 20 MJ/kg. Therefore, it should be reasonable to make subclasses such as herbaceous samples, woody samples, agricultural residues, etc. and each subclass may be evaluated separately.

It is through these literatures, the study focus more precisely in the mathematical analysis that illustrates the optimization problems to be solved, and the results obtained as well as the conclusions during all the analyses.

2 Methodology

This study was divided into four distinct stages (see Fig. 1), each of which involved specific analyses and new methods to achieve the objective. The first step is data collection and analysis, the second is creating the model for minimising model HHV through the regression. The third is maximizing margin separating hyperplane (SVM_svc) and model optimization errors (SVM_svr). The last step is to validate the models after any results obtained.

2.1 Experimental Analysis

The first step was to collect data through experimental analysis at the LabEREE (Laboratoire Energie Renouvelable et Efficacité Energetique). An experimental analysis was carried out on FCW to determine the Moisture Content (Moi), Ash Content (AC) and Volatility Material (VM). The analysis consists in knowing the composition of solid waste in order to know their optimization type by category such as 13 type of waste in type I, 23 type of waste in type II and 43 type of waste in level III. Proximity analysis is carried out in the mixture of food wastes of level I, II and III, and through combustible wastes of level I, II and III (Rojosoa et al., 2024) in order to know the optimum results. All procedures use ISO 11722ASTM D.3173, ISO 1171 ASTM D.3174, ISO 248 ASTM D.3178 Moi, AC and VM respectively, which validate the proximity analysis and allow the data obtained to be processed.

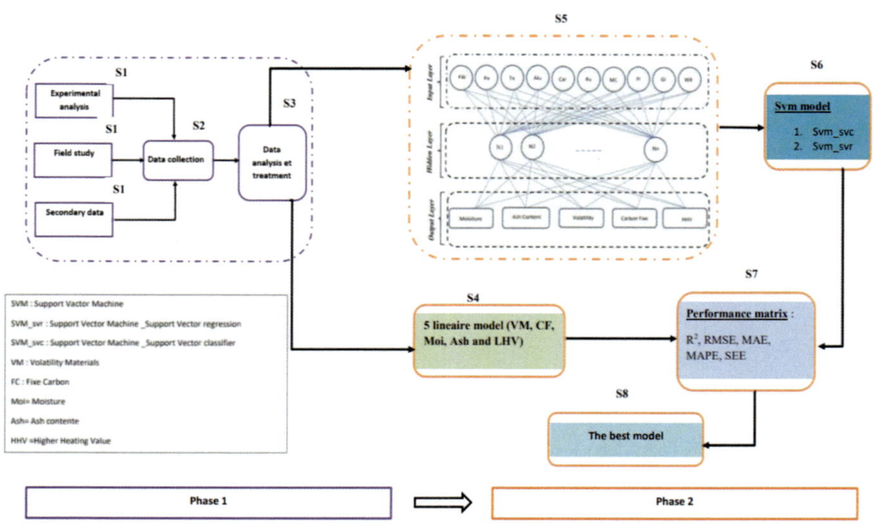

Fig. 1 General methodology of optimization-prediction of higher heating value (HHV) model ($OPH_{hv}M$)

In all cases, the other variables are determined by calculations such as Fixed Carbon (FC) and Higher Heating Value (HHV).

2.1.1 Importance of HHV in Analysis

HHV is the amount of heat released during the combustion is liquid, contrary the Lower Heating Value (LHV) is a vapor. HHV is very important property of fuel of FWC (Kalivodová et al., 2022), whereas, the HHV is very useful of characteristics of thermal energy for combustible waste. The HHV means that all the heat released in the fuels is measured when the products are cooled back to the initial temperature. Some the authors like to use the LHV, however the HHV is used in this study to determine the energy content of FCW. In the practical case, the calorific value is given as 25 °C (McAllister et al., 2011), but if there are different temperatures between the reagents, this could be reduced to 15 °C. To get the most out of the energy released in power plants, the authors need to know the chemical properties of the inputs (waste types), the fuel technology (materials to be used) and the reaction time.

2.1.2 Optimization of the HHV Model Process

The model creation process was carried out in three precise steps. The first step consisted in creating five linear models (Eqs. 1, 2, 3, 4, 5) between the variables coming from the experimental study such as Moi, VM and AC, while the other variables were analysed through mathematical studies such as HHV and FC.

$$HHV = ɥfc + ƛVM + KMoi + ƿAC + Q \tag{1}$$

$$HHV = ƛVM + KMoi + ƿAC + Q \tag{2}$$

$$HHV = ɥfc + ƛVM + KMoi + Q \tag{3}$$

$$HHV = ɥfc + ƛMoi + ƿAC + Q \tag{4}$$

$$HHV = ɥfc + ƛVM + ƿAC + Q \tag{5}$$

In each model, the authors set HHV as the endogenous variable and Q as the intercept in all models. Model 1 (Eq. 1) consists of exogenous variables varied between VM, Moi, FC and AC while model 2 (Eq. 2) represents explanatory variables such as VM, Moi, AC, model 3 (Eq. 3) varies the variables FC, VM, Moi and Q, model 4 (Eq. 4) consists analysed the variables FC, Moi, AC, and the last one consists explored the FC, VM, AC. The models are judged through six performance matrixes, as follows:

- Mean Square Error (MSE): is an estimator that measures the mean square deviation of the errors, i.e. the mean square difference between the estimated values and the true value of the proximal analysis.
- MAE: is a measure of the average size of the errors in a collection of predictions, without taking their direction into account. In general, if the MAE value is less than 10%, it is excellent, if it is between 10 and 20%, it is still good and if it is greater than 50%, it is unacceptable.
- RMSE: The root mean square error is the standard deviation of the prediction errors. In this case, the RMSE indicates the concentration of data around the line of best fit.
- MAPE: is a relative error measure that uses absolute values to prevent positive and negative errors from cancelling each other out, and uses relative errors to be able to compare the accuracy of forecasts. If the MAPE is less than 5%, the accuracy is acceptable, if it is between 10 and 20%, the accuracy is low but acceptable and if it is greater than 25%, the accuracy is very low.
- Standard Error of the Estimated or sum of squares error (SEE): is the estimate of the accuracy of a prediction.
- R-Squard is the proportion of the variance of the dependent variable that can be explained by the independent variable.

This is the main Eq. 6:

$$\alpha = \text{f(x1, x2, \ldots, xp; ß) f(}= \sum_{\epsilon=0}^{n}\left(W_{j\epsilon}\gamma_{\epsilon}\right) \tag{6}$$

After having created the model, the optimization step consists in choosing the best performing model under constraint of the different performance matrix following the supervised learning solves the following problem (Eq. 7):

$$\begin{cases} \text{Max}\,\alpha = f(= \sum_{\in=0}^{n}\left(W_{j\in}\right)(\gamma_{\in}) + \beta j \\ \textit{Subject to}: \sqrt{\sum_{i=1}^{\eta}\frac{(\hat{y}_i - y_i)^2}{n}} > 0 \\ 0 < \sum_{i=1}^{\eta}(\hat{y}_i - y_i)^2 < 1 \\ \frac{1}{n}\sum_{i=1}^{n}\left|\frac{A_i - F_i}{A_i}\right| * 100 > 0 \\ 0 < 1 - \frac{\Sigma_i(y_i - \hat{y}_i)^2}{\Sigma_i(y_i - \overline{y})^2} < 1 \end{cases} \tag{7}$$

Having chosen the right model by following the optimization processes, the second step is to maximize the distance between the separating boundary and the nearest points of each class by following the maximizing margin separating hyperplane (SVM_svc) process (Eq. 8–9):

$$\mathrm{Min}\frac{1}{2}z^{R}z + A\sum_{i=1}^{n}\xi_i z, \xi \quad (8)$$

$$\text{Subject to } \alpha_i\left(z_0^R\varnothing(\gamma_E) + \beta_i \geq 1 - \varsigma_i\right), \varsigma_i \geq O, \mathrm{i} = 1\ldots\ldots n \quad (9)$$

To solve the margin maximization problem, the first step is to minimize the parameter vector (Eq. 10):

$$\mathrm{Max}\frac{1}{\|\gamma\|} \Longleftrightarrow \min\|\gamma\| \quad (10)$$

Normalisation (Eq. 11):

$$\underset{\gamma,\gamma_0}{min}\|\gamma\| \text{ where } \|\gamma\| = \sqrt{\gamma_1^2 + \cdots + \gamma_i^2} \quad (11)$$

The constraints show that all points are in a good location, more precisely they are on the regression line (Eq. 12–13):

$$\text{Object}: \alpha \times f(x_i) \geq 1, i = 1, \cdots, n \quad (12)$$

$$\text{With} f(x_i) = x_i^R\gamma + \gamma_0 \quad (13)$$

The next process solves the problem using the Lagrangian method (Eq. 14)

$$L_p(\gamma,\gamma_0,\omega) = \|\gamma\| - \sum_{i=1}^{n}\omega_i\left[\alpha_i \times (x_i^R\gamma + \gamma_0) - 1\right] \quad (14)$$
$$where\ \omega i\ are\ the\ Langragien\ of\ multiplicator$$

Using the partial derivation of the three variables (Eq. 15) (γ,γ_0,ω):

$$\begin{cases} \frac{\partial L}{\partial \sigma} = \gamma - \sum_{i=1}^{n}\omega_i y_i x_i = 0 \\ \frac{\partial L}{\partial \gamma_0} = \sum_{i=1}^{n}\omega_i \alpha_i = 0 \\ \frac{\partial L}{\partial \omega} = -\alpha_i x(x_i^R\gamma + \gamma_0) + 1 \leq 0, \forall i \end{cases} \quad (15)$$

The final solution based on the resolution of the Lagragian partial drifts is shown in Eqs. 16 and 17.

$$\mathrm{Max}\ L_D(\omega) = \sum_{i=1}^{n}\omega_i - \frac{1}{2}\sum_{i=1}^{\Omega}\sum_{j=1}^{m}\omega_i\omega_j\alpha_i\alpha_j(x_i,x_j) \quad (16)$$

$$\text{Subject to } \sum_{i=1}^{\Omega}\omega_i\gamma_i = 0 \quad (17)$$
$$0 \leq \omega_i \leq A, \forall i$$

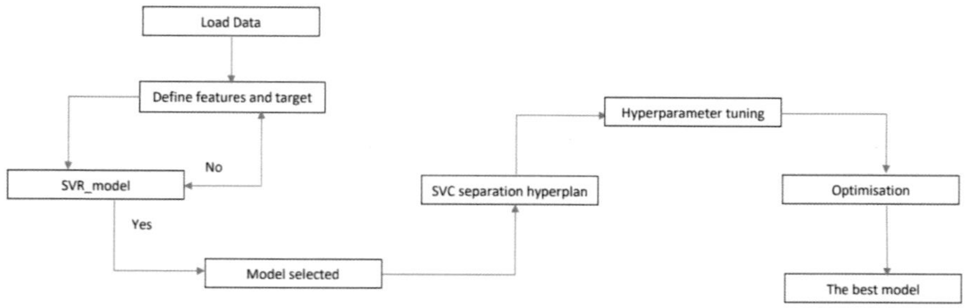

Fig. 2 Creation of the optimization-prediction of higher heating value (HHV) Model ($OPH_{hv}M$) using Python

2.1.3 Optimisation Algorithm

The aim of optimising algorithms is to make teaching easier and to produce results that meet the research objectives (Fig. 2). Here, the mathematical problem and solution have converted into code using python. The code starts by loading data collected to calibrate the model, and then defining the variables in the model to be used, and splitting data into training. The testing sets (70% train, 30% test) is essential before calculating the display regression matrix. The second step is to optimise the use of GridSearchCV for hyperparameter tuning, which consists of finding the best model and retraining the chosen model using the best parameters. Finally, the model evaluation stage ensures that the best optimization prediction model results.

3 Results and Discussion

3.1 Regression Model of Proximate Analysis

The first result in this analysis is shown in Fig. 3, the model of linear regression in the different mixture of variables proximate analysis such us HHV, FC, VM, Moi and AC shows the relationships between variables explained HHV and explanatory variables FC, VM, Moi and AC.

Figure 3 shows the compositional values of experimental data in the FC, Moi, VM, Ash and HHV mixtures

$$\text{HHV} = 10{,}982 - 5{,}5344\text{EMoi} + 0.1136\text{VM} - 0{,}2848\text{AC} \tag{18}$$

A negative significance was observed for Moi and AC, while VM had a positive significance. These results show that Volatility levels have a strong positive impact during the combustion/incineration/gasification processes of FCW, while Moi and AC can reduce the energy released in FCW. These results confirmed by Komilis

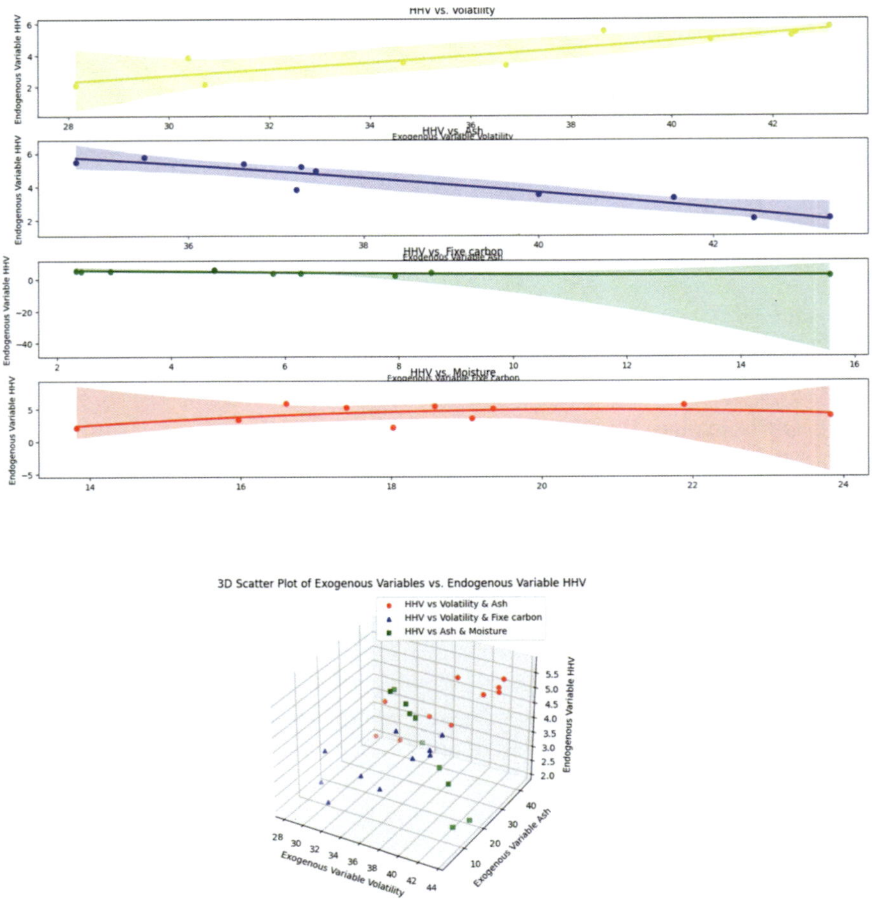

Fig. 3 Model of mixture of moisture, volatility, ash, fixe carbon and HHV and ED scatter plot

et al. (2014), the aim of this study was to experimentally determine the effect of the organic matter and Moi on the calorific value of organic solid wastes. Then the authors show that the Moi at which no ignition was feasible for certain samples. The performance matrices show maximum and minimum value of MSE, MAE, RMSE, MAPE, SEE and R-squared of 0.29, 0.44, 0.53, 9,69%, 0.52 and 0.78 respectively. A better value of r-squared close to 1 is observed, indicating strong independence on the endogenous and exogenous variable. It is from these results, this model is not better enough to predict the energy released from FCW through the HHV equation.

Figure 4 shows the mixtures of exogenous variables such as Moi, VM and AC as well as an endogenous variable HHV.

$$\mathrm{HHV} = 10{,}982 - 5{,}5344^{\mathrm{E}}\mathrm{Moi} + 0{,}1136\mathrm{VM} - 0{,}248\mathrm{AC} \tag{19}$$

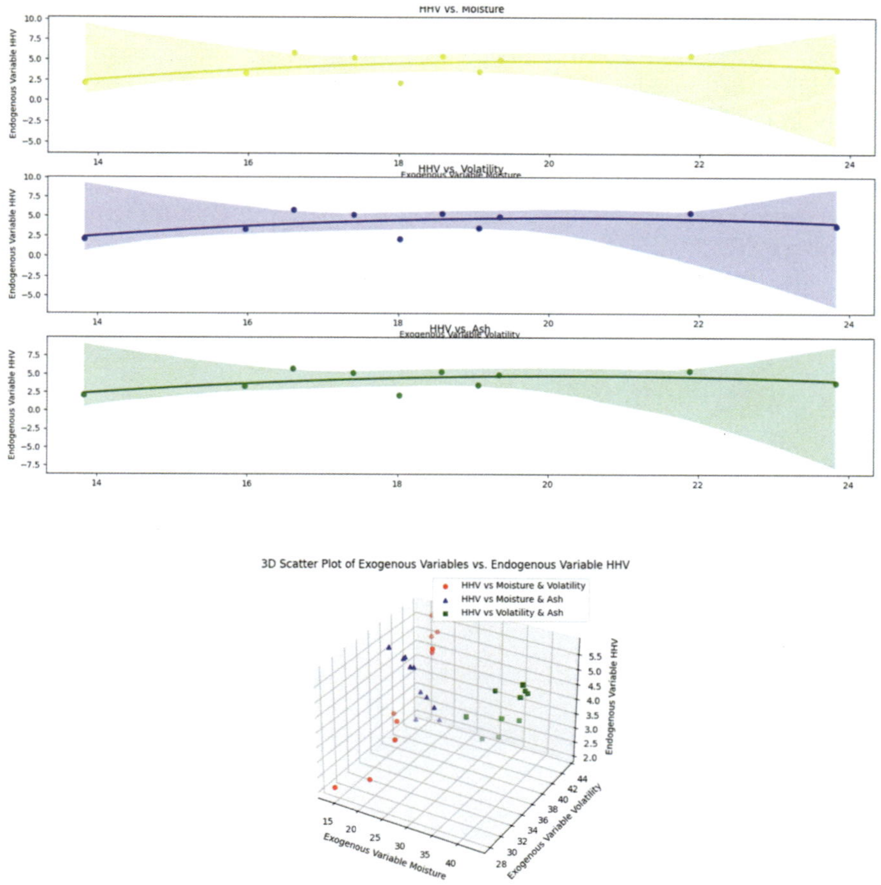

Fig. 4 Model of mixture of moisture, volatility, ash and HHV and ED scatter plot

It was found that the VM of FCW increased with HHV, while Moi and AC had negative effects on the endogenous HHV variables. However, a high VM content does not guarantee a high calorific value, as some VM ingredients are formed from uncombined raw materials (Özyuğuran & Yaman, 2017). The results also show that the AC does not imply a positive effect during energy production, and has a negative effect on the heat produced. The values for the statistical performance of the model on MSE, MAE, RMSE, MAPE, SEE and R-squard are 0.35, 0.44, 0.59, 8.87%, 0.55, 0.73 respectively. This result indicates dependence of HHV through the variables Moi, VM and AC. A low MSE means the model of the HHV model is close of the reality. This means that even if the model's statistical performance parameters are better, this cannot be used to judge the model's quality.

Figure 5 shows the relationship of the endogenous variable HHV through the exogenous variables Moi, VM, FC and HHV.

$$HHV = -17.4989 + 0.2848Moi + 0.3984VM - 0.2848FC \quad (20)$$

In this model, moisture content is highly significant with HHV, as the amount of Moi in the FCW increases, so does the energy release during the combustion process (Lio & Zhou, 2012). This model shows that the volatility parameter has a positive impact on the HHV, while the FC parameter reduces the energy value in the conversion process. This result was validated with (Aich et al., 2020). The statistical parameters MSE, MAE, RMSE, MAPE, SEE and R-squared, are represented as 0.70, 0.80, 0.84, 19.16%, 0.84 and 0.46 respectively. As indicated by those results, there is a strong relationship between the variables, so the model has a good fit data. The MSE here is a little large, perhaps due to the outliers due to the squaring of the errors, which allows a negative effect on the model.

Figure 6 shows the relationship between the Moi, FC and AC variables and HHV using a linear regression model.

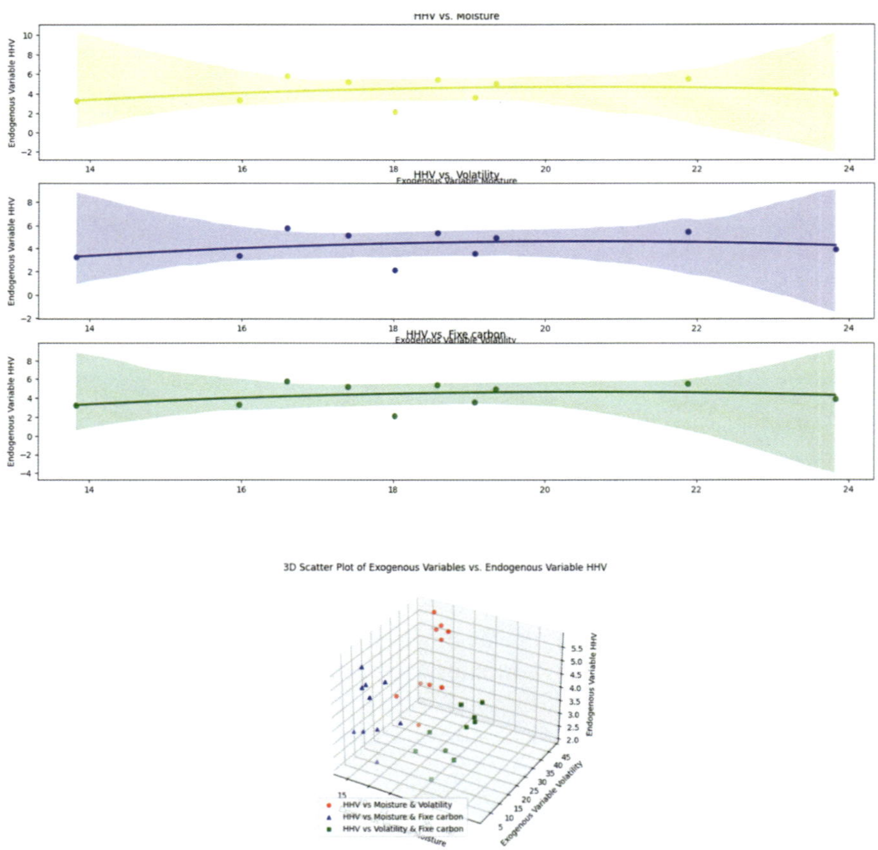

Fig. 5 Model of mixture of volatility, ash, fixe carbon and HHV and ED scatter plot

$$HHV = 22.342 - 0.1136Moi - 0.3984AC - 0.1136FC \tag{21}$$

There was no significant contribution between the endogenous variables HHV and the variable (Moi, FC and AC), which implies that this model cannot be used throughout the rest of the analysis. The performances to qualify this model indicate that the MSE, MAE, RMSE, MAPE, SEE and R-squared, are represented 0.24, 0.42, 0.49, 9.77%, 0.49 and 0.81 respectively. Even if, the results show a strong r-squared which means the good quality of the model for the prediction, it is not enough to judge the model.

The last model in Fig. 7 is the combination of the variables VM, AC, FC and HHV to evaluate its energetic power in food and fuel waste.

$$HHV = 10.982 + 0.1136VM - 0.2848AC + 4.91694FC \tag{22}$$

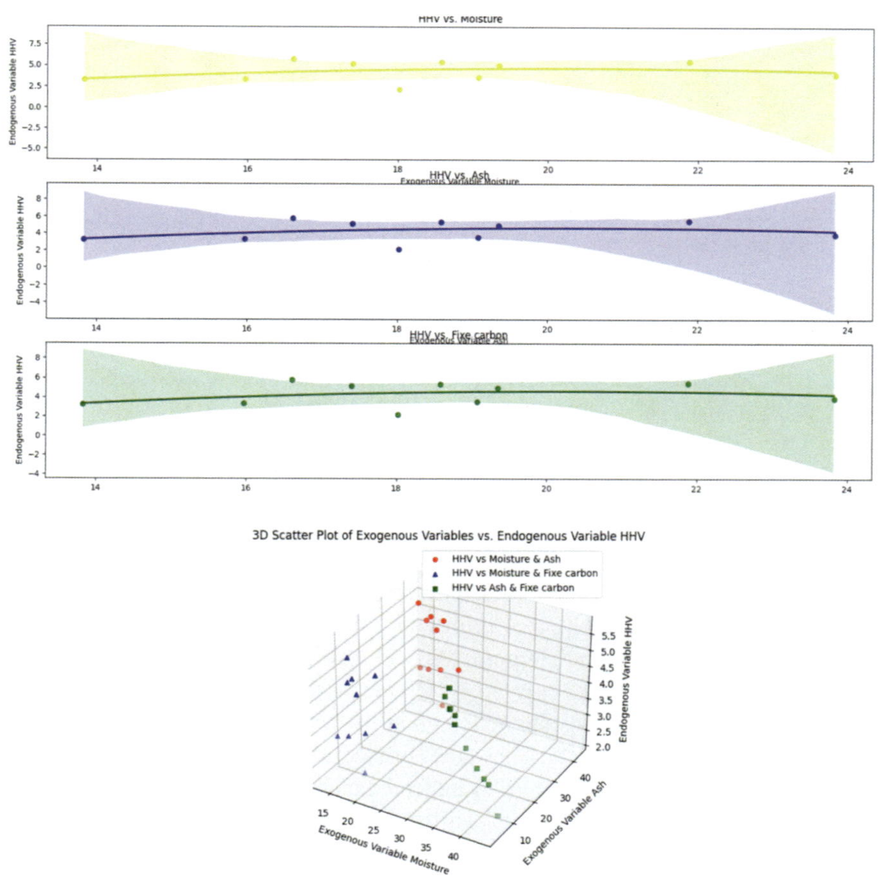

Fig. 6 Model of mixture of moisture, ash, fixe carbon and HHV and ED scatter plot

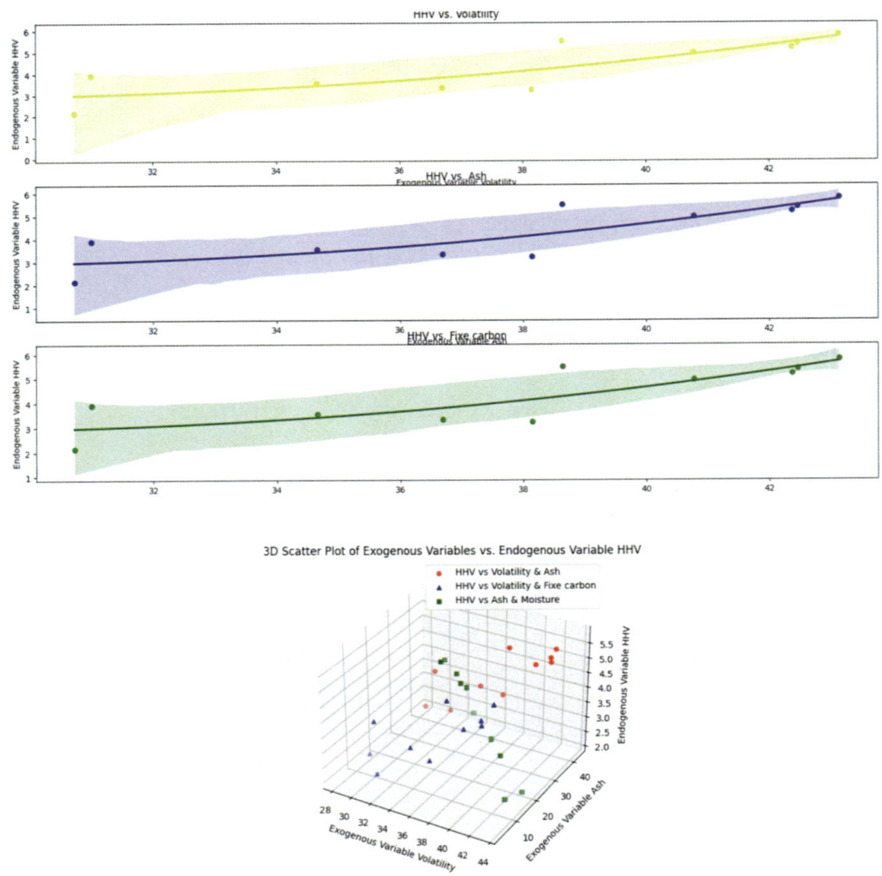

Fig. 7 Model of mixture of volatility, ash, fixe carbon and HHV and ED scatter plot

According to this model, the exogenous variables FC and VM and the endogenous variable HHV are positively significant. This explains why these properties can provide high energy value during the FCW conversion processes. But the effect of the AC property is less obvious and does not affect much the calorific value of the waste (Sygula et al., 2021). Regarding the performance matrixes, the MSE, MAE, RMSE, MAPE, SEE and R-squared, are represented 0.17, 0.40, 0.41, 19.15%, 0.39 and 0.89 respectively. A good model is observed, which means that this model can predict the energy value of waste. The results noticed that the MSE value is close to 0 which means goodness of fit to the data. The values of the MAPE indicate that the model is acceptable, and the value of the SEE is indicated that the model is acceptable also. The authors can interpret that this model is the best comparing by the results with the other models, threfore it confirms with the resolution of the problem of optimizing the choice to minimize the error through performance matrix (Table 1).

Table 1 Comparison of the maximum and minimum value by each model created

Performance matrix	Model 1		Model 2		Model 3		Model 4		Model 5	
	Min	Max	Min	Max	Min	Max	Min	Max	Min	Max
MSE	0.20	0.29	0.35	0.35	0.48	0.70	0.14	0.24	0.12	0.17
MAE	0.29	0.44	0.44	0.51	0.54	0.80	0.35	0.42	0.27	0.40
RMSE	0.44	0.53	0.59	0.59	0.69	0.84	0.37	0.49	0.35	0.41
MAPE (%)	5.58	9.69	8.87	11.60	11.58	19.16	9.50	9.77	6.06	10.15
SEE	0.38	0.52	0.55	0.58	0.66	0.84	0.31	0.49	0.34	0.39
R-Squard	0.85	0.78	0.73	0.73	0.63	0.46	0.89	0.81	0.91	0.87

3.2 Optimization Model Error HHV

This section shows the results of the marginal optimisation of the data classification of the variables of the model 5 (VM, AC, FC and HHV) chosen. The inputs are listed in Table 1 and contain the value obtained through laboratory analysis. What makes this work specific is that most of the studies coming from the literature review take the values all from the laboratory analysis (Amen et al., 2020). However, in this study, in order to obtain accurate results, the VM, Moi and FC analysis are carried out using the classic method and the remaining variables such as FC and HHV are analysed using the statistical and mathematical methods.

The first process indicates the maximisation of the margin between the model and the input variables in order to achieve a linear separation of the variable's Moi, VC, AC, FC and HHV. The optimal hyperplane maximises the distance between the separation boundary and the nearest points of each class (Tao et al., 2006; Berwick, 2008; Nefedov, 2016). The hypothesis is divided into the following two parts:

Representation bias for choosing the form of separation.

Preference bias for choosing the solution from the set of possible solutions.

The results are shown in the form of normalized variable points in the output variables. The output variables are divided into two parts, one at the top of the model and the other at the bottom of the model. The aim is to know the distance of any point of the variables from the frontier, so the points at which the margin lines lie are the vectors and also the minimum margin is (Eq. 23):

$$\alpha = \frac{1}{\|\gamma\|} \tag{23}$$

Where the Maximum margin is $\beta = 2\alpha$.

This step makes it possible to improve the model better than before by suppressing the errors at the level of the five performances that validate the model (Table 1). If the authors take the case by case, the results were found values closer to reality through the optimization method which Maximising margin separating hyperplane (SVM_svc) and model optimisation errors (SVM_svr). The adjustment of the errors (Fig. 8)

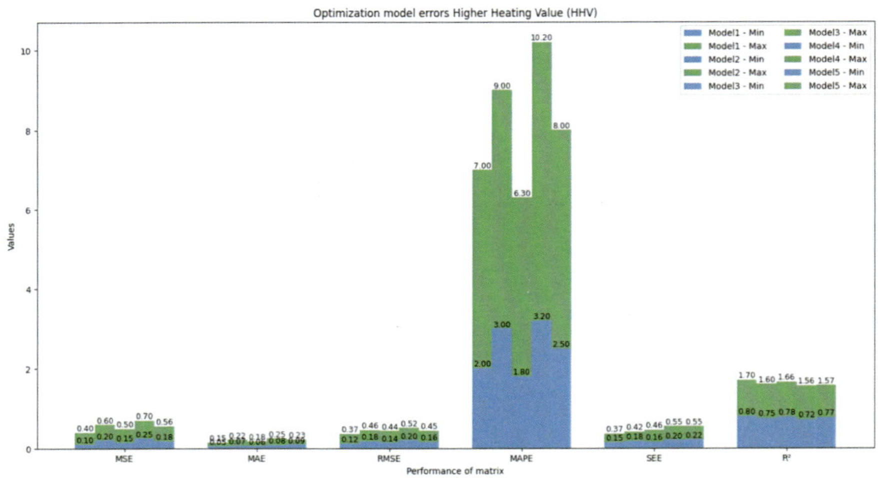

Fig. 8 Optimization-prediction of higher heating value (HHV) model ($OPH_{hv}M$)

in the MSE_SVM parameters in model 1, model 2, model 3, model 4 from 0.29 to 0.29, 0.35 to 0.35, 0.48 to 0.70, 0.14 to 0.24 and 0.12 to 0.17 respectively, allows to reduce the errors.

The authors have noticed the minimum value of the MSE-SVM of 0.12 if it was previously the value of 0.17. The lower MSE_SVM indicates that the predicted energy output of HHV is closer to the actual values, which means that the model created provides better heat performance in the combustion of FCW. The MAE-SVM parameter indicates the maximum and minimum values in model 1, model 2, model 3 and model 4 from 0.29 to 0.44, 0.44 to 0.51, 0.54 to 0.80, 0.35 to 0.42 and 0.27 to 0.40 respectively.

It can be seen that model 5 represents the minimum value of 0.27, indicating that the information is more accurate in predicting HHV among the other variables. The MAE-SVM parameter indicates the maximum and minimum values in model 1, model 2, model 3 and model 4 as 0.29 to 0.44, 0.44 to 0.51, 0.54 to 0.80, 0.35 to 0.42 and 0.27 to 0.40 respectively (Fig. 8).

It can be seen that model 5 represents the minimum value of 0.27, indicating that the information is more accurate in predicting HHV among the other variables. The RMSE-SVM parameter indicates the accuracy of the model across data variables. The minimum and maximums values for model 1, model 2, model 3, model 4 and model 5 from 0.44 to 0.53, 0.59 to 0.59, 0.69 to 0.84, 0.37 to 0.49 and 0.35 to 0.41 respectively. The minimum value was found in model 5 which confirms a better model for predicting the amount of energy during the thermal process of FWC. The minimum and maximums values of the SEE-SVM parameter in model 1, model 2, model 3, model 4 and model 5 from 0.38 to 0.52, 0.55 to 0.58, 0.66 to 0.84, 0.31 to 0.49 and 0.34 to 0.39 respectively. The minimum value in model 4 is 0.31, indicating that the regression line depreciates the total squared deviation of the prediction.

In general, MAPE_SVM parameters measure the precision of the sample mean relative to the variable mean. This ensures that the model 4 in the mixture of Moi, AC and FC variables indicates a good characteristic. The other parameter is MAPE-SVM such that the minimum and maximums values of model 1, model 2, model 3, model 4 and model 5 from 5.58% to 9.69%, 8.87 to 11.60%, 11.58% to 19.16%, 9.50% to 9.77%, 6.06% to 10.15% respectively (Fig. 8). This indicator measures the accuracy of the model between the current and future values. In this case, the results show the minimum value in model 1 of 5.58% which indicates that this model can predict an accuracy of current and future data in the conversion of waste to energy by the combustion or incineration procedure.

The last parameter is the R-squared-SVM, with minimum and maximums values for model 1, model 2, model 3, model 4 and model 5 of 0.85 to 0.78, 0.73 to 0.73, 0.63 to 0.46, 0.89 to 0.81, and 0.91 to 0.87 respectively (Fig. 8). According to these results, the results show in model 3 the minimum value of 0.63, which means that the mixture of moisture, volatile and fix carbon variables is significantly positive on the HHV variable. The performance of the HHV model was corrected in this parameter through errors in the calibration of the data in the model.

The model 5 (Eq. 14) is qualified as the best model of $OPH_{hv}M$ with the lowest performance parameters in MSE_SVM, MAE_SVM and RMSE_SVM of 0.12, 0.25 and 0. 37 respectively (Fig. 8) after the three process of optimisation methods such as Minimising model HHV through the regression with the supervised learning problem, Maximising margin separating hyperplane (SVM_svc) and model optimisation errors (SVM_svr). Additionally, the Fig. 8 shows a strong relationship between the exogenous variables Volatile, Ash and fixed carbon and the endogenous variable HHV. All the points form a line and a clean curve, indicating a straight relationship, which suggests a significant relationship between the endogenous and exogenous variables. Unusual points or contours in the graphs may indicate special cases or errors in the data.

In order to confirm the validation of the model, a comparison was made with the different authors of the same topic, which showed a low MAPE value 0.018% (Fig. 9) in the current study.

4 Conclusion

This work is focused on mathematical and statistical analysis involving the science and engineering of FCW analysis. In addition, the data are gathered through fieldwork in the collection of quantitative and qualitative data on waste typology. Other data are collected during experimental analysis in the laboratory. Several studies were grouped from this analysis in order to specify the research problem through literature reviews. The creation of the prediction-optimization model solves the choice optimization from the linear model created in the mixture of proximate analysis variables such as moisture, volatile, fixed carbon, ash content and HHV. Another optimization is the second step to classify the variables and the model through the maximising

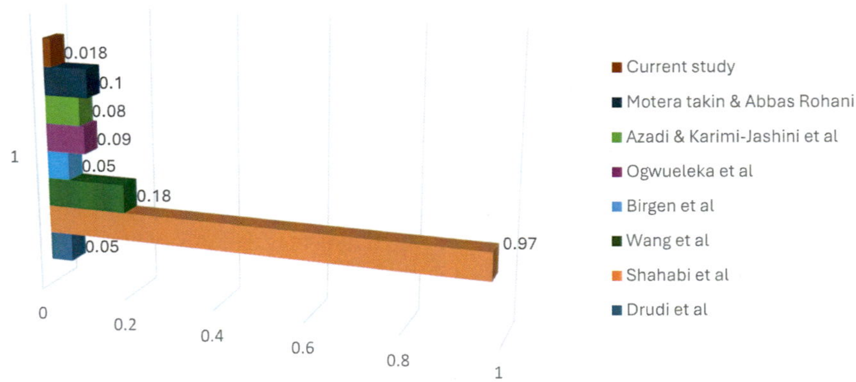

Fig. 9 Comparison of the results MAP on the optimization-prediction of higher heating value (HHV) model ($OPH_{hv}M$)

margin separating hyperplane (SVM_svc) methods. A final optimization on the errors in the six model performance indicators such as coefficient of determination (R2), Mean Squared Error (MSE), Root Mean Squared Error (RMSE), Mean Absolute Error (MAE), Mean Absolute Percentage Error (MAPE) and Standard Error of the Estimate (SEE) in order to minimize the model error. The results show that the variable HHV is highly dependent on the exogenous variables volatile: ash, fixed carbon. The authors suggest a high accuracy of the results on the HHV model created both chosen after following the optimization steps. Finally, the results in all analysis can be used a tool in the optimisation-prediction of HHV of volatile, ash and carbon fix.

References

Aich, S., Behera, D., Nandi, B. K., & Bhattacharya, S. (2020). Relationship between proximate analysis parameters and combustion behaviour of high ash Indian coal. *International Journal of Coal Science & Technology, 7*(4), 766–777. https://doi.org/10.1007/s40789-020-00312-5

Amen, R. et al., (2020). Modeling the higher heating value of municipal solid waste for assessment of waste-to-energy potential: A sustainable case study. Request PDF. https://www.researchgate.net/publication/347444719_Modeling_the_Higher_Heating_Value_of_Municipal_Solid_Waste_for_Assessment_of_Waste-To-Energy_Potential_A_Sustainable_Case_Study

Berwick, R. (2008). An idiot's guide to support vector machines (SVMs). https://web.mit.edu/6.034/wwwbob/svm.pdf

IRENA. (2023). World Energy Transitions Outlook 2023: 1.5°C Pathway, International Renewable Energy Agency, Abu Dhabi. Available for download: https://www.irena.org/publications

Iwasaki, E. (2022). Preparatory survey on the project for the improvement of solid waste management equipment for clean city in antananarivo final report. Retrieved 30 October 2024, from https://openjicareport.jica.go.jp/pdf/12364139.pdf

Jameel, M. K., Mustafa, M. A., Ahmed, H. S., Mohammed, A. jassim, Ghazy, H., Shakir, M. N., Lawas, A. M., Mohammed, S. khudhur, Idan, A. H., Mahmoud, Z. H., Sayadi, H., & Kianfar, E. (2024). Biogas: Production, properties, applications, economic and challenges: A review. Results in Chemistry, 7, 101549. https://doi.org/10.1016/j.rechem.2024.101549

Jiang, P., Li, L., Lin, H., Ji, T., Mu, L., Ji, Y., Lu, X., & Zhu, J. (2024). Establishing a generalized model for accurate prediction of higher heating values of substances with large ash fractions. *Green Chemical Engineering*. https://doi.org/10.1016/j.gce.2024.08.002

Kalivodová, M., Balas, M., Milčák, P., Lisá, H., Lisý, M., Lachman, J., Kracík, P., Križan, P., & Vejrazka, K. (2022). Determination of higher heating value by calculation based on elemental analysis. *Paliva,* 8–20. https://doi.org/10.35933/paliva.2022.01.02

Kalogirou, E. N. (2017). *Waste-to-Energy technologies and global applications.* International Standard Book Number-13: 978-1-138-03520-1 (Hardback). Retrieved 30 October 2024, from https://wtert.org/wp-content/uploads/2022/06/Waste-to-Energy-Technologies-and-Global-Applications-by-Efstratios-N.-Kalogirou.pdf

Klemeš, J. J., Bokhari, A., & Asif, S. (2021). Modelling the higher heating value of municipal solid waste for assessment of waste-to-energy potential: A sustainable case study. *Journal of Cleaner Production, 287*, 125575. https://doi.org/10.1016/j.jclepro.2020.125575

Komilis, D., Kissas, K., & Symeonidis, A. (2014). Effect of organic matter and moisture on the calorific value of solid wastes: An update of the Tanner diagram. *Waste Management, 34*(2), 249–255. https://doi.org/10.1016/j.wasman.2013.09.023

Luo, Z., & Zhou, J. (2012). Thermal conversion of biomass. In W.-Y. Chen, J. Seiner, T. Suzuki, & M. Lackner (Eds.), *Handbook of climate change mitigation* (pp. 1001–1042). Springer US. https://doi.org/10.1007/978-1-4419-7991-9_27

McAllister, S., Chen, J.-Y., & Fernandez-Pello, A. C. (2011). Thermodynamics of combustion. In S. McAllister, J.-Y. Chen, & A. C. Fernandez-Pello (Eds.), *Fundamentals of combustion processes* (pp. 15–47). Springer New York. https://doi.org/10.1007/978-1-4419-7943-8_2

Nefedov, A. (2016). Support vector machines: A simple tutorial. https://sustech-cs-courses.github.io/IDA/materials/Classification/SVM_tutorial.pdf

Olatunji, O., Akinlabi, S., Madushele, N., Adedeji, P., & Ishola, F. (2019). Multilayer perceptron artificial neural network for the prediction of heating value of municipal solid waste. *AIMS Energy, 7*, 944–956. https://doi.org/10.3934/energy.2019.6.944

Özyuğuran, A., & Yaman, S. (2017). Prediction of calorific value of biomass from proximate analysis. *Energy Procedia, 107*, 130–136. https://doi.org/10.1016/j.egypro.2016.12.149

Rijsberman, F. (2002). GGGI-Cote-dIvoire-CPF-02.11.22.pdf. Retrieved 30 October 2024, from https://gggi.org/wp-content/uploads/2022/02/GGGI-Cote-dIvoire-CPF-02.11.22.pdf

Rojosoa, H., Sidibe, S., Sawadogo, M., Di Maria, F., & Ouimenga, S. (2024). Physical composition of household and industrial solid waste under a rigorous statistical and case study analysis approach based on Python.

Syguła, E., Świechowski, K., Stępień, P., Koziel, J., & Bialowiec, A. (2021). The prediction of calorific value of carbonized solid fuel produced from refuse-derived fuel in the low-temperature pyrolysis in CO2. *Materials, 14*, 1–17. https://doi.org/10.3390/ma14010049

Tao, D., Tang, X., Li, X., & Wu, X. (2006). Asymmetric bagging and random subspace for support vector machines-based relevance feedback in image retrieval. *IEEE Transactions on Pattern Analysis and Machine Intelligence, 28*(7), 1088–1099. ISSN 0162-8828. Downloaded from: https://eprints.bbk.ac.uk/id/eprint/443/

World Bank. (2015). Energy_Profile_Burkina_Faso.pdf. Retrieved 30 October 2024, from https://www.zvei.org/fileadmin/user_upload/Themen/Maerkte_Recht/Aussenwirtschaft/Afrika/Burkina-Faso/Burkina-Faso-EZ-Elektro-EnergieRohstoffe/Energy_Profile_Burkina_Faso.pdf

Predictive Modelling of Electricity Consumption in Togo Using Deep Learning: A Comparative Approach

Zourkalaini Boubakar, Belko Abdoul Aziz Diallo, Nathan Williams, and Yendoubé Lare

Abstract Electrical energy plays a vital role in daily life. Since the adoption in 2015 of the new United Nations (UN) Sustainable Development Goals (SDGs), a lot of attention has been paid to the energy sector in particular. The SGD number 7 aims to ensure by 2030, access to affordable, reliable, sustainable, and modern energy for all. Thus, understanding electricity consumption growth remains a fundamental aspect of achieving universal access to energy. This research aimed to investigate the multivariate modelling of yearly residual electricity consumption in Togo by conducting a comparative approach between different deep learning algorithms. The analysis uses a data-driven approach based on postpaid residential electricity consumption data from Togo Electricity Company, aggregated at the regional level, from 1994 to 2020. Alongside this consumption data, additional covariates were selected through a feature selection procedure. The data were modelled using deep learning algorithms. Three algorithms were used namely Multi-Layer Perceptron (MLP), Convolutional Neural Network (CNN), Long Short-Term Memory (LSTM), and a hybrid CNN-LSTM model. The model's performances were assessed using the Mean Absolute Error (MAE) metric. As a result, from this study, the CNN-LSTM hybrid model performed the best with an MAE of 6.2 GWh on the test set while MLP, CNN, and LSTM performed with an MAE of 9.1 GWh, 14.4 GWh, and 14.8 GWh respectively. These results revealed the potential of deep learning techniques in modelling yearly residential electricity consumption in Togo. Thus, the utility company can rely on a such deep learning model to right-size its residential electricity supply for a more sustainable energy use and a fair and cost-efficient green energy transition.

Z. Boubakar (✉) · B. A. A. Diallo
Department of Data Management, WASCAL Competence Center, WASCAL, Ouagadougou, Burkina Faso
e-mail: boubakar.z@edu.wascal.org

N. Williams
Department of Sustainability, Golisano Institute for Sustainability, Rochester Institute of Technology, Rochester, USA

Y. Lare
Centre d'Excellence Rgional pour la Maitrise de l'Electricité (CERME), Department of Physics, University of Lomé, Lomé, Togo

M.-M. Narra and S. Narra (eds.), *African Green Transition Through Innovative Pathways*, World Sustainability Series, https://doi.org/10.1007/978-3-031-87043-9_17

Keywords Electricity consumption · Growth · Modelling · Deep learning · Togo

Nomenclature

AI	Artificial intelligence
ANN	Artificial Neural network
DL	Deep Learning
GWh	Giga Watts hours
KWh	Kilo Watts hours
ML	Machine Learning
MLP	Multi-Layer Perceptron
RNN	Recurrent Neural Network

1 Introduction

Every advanced economy can attribute its development and success to reliable access to modern energy (International Energy Agency, 2014). Since the adoption of the new United Nations (UN) Sustainable Development Goals (SDGs) in 2015, a lot of attention has been paid to the energy sector in particular. "The energy sector has been long recognized as essential for humanity to develop and thrive, it is also crucial to the achievement of many other SDGs, including those concerned with gender equality, poverty reduction, and improvements in health" (International Energy Agency, 2017). Particularly, the SDG number 7 aims to "ensure access to affordable, reliable, sustainable and modern energy for all" (Johnston, 2016). As defines by Moss (2021), Modern Energy Minimum as a higher, more inclusive level of electricity consumption at 1,000 KWh per person per year, with at least 300 KWh at home and 700 KWh consumed in the wider economy. However, electricity demand is set to increase further because of population growth, rising household incomes, growing demand for digital connected devices, and environmental concerns. Therefore, policymakers need to understand energy demand and how that demand will evolve with time.

Analysis of electricity consumption in a meaningful manner can yield useful information for understanding and forecasting demand. Such analysis can give insights to policymakers regarding the decision-making of a country's energy sector, especially for developing countries where strategies on how resources should be allocated to increase access to electricity are often made with little evidence (Fobi et al., 2018). In 2018, Togo established its electrification strategy. Several approaches and sources of information were gathered to define the electrification strategy in light of SDG number 7 (AT2ER, 2018). However, the strategy does not consider the differences in consumption growth. Thus, the electrification strategy might underestimate the likely near-term energy demand.

Advances in technologies characterized by large amounts of data create enormous prospects for all activity sectors, and as a result, businesses are increasingly turning to Machine Learning (ML) tools to tap into that potential.

ML includes several tools for extracting important insights from raw data, which can then be utilized to help enterprises improve their operations and strategic decisions. ML encompasses a wide range of tools including Deep Learning (DL). DL algorithms can automatically learn arbitrary complex mappings from inputs to outputs and support multiple inputs and outputs, offering opportunities for time series forecasting, especially on data with complex nonlinear dependencies, multivariate inputs, and multi-step forecasting (Brownlee, 2021). In the energy sector, ML is now well known. The work presented in (Narciso & Martins, 2020) emphasizes the application of ML tools for energy-related data analysis. Exploring the potential of such a new advanced data analysis tool, the study addresses this research question: What is the most effective DL model for predicting yearly residential electricity consumption? To answer this question a comparative study has been conducted on different DL architectures.

The study is structured as follows: The next section presents previous works, followed by data and the methodology used to conduct the study. Then, predictive modelling is presented, followed by results and their discussion. A conclusion is presented at the end, recalling the takeaway of this study and how the work can be deepened.

1.1 Previous Works

Energy demand and supply forecasting has received a lot of attention in recent years due to the fluctuation of both variable generation and demand. Different techniques are used to forecast energy demand over different geographic extend, temporal resolutions, and horizons. Traditionally, these techniques are classified into Short-term energy consumption forecasting and long-term energy demand projections. The former is used to inform real-time electricity scheduling (hours up to a month) while the latter, comes in longer-term system planning (at least one year ahead). Short-term forecasts can assist customers in more efficient use of energy. Thus, they can save money as well as proactively reduce energy use impacts on the environment.

Better long-term energy forecasting can help utilities, governments, and investors determine where, when, and how to supply energy. Hence, forecasts with high temporal and spatial resolution are important in energy demand modelling. ML techniques such as artificial neural networks (ANNs) are the most common forecasting techniques used in residential electricity consumption modelling (Verwiebe et al., 2021). A review of electrical load forecasting models conducted by (Kuster et al., 2017) confirmed the wide use of ANNs that can capture recurrent patterns. In their systematic review of electricity demand forecasting using ANNs-based machine learning algorithms, (Román-Portabales et al., 2021) discovered that the Long Short Term Memory (LSTM) model, which is the most commonly used algorithm in the reviewed papers, has been shown to achieve very good results in aggregated load forecasting, and that their predictions typically get more accurate as the number of electricity consumers grows. However, a mix of multiple ANN-based techniques could be used to create more accurate models at the trade-off of increased complexity (Bashari & Rahimi-Kian, 2020).

There are fewer studies on residential electricity demand modelling using ANNs in developing Africa. Several studies used a bottom-up approach to model electricity demand (Castellano, 2015). Adeoye and Spataru (2019) modelled electricity demand for WAPP (West African Power Pool) country members using multiple linear regression. Adjamagbo et al. (2011) Used parametric models to model Togo's electricity demand. A recent study (Amega, 2018) makes use of Stella (a system dynamic model software) to predict Lomé (Togo) residential electricity consumption. Unlike these studies, the present work models Togo residential electricity demand using ANNs by examining the historical electricity consumption of Togo Electricity Company residential customers.

2 Data and Methodology

2.1 Data Sources and Data Preparation

Two main datasets were used to conduct this study: Electricity consumption and population size. The former is provided by Togo Electricity Company while the latter is a public dataset. The population size datasets were collected from the National Institute of Statistics (INSEED, 2011).

Table 1 describes the different datasets and data sources.

The study was conducted in Togo. The Republic is located in West Africa between latitudes 6° and 11° North and longitudes 0° and 1°40 East. The country is divided into five (5) administrative regions (each one counts several prefectures) namely

Table 1 Datasets and data sources

Data source	Dataset	Spatial resolution	Temporal resolution	Years
Togo National Institute of Statistics (INSEED)	Population size	Prefectural level	Yearly	1990–2020
Togo Electricity Company (CEET)	Postpaid Residential electricity consumption (KWh) & number of customers	Agency scale	Monthly	1994–2020

(from South to North) Maritime, Plateaux, Centrale (Centre), Kara, and Savanes. The utility counts a total number of 31 operational agencies across the country. The raw data from the utility were available in Excel Open XML Spreadsheet (XLSX). After the data cleaning procedure, the raw data were transformed into Common Separated Values (CSV) format. The remaining raw data were available in CSV. Next to data transformation, filtering proceeded to discard incomplete records that cannot be used in the analysis. Finally, for analysis purposes, different datasets were aggregated at regional spatial resolution and yearly temporal resolution (from 1994 through 2020). Therefore, the analysis was built upon complete datasets instead of sampling data, as is done in some statistical approaches.

2.2 *Methods*

2.2.1 Deep Learning for Time Series

Deep Learning (DL) is a subset of ML (See Annexe for more about DL). Recent years have seen a proliferation of DL applications in many domains. Many DL algorithms have been applied more recently to time series, both with univariate and multivariate time series. DL algorithms applied to forecasting could overcome many of the challenges faced by classical approaches most importantly non-linear dynamics, usually neglected by traditional methods. These algorithms encompass Convolutional Neural Networks (CNNs) (LeCun et al., 1998) and Recurrent Neural Networks (RNNs) (Rumelhart et al., 1988).

The residential electricity consumption forecasting was performed using Tensorflow version 2.8.0. TensorFlow is a Python end-to-end open-source package designed for deep learning.

2.2.2 Problem Framing

The problem to be modelled can be expressed as follows: given 5 devices measuring the yearly residential energy consumption of the different 5 regions, build a model to forecast the next residential energy consumption of each region. This corresponds to forecasting the next multiple time steps.

2.2.3 Feature Selection

A set of features were used as ingredients for consumption growth. These variables are Residential consumption (RCons) data (100 GWh); Residential number of customers (RCust); Total population (TPop); Electrification rate (ER, %); Medium voltage consumption (MVolt) data (GWh); and Medium voltage number of customers (MCust). Residential consumption data is the independent variable (predictant).

Table 2 Pearson correlation coefficients

	RCons	RCust	MVolt	MCust	TPop	ER
RCons	1.00	0.98	0.99	0.96	0.91	0.94
RCust	0.98	1.00	0.95	0.92	0.88	0.96
MCons	0.99	0.95	1.00	0.99	0.91	0.92
MCust	0.96	0.92	0.99	1.00	0.93	0.89
TPop	0.91	0.88	0.91	0.93	1.00	0.86
ER	0.94	0.96	0.92	0.89	0.86	1.00

Medium voltage consumption and the number of customers are used as economic factors and the Electrification rate is derived from population size, number of residential customers, and household size. The variables were selected through a correlation check (see Table 2).

2.2.4 Data Normalization

Each time series in the data is on a different scale. Thus, each feature was normalized independently (to have a mean of 0 and a standard deviation of 1) so that they all take small values on a similar scale. Similar to input data, target data were also normalized because of their wide range of values. Data were normalized using formula (1).

$$X^k = \frac{x^k - mean}{std} \tag{1}$$

In this formula, X^k is the normalized value of series k, x^k is the original input data value of series k, mean is the mean of the input data value of series k and std is the standard deviation of the input data of series k.

2.2.5 Modelling

The experiment was run on four (4) different models: MLP, CNN, LSTM, and CNN-LSTM ensemble model. Figure 1 illustrates the architectural design of each model. All models use an RMSprop optimizer with Mean Square Error (MSE) as an objective function (loss). The mean absolute error (MAE) is used as a metric to evaluate the performance of models. MAE evaluates the absolute distance of the observation to the prediction. MAE takes the average of absolute errors for a group of predictions and observations as a measurement of the magnitude of error for the entire group. A small value of MAE indicates good performance.

Evaluating a model always boils down to splitting the available data into three (3) sets training, validation, and test. The model is trained on the training data while being evaluated on the validation data. Once the model is ready, it's tested on the test data.

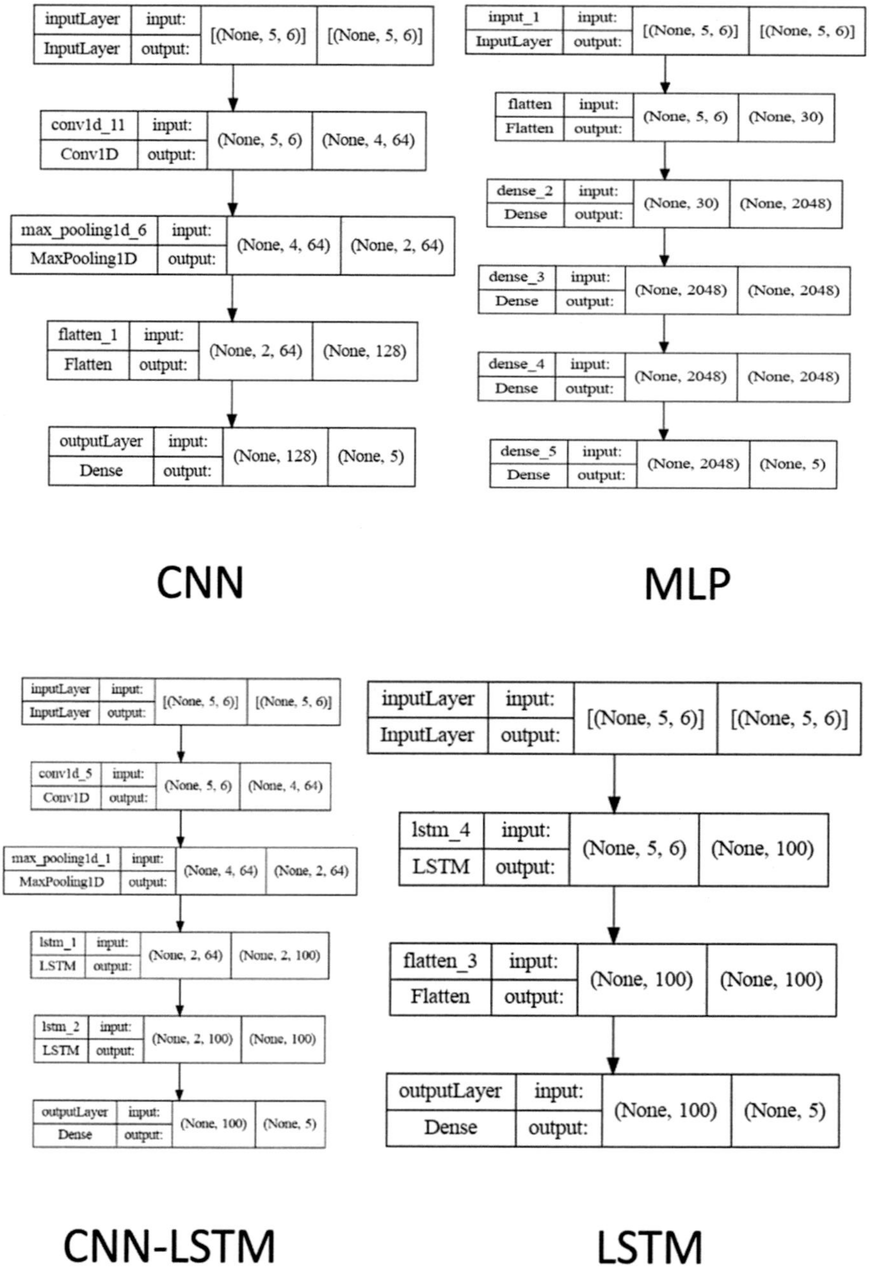

Fig. 1 Model architecture

The reason for validating a model is that developing a model always involves tuning its configuration (called hyper-parameters). There is no universal rule regarding the computation of the number of samples for each data split. However, data representativeness, the arrow of time, and data availability are used as indicators for data splitting.

For this forecasting problem, the dataset was split as follows:

Training sample (80%): Dataset from 1994 to 2014.

Validation sample (15%): Dataset from 2015 to 2018.

Test sample (5%): Dataset from 2019 to 2020

$$MAE = \frac{1}{N}\sum_{1}^{N}|yi - \tilde{y}| \tag{2}$$

$$MSE = \frac{1}{N}\sum_{1}^{N}(yi - \tilde{y})^2 \tag{3}$$

3 Results and Discussion

All models were trained and evaluated using the same training, evaluation, and testing data. Models performed well on the validation and testing sets as shown in Fig. 2. These results were obtained with the best hyperparametrizations. CNN and LSTM have similar performance on the validation set. These two models have the same number of parameters and the same learning rate. However, CNN requires more epochs (250 epochs) to get this performance than LSTM which requires 70 epochs, indicating the ability of LSTM to model the temporal dependency in data. CNN-LSTM model (ensemble model) combines both CNN and LSTM properties. On the one hand, the feature extraction process is performed at the CNN level (spatial information), and on the other hand, the extracted features are fed to the LSTM model that has a memory (temporal information). Thus, the Spatio-temporal CNN-LSTM model will require more parameters to perform the forecasting. It has achieved the best performance on the validation set with a few epochs (50 epochs). The experiment was also carried out with the MLP model that is used as a baseline. MLP unlike the precedent models is a feed-forward model without spatial information (as opposed to CNN) and memory (as opposed to RNN). MLP processed the temporal series as a single data point. Thus, the temporal and spatial dependence are ignored. The best hyper-parameters (ending up having the lowest validation performance) for this model were found with huge computational resources with more than 8 Million parameters. After the training process, the final models were evaluated using test sets. Unlike validation sets, test sets are data that models have never seen before. Such data sets are used to evaluate the performance of a trained model.

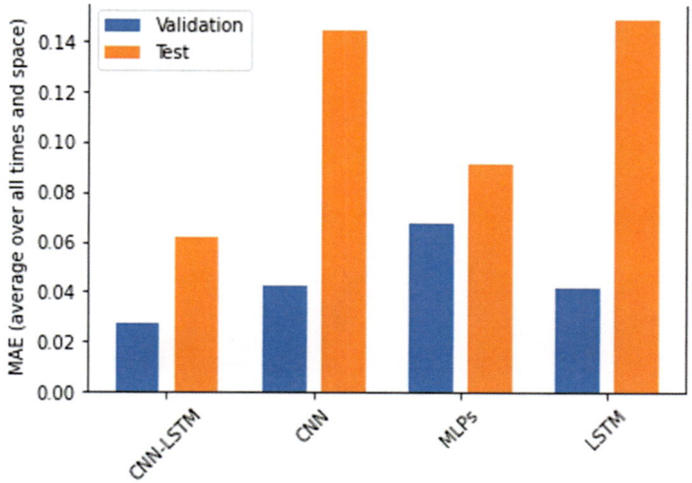

Fig. 2 Model evaluation

Table 3 Testing performance

Models	MAE (100 GWh)
MLP	0.091
CNN	0.144
LSTM	0.148
CNN-LSTM	0.062

Table 3 lists the models' performances on testing sets. MAE calculations are carried out both at the training level and testing level as well. The MLP performed better on unseen data than CNN and LSTM models that performed similarly.

All criteria (hyperparameters and performance on unseen set) taken into account, the ensemble CNN-LSTM model ranked the best. Figure 3 shows actuals versus the ensemble model forecasts in the five different regions of Togo country. As a recall, models were trained with datasets covering the 1994 to 2014 period. It can be noticed that globally CNN-LSTM model is good at reproducing different electricity consumption from different regions. The model reaches high accuracy for the Kara region with an MAE of 1.32 GWh while the highest MAE value of 28.03 GWh is observed for the Lome Region where consumption is very high.

Fig. 3 CNN-LSTM 's forecasts versus actual values

4 Conclusion and Perspectives

This paper investigates the modelling of yearly electricity consumption in Togo using a deep learning model. As a result of this research, it was found that the multivariate and multi-step ensemble CNN-LSTM model is the most effective Deep learning architecture for predicting the yearly electricity consumption of Togo's different regions, among the most common DL architecture for forecasting tasks: MLP, CNN, and LSTM. The experiment carried out on these models uses postpaid electricity consumption data (from CEET) and was evaluated using the MAE.

Electrical energy plays a vital role in daily life. Understanding energy consumption remains a fundamental aspect of energy planning. This work contributes to understanding consumption dynamics and gives insights to decision-makers. The outcome of this work can be considered a first step in the application of ANN in electrical energy modelling in Togo. The multivariate and multi-step CNN-LSTM ensemble model can forecast electricity consumption in the five regions simultaneously without the need to build a specific model for each region. Deep learning methods present undeniable advantages compared to traditional forecasting methods. The CEET could use these forecast techniques to right-size resource allocation regarding grid extension.

This study was built upon a postpaid electricity consumption dataset. However, Togo Electricity Company counts postpaid as well as prepaid customers. Future work will be needed to extend our understanding of consumption growth among prepaid customers. Findings indicate that ANN models are good at forecasting yearly residential electricity consumption. However, from the experiment, it can be seen that the model accuracy is influenced by consumption growth ingredients. Further work on ensemble feature selection for modelling residential electricity consumption could be an interesting step in increasing model reliability.

Acknowledgements The authors gratefully acknowledge the support of Togo Electricity Company for raw data collection. The authors would like also to express their gratitude to WASCAL for funding this project.

Annex

Deep Learning Overview

DL is a specific subset of ML that learns patterns from data using successive layers (stacked on top of each other) of increasingly meaningful representations of data via Neural Networks (NNs). The basic function of neurons was formalized by Frank Rosenblatt in 1958 as the perceptron (see Fig. 4), a model that contains the essentials of most modern deep learning. In the perceptron model, a neuron (node) receives input from other neurons. These inputs get integrated by summing them up. In this

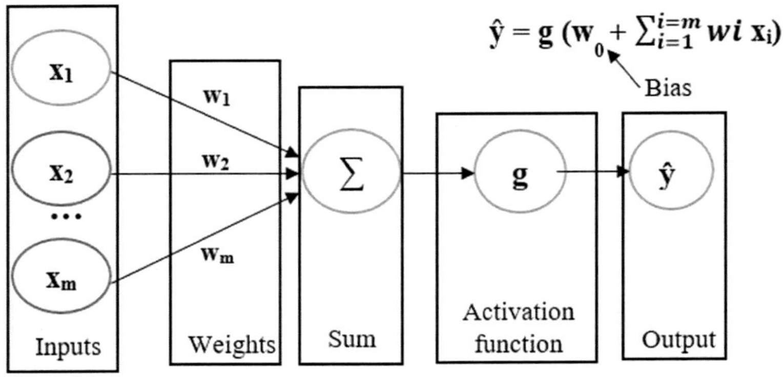

Fig. 4 The perceptron model

sum, each input from a neuron comes with its weight that marks its importance. The integrated input can then lead to a neural activation that yields the output. This process is called feedforward propagation. In modern Neural Networks (NNs), activation functions are non-linear functions (sigmoid function, hyperbolic, rectified linear unit …). The purpose of activation functions is to introduce non-linearity into the network. Linear activation functions produce linear decisions no matter the network size. In contrast, non-linearity allows for the approximation of arbitrarily complex functions.

Deep NNs, called Multi-Layer Perceptron (MLP) are NNs that contain multiple hidden layers. These NNs can be trained through a backpropagation algorithm as illustrated in Fig. 5.

Convolutional Neural Networks

CNN is a type of deep model used in computer vision applications. The difference between densely connected layers and convolution layers is that the latter learn local patterns in their input feature space whereas the former learn global patterns. CNN can automatically extract features from high-dimensional raw data with a grid topology, such as the pixels of an image, without the need for any feature engineering. The patterns they learn are translation-invariant, which means that the features are extracted regardless of where they are in the data as shown in Fig. 6. These characteristics make CNN suitable for dealing with one-dimensional data such as time series. Sequence data can be seen as a one-dimensional image from where the convolutional operation can extract features. In a fully connected layer, each neuron receives information from all neurons of the previous layer.

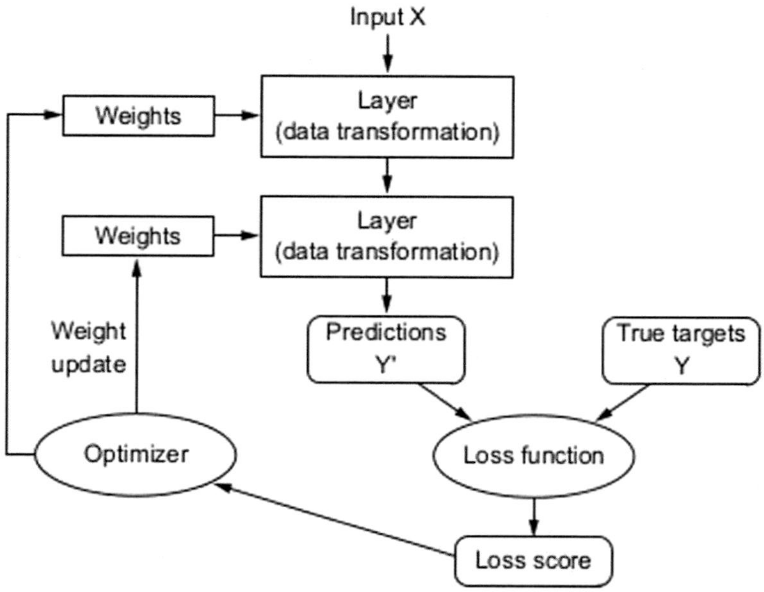

Fig. 5 Backpropagation flowchart

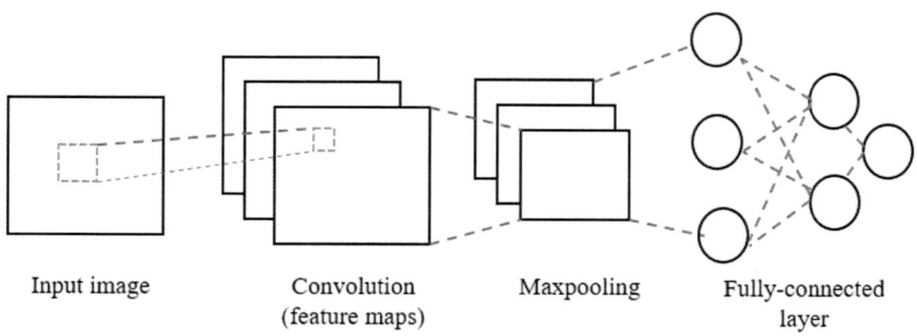

Fig. 6 Convolutional neural network

Recurrent Neural Networks

Densely connected (fully connected) networks and convolutional networks are feed-forward networks. Feedforward networks have no memory; each input is processed independently. To process a temporal series of data points, these networks request the entire sequence at once. However, RNNs process sequences by iterating through the sequence elements and maintaining a state (h) that contains information relative to what it has seen so far (Fig. 7). A data point is no longer processed in a single step, rather the network internally loops over sequence elements. RNNs are used in sequence.

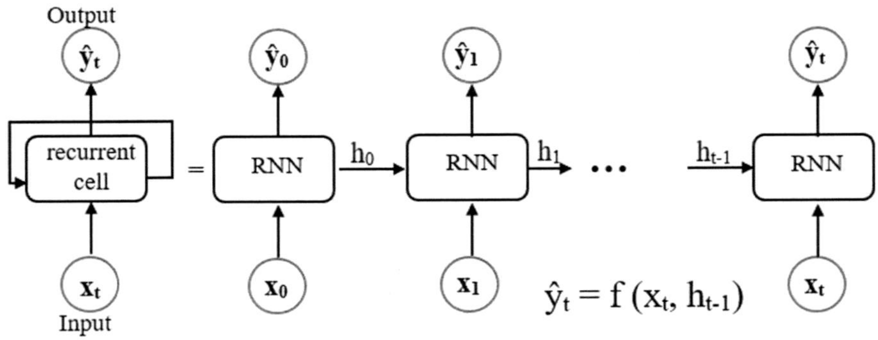

Fig. 7 Recurrent neural network

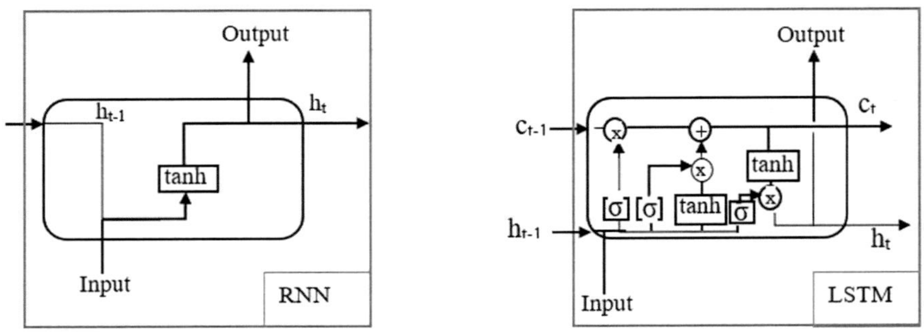

Fig. 8 Anatomy of RNN cell versus anatomy of LSTM cell

Sequence modelling requires the handling of variable-length sequences, tracking of long-term dependencies and information about the order, and parameters sharing across the sequence. RNNs do take these requirements into account. However, long-term dependencies are impossible to learn in practice. To address this issue, LSTM and Gated Recurrent Unit (GRU) were designed. LSTM is a family of RNNs that saves information for later, preventing older signals from gradually vanishing during processing. Figure 8 illustrates a comparison between the RNN cell and the LSTM cell (tanh and σ are respectively hyperbolic and sigmoid activation functions). GRU is very similar to LSTM. GRU is a slightly simpler, streamlined version of the LSTM architecture. LSTMs have been applied a lot to multivariate electricity consumption forecasting as the state of the (Makridakis et al., 2020).

References

Adeoye, O., & Spataru, C. (2019). Modeling and forecasting hourly electricity demand in West African countries. *Applied Energy, 242*, 311–333. https://doi.org/10.1016/j.apenergy.2019.03.057.

Adjamagbo, C., Ngae, P., & Vigneron, A. V. V. (2011). *Modélisation De La Demande En Énergie Électrique Au Togo, 14*, 67–83.

Amega, K. (2018). Master research program on climate change and energy energy efficiency potential and its impact on long-term. Universite Abdou Moumouni.

AT2ER. (2018). Rapport annuel D ' activités 2018. 83.

Bashari, M., & Rahimi-Kian, A. (2020). *Forecasting electric load by aggregating meteorological and history-based deep learning modules.* IEEE Power and energy society general meeting, 2020-Augus, pp 0–4. https://doi.org/10.1109/PESGM41954.2020.9282124.

Brownlee, J. (2021). *Deep learning for time series forecasting*, 574. https://doi.org/10.1007/978-3-030-64777-3_9.

Castellano, A. (2015). Brighter Africa. February. *Journal of Molecular and Cellular Cardiology, 4*(1), 144. https://doi.org/10.1016/0022-2828(72)90097-1.

Fobi, S., Deshpande, V., Ondiek, S., Modi, V., & Taneja, J. (2018). A longitudinal study of electricity consumption growth in Kenya. *Energy Policy, 123*(August), 569–578. https://doi.org/10.1016/j.enpol.2018.08.065.

INSEED. (2011). Togo fourth census 2010.

International Energy Agency. (2014). Africa energy outlook. A focus on the energy prospects in sub-Saharan Africa. World Energy Outlook Special Report. *International Energy Agency Publication*, 1–237.

International Energy Agency. (2017). Energy access outlook from poverty to prosperity.

Johnston, R. B. (2016). Arsenic and the 2030 Agenda for sustainable development. Arsenic research and global sustainability. In *Proceedings of the 6th International Congress on Arsenic in the Environment, AS 2016* (pp 12–14). https://doi.org/10.1201/b20466-7.

Kuster, C., Rezgui, Y., & Mourshed, M. (2017). Electrical load forecasting models: A critical systematic review. *Sustainable Cities and Society, 35*, 257–270. https://doi.org/10.1016/j.scs.2017.08.009.

LeCun, Y., Bengio, Y., & Laboratories, T. B. (1998). Convolutional networks for images, speech, and time-series.

Makridakis, S., Spiliotis, E., & Assimakopoulos, V. (2020). The M4 competition: 100,000-time series and 61 forecasting methods. *International Journal of Forecasting, 36*(1), 54–74. https://doi.org/10.1016/j.ijforecast.2019.04.014.

Moss, T. (2021). What is the modern energy minimum? Energy for growth hub. https://www.energyforgrowth.org/multimedia/what-is-the-modern-energy-minimum/.

Narciso, D. A. C., & Martins, F. G. (2020). Application of machine learning tools for energy efficiency in industry: A review. *Energy Reports, 6*, 1181–1199. https://doi.org/10.1016/j.egyr.2020.04.035.

Román-Portabales, A., López-Nores, M., & Pazos-Arias, J. J. (2021). A systematic review of electricity demand forecast using ann-based machine learning algorithms. *Sensors, 21*(13), 1–23. https://doi.org/10.3390/s21134544.

Rumelhart, D. E., Hinton, G. E., & Williams, R. J. (1988). Learning internal representations by error propagation. In *Readings in Cognitive Science* (pp. 399–421). Elsevier. https://doi.org/10.1016/B978-1-4832-1446-7.50035-2.

Verwiebe, P. A., Seim, S., Burges, S., Schulz, L., & Müller-Kirchenbauer, J. (2021). Modeling energy demand—A systematic literature review. *Energies, 14*(23), 1–58. https://doi.org/10.3390/en14237859.

Correction to: Assessment of Energy Potential of Residential Solid Waste in the Asokore Mampong Municipality of the Ashanti Region of Ghana

Grant Degbe Dzitse, Helen M. K. Essandoh, and Lawrence Darkwah

Correction to:
Chapter "Assessment of Energy Potential of Residential Solid Waste in the Asokore Mampong Municipality of the Ashanti Region of Ghana" in: M.-M. Narra and S. Narra (eds.), *African Green Transition Through Innovative Pathways*, World Sustainability Series, https://doi.org/10.1007/978-3-031-87043-9_1

The original version of Chapter 1 was updated with the following corrections post publication:

The chapter's title has been changed to "Assessment of Energy Potential of Residential Solid Waste in the Asokore Mampong Municipality of the Ashanti Region of Ghana".

Changes in the authors' affiliations:

Previous affiliations:

Grant Degbe Dzitse, Regional Water and Environmental Sanitation Centre, Kumasi (RWESCK), College of Civil Engineering, Kwame Nkrumah University of Science and Technology, Kumasi, Ghana; Helen M. K. Essandoh, Regional Water and Environmental Sanitation Centre, Kumasi (RWESCK), College of Civil Engineering, Kwame Nkrumah University of Science and Technology, Kumasi, Ghana; Lawrence Darkwah, College of Chemical Engineering, Kwame Nkrumah University of Science and Technology, Kumasi, Ghana.

Current affiliations:

The updated version of this chapter can be found at
https://doi.org/10.1007/978-3-031-87043-9_1

M.-M. Narra and S. Narra (eds.), *African Green Transition Through Innovative Pathways*, World Sustainability Series, https://doi.org/10.1007/978-3-031-87043-9_18

Grant Degbe Dzitse, Regional Water and Environmental Sanitation Centre, Kumasi (RWESCK), Department of Civil Engineering, College of Engineering, Kwame Nkrumah University of Science and Technology, Kumasi, Ghana; Helen M. K. Essandoh, Regional Water and Environmental Sanitation Centre, Kumasi (RWESCK), Department of Civil Engineering, College of Engineering, Kwame Nkrumah University of Science and Technology, Kumasi, Ghana; Lawrence Darkwah, Department of Chemical Engineering, College of Engineering, Kwame Nkrumah University of Science and Technology, Kumasi, Ghana.

Changes in the table: Table 7 has been deleted as it is the same as Table 5.

Errors that have the potential to mislead, distort the scientific record, or lead to consequential errors.

Insertion of the references.

The chapter has been updated with the changes.

MIX

Papier | Fördert gute Waldnutzung

FSC® C083411

Kolibri 360 GmbH
Ferdinand-Jühlke-Straße 7
99095 Erfurt, Deutschland
produktsicherheit@kolibri360.de